D1610173

Anaerobic Bacteriology:

Clinical and Laboratory Practice

To Sir James Howie

Anaerobic Bacteriology: Clinical and Laboratory Practice

Third edition

A. TREVOR WILLIS
DSc, MD, FRACP, PhD, FRCPath, FRCPA

Director, Public Health Laboratory, Luton
Consultant Microbiologist, Luton and Dunstable Hospital

BUTTERWORTHS
LONDON - BOSTON
Sydney - Wellington - Durban - Toronto

THE BUTTERWORTH GROUP

United Kingdom — Butterworth & Co (Publishers) Ltd
London: 88 Kingsway, WC2B 6AB

Australia — Butterworth Pty Ltd
Sydney: 586 Pacific Highway, Chatswood NSW 2067
Also at Melbourne, Adelaide and Perth

Canada — Butterworth & Co (Canada) Ltd
Toronto: 2265 Midland Avenue, Scarborough, Ontario, M1P 4S1

New Zealand — Butterworths of New Zealand Ltd
Wellington: 26–28 Waring Taylor Street, 1

South Africa — Butterworth & Co (South Africa) (Pty) Ltd
Durban: 152–154 Gale Street

USA — Butterworth (Publishers) Inc
Boston: 19 Cummings Park, Woburn, Mass. 01801

First published 1960

Second edition 1964
Third edition 1977

ISBN 0 407 00081 X

British Library Cataloguing in Publication Data

Willis, Allan Trevor
Anaerobic bacteriology. – 3rd ed.
1. Bacteriology, Medical 2. Bacteria, Anaerobic
I. Title
616.01'4 QR46 77–30019

ISBN 0–407–00081–X

Type set by Butterworths Litho Preparation Department
Printed in Great Britain by Butler & Tanner Ltd, London & Frome

Preface

In the twelve years since the second edition appeared there has been a great resurgence of interest in anaerobic bacteriology among both medical microbiologists and their clinical colleagues. This is due largely to an increasing recognition of the importance of the non-sporing anaerobic bacteria as significant causes of infection in man. Although it has been suggested by some that the non-clostridial anaerobes, like the mycoplasmas and 'L-forms' before them, are simply another band-wagon that commands fashionable attention, it seems unlikely that the importance attached to their role in human and animal pathology will be so evanescent.

These developments have led to major additions of detail and to changes in emphasis in our knowledge on many subjects, especially in anaerobic methodology and in the systematics and the ecological and pathogenetic associations of the non-sporing anaerobes. These additions and changes are reflected in the present text, which now includes separate chapters devoted to the descriptive bacteriology of the non-clostridial anaerobes, and to the clinical syndromes produced by them in man. The remainder of the text has been thoroughly revised and much of it rewritten, and appropriate additions made to the bibliography.

My grateful thanks are due to Professor R. A. Willis for literary criticism and advice, to Miss B. H. Whyte and her staff at the Central Public Health Laboratory Library at Colindale for their help in obtaining many of the original publications, and to my colleagues at the Luton Public Health Laboratory and at the Luton and Dunstable Hospital for many helpful and stimulating discussions. During the preparation of this new edition I was fortunate in working first under Sir James Howie and subsequently under Sir Robert Williams in the Public Health Laboratory Service, to both of whom I owe a considerable debt of gratitude for their interest and support.

I am deeply grateful to my wife for her constant encouragement and for her help in preparing the manuscript, in reading proofs and in checking references.

Finally, my thanks are due to my Publishers for their help during the production of this book.

Public Health Laboratory, and
Luton and Dunstable Hospital,
Luton

A. Trevor Willis

Contents

1

Methods of Growing Anaerobes

A variety of methods is available for the culture of anaerobic organisms in the clinical laboratory. Exclusion of oxygen from part of the medium is the simplest method, and is effected by growing the organism *within* the culture medium as a shake or fluid culture.

When an oxygen-free or anaerobic atmosphere is required for obtaining surface growths of anaerobes, anaerobic jars provide the method of choice. Some of the methods for obtaining anaerobiosis in a jar are also applicable to single tube or plate cultures, but for laboratories that intend to undertake anaerobic bacteriology seriously, anaerobic jars are essential.

More sophisticated methods for the surface culture of anaerobes are the pre-reduced anaerobically sterilized (PRAS) roll-tube technique of Hungate (1950), and use of the anaerobic cabinet or glove box. These two techniques utilize complex and expensive equipment, are time consuming, and require specially trained staff and specialized medium preparation facilities. There is no doubt that these complex techniques provide the most meticulous anaerobic conditions, and are appropriately used for the isolation and study of exacting anaerobic species that are highly sensitive to oxygen (Spears and Freter, 1967; Barnes and Impey, 1970). They are, however, too demanding in time, space and expense for routine use in the clinical laboratory where anaerobic work is but one facet of the diagnostic routine. Moreover, comparative studies have shown that these methods are not superior to the anaerobic jar for the recovery of clinically significant anaerobes from pathological specimens (Killgore *et al.*, 1972, 1973; Rosenblatt *et al.*, 1972, 1973, 1974; Dowell, 1972; Cadogan-Cowper and Wilkinson,

1974; Spaulding *et al.*, 1974; Starr, 1974; Watt, Hoare and Brown 1974). Although Peach and Hayek (1974) experienced a three-fold increase in the isolation rate of anaerobes from clinical specimens when an anaerobic cabinet was used instead of an anaerobic jar, it seems likely from the data they published that their conventional anaerobic technique was faulty.

THE ANAEROBIC JAR

The modern anaerobic jar is based, both in design and in operation, on that originally described by McIntosh and Fildes (1916). In their anaerobic jar, hot platinum was used to catalyse the combination of oxygen with hydrogen to form water—a principle first applied to the culture of anaerobes by Laidlaw (1915). In the original jar, the catalyst capsule was composed of asbestos wool impregnated with finely divided palladium and enclosed in a fine-mesh brass or copper gauze envelope. The palladium was activated by heating it in a bunsen burner flame immediately before setting up the jar, and the hydrogen was supplied to the jar from a Kipp's apparatus or from a cylinder of the gas.

Electrical heating of the catalyst was introduced by Smillie (1917), and was adopted by Fildes and McIntosh (1921) and Brown (1921, 1922). Subsequently various minor modifications to jar design were made by Brewer (1938–39) and Evans, Carlquist and Brewer (1948).

These early forms of anaerobic jar have now been replaced by cold catalyst jars; indeed, electrically operated jars are no longer manufactured. The room temperature catalyst (Heller, 1954), which is manufactured under patent as Deoxo pellets by Engelhard Industries Ltd, London, consists of pellets of alumina coated with finely divided palladium. Baird and Tatlock (London) developed the first commercially available jar (the BTL or Torbal jar) which utilized this principle of catalysis (Laboratory Equipment Test Report, 1958). Subsequently Becton, Dickinson, in association with Baltimore Biological Laboratories in America introduced the Gaspak anaerobic system (Brewer and Allgeier, 1966), and more recently Don Whitley Scientific developed the Gaskit system for use with their Schrader valve vented jars.

Figure 1.1 depicts the BTL jar. Anaerobic jars are cylindrical vessels made of metal, glass or plastic, flanged at the top to carry an air-tight lid which is held firmly in place by a clamp or lug closure. The lid carries on its undersurface the room temperature catalyst capsule. The essential difference between the BTL and the Gaspak jars is that in the former the lid is provided with two valves through which air can be withdrawn and hydrogen introduced. The lid of the standard Gaspak jar is not vented, because the jar is specifically designed for use with an internal disposable hydrogen–carbon dioxide generator. Important

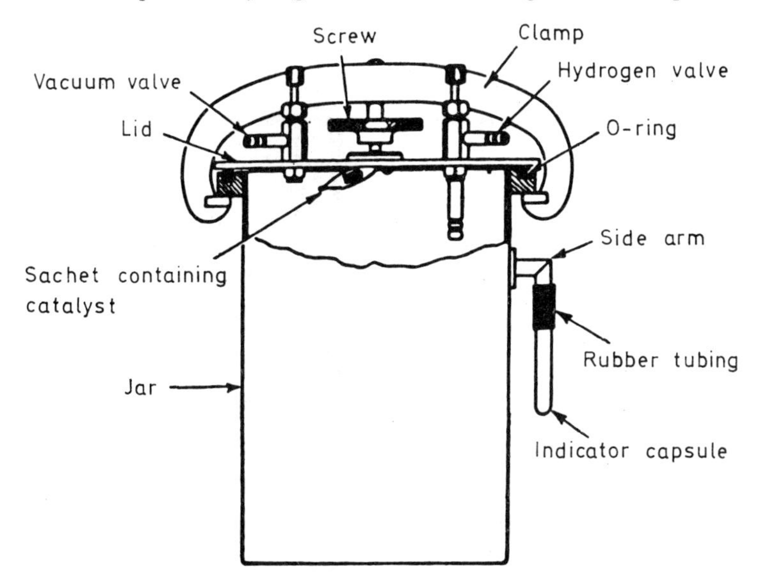

Figure 1.1. BTL anaerobic jar. (Reproduced by courtesy of Baird and Tatlock (London) Ltd)

features of the Whitley anaerobic jar (Burt and Whitley, in press) include venting by special Schrader valves, and a quick release 4-g cold catalyst capsule of large surface area.

The BTL and Whitley anaerobic jars

The hydrogen source

Hydrogen is obtained from a cylinder of the compressed gas. The use of coal gas as a source of hydrogen should be avoided since it contains carbon monoxide which is inhibitory to some organisms. The use of illuminating gas or natural gas is inappropriate, since it consists mainly of methane, and is usually deficient in hydrogen. Since it is important that the hydrogen should be supplied to the jar at low pressure, the hydrogen cylinder should be fitted with a reducing valve and the gas delivered at a pressure of not more than 0.5 psi. If a reducing valve is

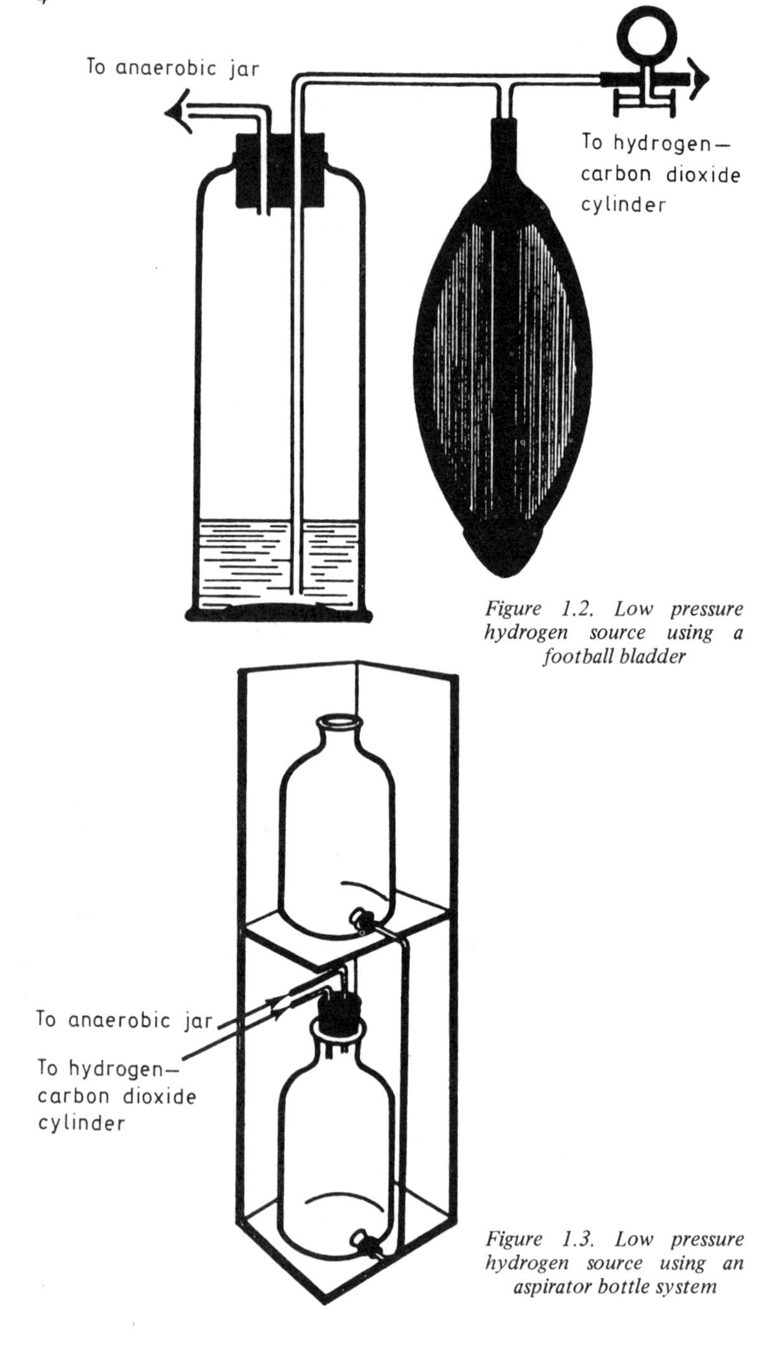

Figure 1.2. Low pressure hydrogen source using a football bladder

Figure 1.3. Low pressure hydrogen source using an aspirator bottle system

not available, a convenient low pressure source is either a football bladder or anaesthetic bag (*Figure 1.2*), or an aspirator bottle system (*Figure 1.3*) filled from the cylinder. The pressure reducing mechanism is attached to a gas washing bottle, which acts as a flow meter, and this in turn is connected to the anaerobic jar.

The carbon dioxide source

Since the growth of many anaerobes is improved by the addition to the jar of 10% carbon dioxide, and since no anaerobes are adversely affected by this concentration of the gas, it is important to have available a carbon dioxide source. This may be provided from a separate cylinder of the compressed gas, but it is much more convenient to use a hydrogen–carbon dioxide mixture. A suitable mixture is 90% hydrogen with 10% carbon dioxide, and is available from BOC Ltd. Dowell and Hawkins (1968) recommended a gas mixture containing 80% nitrogen, 10% hydrogen and 10% carbon dioxide.

In the interests of safety, it is important that compressed gases, and especially explosive gases, should be housed outside the building (Department of Health and Social Security, 1972), and the gases led into the laboratory through copper tubing. Cylinders containing mixtures of hydrogen with carbon dioxide must be stored horizontally, since the gases tend to 'layer off'. When this happens the cylinder is likely to supply mainly hydrogen early in its life, and mainly carbon dioxide as it becomes empty.

The catalyst

The Deoxo pellet catalyst (grade 0.5% Pd) is contained in a wire gauze capsule attached to the underside of the lid of the jar. The sachet contains pellets of alumina coated with finely divided palladium; the sachet in the BTL jar contains about 1 g of catalyst, that in the Whitley jar about 4 g. Certain gases such as chlorine, sulphur dioxide, carbon monoxide and hydrogen sulphide poison the catalyst, as does oil, the vapour of some organic solvents and strong acids. Inactivation by hydrogen sulphide is especially relevant to the microbiologist, since many anaerobes give off appreciable amounts of this gas, especially from fluid cultures. The catalyst is also inactivated by moisture, which is plentiful in the anaerobic jar, but it is readily reactivated by heating in a hot air oven (*see below*).

The use of multiple catalyst sachets in the BTL anaerobic jar speeds the development of anaerobiosis (Moore, 1966; Walker, Harris and Moore, 1971), since the rate of catalysis is directly proportional to the amount of catalyst present. Multiple sachets are thus of value when organisms that are intolerant of oxygen are cultured, but it seems that no additional cultural advantage is gained by the use of more than three sachets in the standard sized jar (Watt, Hoare and Collee, 1973). The use of multiple sachets is not required with Whitley anaerobic jars because of the larger size of the catalyst capsule; indeed, Don Whitley Scientific supply a 4-g catalyst conversion kit which can be fitted to most other types of anaerobic jar (*see also* Baldwin, 1975).

In the event that cold catalyst is not available, Wright's capsule (Wright, 1943), which consists of asbestos wool impregnated with finely divided palladium can be prepared in the laboratory (*see* Appendix, p. 26).

The anaerobic indicator

For all types of anaerobic jar it is a common practice to include some system which serves as a check on the development of anaerobic conditions. A faulty capsule or a leaking jar may well prevent the development or maintenance of complete anaerobiosis, although if the BTL and Whitley jars are properly tested before each use, failure due to these causes will not occur.

The BTL jar is fitted with an external indicator tube attachment containing Lucas semi-solid indicator (*see below*), which is obtainable from the manufacturer in sealed tubes ready for use. As the jar becomes anaerobic, the indicator dye (methylene blue) becomes colourless. With continued exposure the surface of the indicator begins to dry and it becomes ineffective. It can easily be rejuvenated by heating the top layer in a bunsen burner flame, pouring it off, and leaving the moist part below exposed.

Setting up the jar

1. Place cultures in the jar so that the medium in the petri dishes is uppermost, and secure the lid. Attach the jar through one valve to a vacuum pump, and through the other to a mercury manometer; evacuate the air until there is a 'negative pressure' in the jar of about 300 mmHg.

2. Close the valve connecting the jar to the vacuum pump, disconnect the pump and replace it by the low pressure hydrogen source.
3. During this changeover the jar remains connected to the manometer which shows if the jar is holding the vacuum. If there is any sign that the vacuum is not being held, the jar is probably leaking and must not be used until it has been thoroughly tested.
4. Open the valve to admit the hydrogen, which is drawn in rapidly as the jar loses its partial vacuum. As soon as this inrush of gas has ceased, close both valves on the jar and disconnect the manometer.
5. After about five minutes again open the valve to the hydrogen supply; there should again be an immediate inrush of gas, since catalysis creates a reduced pressure within the jar due to decrease in the amount of gases; at this time the top of the jar adjacent to the catalyst will feel warm due to the exothermic catalysis. If there is no inrush of gas, the catalyst is inactive and must be replaced.
6. Leave the jar connected to the hydrogen supply for about five minutes, then close the valve and place the jar in the incubator, where catalysis will continue until all the oxygen in the jar has been used up.

If carbon dioxide is being added from a separate cylinder, this is run into the jar after evacuation but before the hydrogen is admitted. Evacuate the jar to – 300 mmHg; then, with the manometer still attached to the jar, run in the carbon dioxide until the 'negative pressure' is reduced by approximately 75 mmHg. If atmospheric pressure is assumed to be 760 mmHg, the jar will now contain the equivalent of 10% of carbon dioxide.

Additional notes on setting up the BTL and Whitley jars

EVACUATION OF THE JAR

There are several different recommendations for the degree of evacuation involving in setting up the BTL and Whitley jars. Some operating instructions recommend that six-sevenths of the air should be removed before hydrogen is admitted, a procedure that is also recommended by Collee, Rutter and Watt (1971), Futter and Richardson (1971) and Jarvis and Bruten (1972). The chief advantages of drawing off most of the air are that anaerobiosis may be achieved more rapidly, that less moisture will accumulate in the jar and that an explosion will be less forceful. The main disadvantages are that fluid cultures are likely to boil over as the higher vacuum is drawn, and that the medium in plate cultures sometimes becomes detached from the petri dish due to the expansion of small locules of air trapped between the dish and the medium, and released from solution. Another disadvantage is that the activity of the catalyst is not so easily assessable, although this can be monitored manometrically. Finally, unless care is taken to ensure that the rims and lids of petri dishes are quite dry, a film of water between them acts as a non-return valve; air is removed from within the petri dish as the higher vacuum is applied, then, as hydrogen is admitted to the jar, the base and lid of the petri dish are forced together by the increasing pressure. Such plates have their own 'sealed-in' atmosphere and do not become anaerobic. This may be averted by the use of vented petri dishes.

THE HYDROGEN SOURCE

If hydrogen compressed in cylinders is not available, the Gaspak or Gaskit disposable hydrogen–carbon dioxide generator (*see* p. 13) may be used conveniently with the BTL and Whitley jars. If conditions are such that the laboratory must prepare its own hydrogen, the gas is obtained from a Kipp's or similar apparatus in which 25% sulphuric acid is allowed to act on zinc. In this case the gas must be passed through a 10% solution of lead acetate to remove hydrogen sulphide, and then through a 10% solution of silver nitrate to remove arsine, since the catalyst is poisoned by both arsine and hydrogen sulphide.

Futter and Richardson (1971) recommended that jars should be filled with hydrogen from a bladder, and that the bladder should be left attached to the jar for the first hour or so of incubation. This ensures that a partial vacuum is not developed in the jar, which might otherwise result in air being drawn in through a faulty seal in the lid of the jar. They pointed out that a partial vacuum is especially prone to develop if the jar contains a large volume of medium in which oxygen is dissolved.

THE ADDITION OF CARBON DIOXIDE

It has been known for many years that carbon dioxide greatly improves the growth of many aerobic and anaerobic organisms, and that amongst the latter it is especially favourable to the growth of non-spore formers

(Khairat, 1940; Smith, 1955; Willis, 1964; Wilson and Miles, 1964; Barnes, Impey and Goldberg, 1966; Smith and Holdeman, 1968; Dehority, 1971). This has been reconfirmed by Watt (1973) and Stalons, Thornsberry and Dowell (1974). It is also clear that the germination of spores of some clostridia is greatly stimulated by the presence of carbon dioxide (Roberts and Hobbs, 1968; Holland, Barker and Wolf, 1970). The well known stimulatory effect of carbon dioxide on the growth of some strains of streptococci may lead to their mistaken identification as 'anaerobic cocci'; carbon dioxide dependence can be shown by culturing these organisms in parallel in air, in air plus 10% carbon dioxide and under anaerobic conditions.

EXPLOSIONS

I have not seen a record of a significant explosion occurring with the BTL or Whitley jars. Explosions in anaerobic jars are due to ignition of the hydrogen and oxygen when these gases are present in the proportion of about 2 to 1, and this ratio exists for a short time when the first inrush of hydrogen occurs. This explosive phase soon passes as gaseous diffusion occurs in the jar. Since the room temperature catalyst is active before the jar is sealed, theoretically an explosion can occur when the hydrogen is first admitted. It will be noted that the hydrogen valve on the BTL jar is provided with a lipped nozzle on the underside of the lid. This is for the attachment of rubber or plastic tubing which directs inflowing hydrogen away from the catalyst capsule and materially aids rapid diffusion of gases in the jar.

Occasional minor explosions are encountered with cold catalyst jars when a jar, recently put up, is opened while active catalysis is proceeding.

ORIENTATION OF CULTURE PLATES IN THE JAR

It is common practice to place petri dishes 'upside down' in the anaerobic jar, i.e. with the inoculated surface of the agar facing upwards. This is inappropriate because moisture is likely to gravitate onto the agar surfaces and spoil the cultures either by gross flooding, or by facilitating swarming growth of organisms. The practice probably derives from the fact that the agar in properly placed petri dishes sometimes becomes detached from the dish when it is subjected to a high vacuum, and it 'lands butter-side down' on the petri dish lid. This is avoided by ensuring that an unnecessarily high vacuum is not applied to the jar (*see above*).

THE CATALYST

One of the commonest causes of failure in the anaerobic jar is use of an inactive catalyst. The catalyst may fail quite suddenly (rarely is there a noticeable gradual 'wearing out'), so that testing it each time the jar is used is essential. Since palladium catalysts are inactivated by moisture, it is important that the capsule should be kept dry at all times. Rejuvenation of the catalyst is effected by heating it in a hot air oven at 160 °C for 1.5 h. It is now widely recommended that the catalyst in the lid of the jar is replaced each time the jar is used (Dowell, 1972; Holdeman and Moore, 1975), the used catalyst being sent for reactivation. In the BTL jar this manoeuvre is greatly simplified if a small electrical alligator clip is screwed into the hole on the underside of the lid that normally holds the catalyst sachet (Warren, 1973); in the Whitley jar the catalyst is held in place by a quick release finger spring. Reactivated catalyst is stored in sealed screw capped jars in a warm dry place.

INDICATORS OF ANAEROBIOSIS

The most commonly used chemical methods for indicating anaerobiosis are those recommended by Fildes and McIntosh (1921), Lucas (*see* Stokes, 1968), and Brewer, Allgeier and McLaughlin (1966). They depend on the fact that when methylene blue is placed in an anaerobic environment it is reduced from its coloured oxidized form to a colourless reduced leuco-compound.

Resazurin has also been used as an Eh indicator in thioglycollate media (Pittman, 1946; Mossel, Golstein Brouwers and de Bruin, 1959). The single solution indicator of Ulrich and Larsen (1948) and Parker's modification of the Fildes and McIntosh indicator (Parker, 1955) do not appear to be superior to those about to be described. Indicators for use with anaerobic jars have been briefly reviewed by Drollette (1969).

Lucas semisolid indicator. The following formula is taken from the BTL anaerobic jar instruction sheet: Add 12 drops of 9% thioglycollic acid and 2 drops of phenol red to 5 ml of a 2% borax solution in a boiling tube. Then add 10 ml of methylene blue solution (one BDH methylene blue tablet to 200 ml water). Add 10 ml of previously melted sloppy agar. Heat the blue mixture in a boiling water bath until it is colourless. Transfer it with a warmed pasteur pipette into a series of warm ampoules, and seal the ampoules immediately. When required, open an ampoule and either place it inside the anaerobic jar or attach it with rubber tubing to the side arm of the jar. In the presence of oxygen the indicator becomes blue to a depth of about 5 mm.

The indicator of Brewer, Allgeier and McLaughlin. Brewer, Allgeier and McLaughlin (1966) described an anaerobic indicator sachet for inclusion inside the anaerobic jar; this was the forerunner of the presently available Gaspak disposable anaerobic indicator (*see* p. 13). In the original sachet, the indicator solution, which contained equal parts of 60% tris (hydroxymethyl) aminomethone, 4% dextrose and 0.02% methylene blue, was contained in an oxygen-permeable teflon bag. The Gaspak disposable anaerobic indicator consists of a foil envelope containing a pad saturated with the methylene blue solution, and is available from the manufacturers, Becton, Dickinson UK Ltd.

Fildes and McIntosh indicator. The preparation of this indicator is described in the Appendix, p. 26.

Bacteriological indicator. In addition to, or instead of, using a chemical indicator, some workers include in the jar a plate culture of a known strict anaerobe such as *Cl. tetani* or *B. fragilis,* and of a strict aerobe, such as *Ps. aeruginosa.* This method is quite reliable if the indicator anaerobe grows and the aerobe does not. Failure of the anaerobe to grow, however, does not necessarily mean that that anaerobic equipment is at fault; it may be that the medium on which the organism was plated is unsuitable (this is unlikely if quality control of media is practised), or that the organism was not viable when the jar was set up. Growth of the aerobe indicates failure to achieve anaerobiosis.

The fact that indicators are necessary in anaerobic work draws attention to the importance of examining the anaerobic apparatus before use to ensure that it is in proper working order. Five minutes spent in testing the equipment may save many hours of futile incubation.

Opening the BTL and Whitley jars

The external indicator, if fitted, is checked to ensure that it is colourless. Such an observation should be made daily when incubation is continued for a number of days. One of the valves in the lid of the jar is opened; this may be accompanied by a hiss of inrushing air – evidence of an air-tight jar, but not necessarily indicative that complete anaerobiosis has been achieved. Indeed, the hiss is sometimes due to the escape of gases from the jar, especially if bacterial fermentation reactions have occurred during incubation. Similarly, the slight discoloration (a blueing) of the blood in fresh blood agar plates, which commonly occurs under anaerobic conditions, is not indicative of complete anaerobiosis.

After removal of cultures the jar is cleaned and dried, particular attention being paid to the lid of the jar where condensate from catalysis is present. The jar is stored open with the lid inverted, in a warm dry place. Since the palladium catalyst is inactivated by moisture, the practice of storing jars set up and sealed is not recommended. Indeed, the life of the cold catalyst may be appreciably shortened by this procedure. The best practice is to remove the catalyst capsule for heat-reactivation, replacing it with a new or reactivated one immediately before the jar is set up again.

Cleaning and servicing the BTL and Whitley jars

The jar may be cleaned with a soft rag and methylated spirit. Care must be taken to ensure that solvents or grease do not come into contact with the catalyst sachet or the 'O' ring, and that the sachet does not become wet. Leaking needle valves on the BTL jar are usually remedied by unscrewing them and lightly greasing their points; damaged needle valves are expensive to regrind or replace. Leaking Schrader valves on the Whitley jar are remedied by replacing the valve core. A common cause of a leaking BTL jar is perished tubing on the indicator side arm. Spare capsules, 'O' rings and valves are available from the respective manufacturers.

It is a simple and inexpensive matter for the laboratory to prepare its own catalyst sachets. The envelope is made from fine mesh copper, brass or Monel-metal gauze, and is filled with about 4 g of Deoxo catalyst (Engelhart Industries). Large sachets containing more pellets produce more rapid catalysis than small ones; the pellets should be evenly distributed in the sachet and not bunched together, and care must be taken not to crush the catalyst.

In laboratories where more than a few anaerobic jars are in use, it is convenient to permanently number each jar and its lid. This helps to keep track of vagrant jars, and facilitates spot testing and servicing.

The Gaspak anaerobic system

The remarkable ingenuity of Brewer and his colleagues (Brewer, Heer and McLaughlin, 1955; Brewer and Allgeier, 1965; Brewer, Allgeier and

McLaughlin, 1966; Brewer and Allgeier, 1966) led to the development of the Gaspak anaerobic system, which is marketed in the United Kingdom by Becton, Dickinson UK Ltd. The Gaspak jar, which is available in two sizes (to hold 12 and 36 dishes), is made of transparent polycarbonate plastic. It utilizes the Deoxo room temperature catalyst, and has been designed specifically for use with a disposable hydrogen–carbon dioxide generator and a disposable methylene blue anaerobic indicator.

The hydrogen–carbon dioxide generator

This consists of a sealed foil sachet containing two tablets – one composed of citric acid and sodium bicarbonate, the other of sodium borohydride and cobalt chloride. When water is introduced into the sachet, the following reactions take place:

1. $C_6H_8O_7 + 3NaHCO_3 \rightarrow Na_3(C_6H_5O_7) + 3H_2O + 3CO_2$
2. $NaBH_4 + 2H_2O \rightarrow NaBO_2 + 4H_2$

The generator is activated with water and placed inside the jar immediately before it is sealed. Thus, the gases are produced inside the closed jar, so that facilities for evacuation are not required.

Recently, Don Whitley Scientific have produced a new hydrogen–carbon dioxide generator for use with anaerobic jars. It is marketed under the name of Gaskit (Ferguson, Phillips and Willis, 1976).

Other features of the Gaspak jar

The Gaspak jar utilizes the room temperature catalyst (p. 5), and a disposable anaerobic indicator (p. 11) which is placed inside the jar. Many of the notes made earlier about the BTL and Whitley anaerobic jars also apply to the Gaspak jar. The Gaspak anaerobic jar, spare parts, disposables and accessories are available from Becton, Dickinson UK Ltd.

Setting up the Gaspak jar

1. Place the cultures in the jar.
2. Open a disposable anaerobic indicator sachet and place it in the jar so that the indicator wick is visible. The indicator rapidly

turns blue on exposure to air, but becomes colourless again under anaerobic conditions. Decolorization may take several hours.

3. Open a disposable hydrogen–carbon dioxide generator (Gaspak or Gaskit) (use three generators for the large 36–40 plate jars), activate the generator and place it upright in the jar.
4. Secure the lid of the jar immediately, and place the jar in the incubator.

Which is the anaerobic jar of choice?

The BTL and Whitley jars on the one hand, and the Gaspak system on the other have relative advantages and disadvantages, but both provide effective means of producing an anaerobic environment. The outstanding feature of the Gaspak system is the disposable gas generator envelope, which does away with the need for a vacuum pump, manometer and cylinders of compressed gas; operation of the jar is consequently very quick and simple. On the other hand, the internal production of hydrogen makes it virtually impossible to test the apparatus for catalyst activity, especially as the standard Gaspak jar is not vented. Further, as the standard Gaspak jar is not evacuated before use a relatively large volume of water is formed during catalysis.

In the Gaspak gas generator the reaction of borohydride with water results in a highly alkaline residue which commonly prevents activation of the bicarbonate tablet, so that only a fraction of the available carbon dioxide is released. Moreover, the alkaline residue is likely to absorb any carbon dioxide that is present in the atmosphere of the jar, so that during incubation there is a decreasing amount of carbon dioxide present. The proportion of carbon dioxide in the Gaspak jar under these circumstances must therefore be regarded as random, variable and unknown (Ferguson, Phillips and Tearle, 1975).

The Gaskit generator (Don Whitley Scientific) does not have these disadvantages, since its activation by citric acid ensures that poisoning of the carbon dioxide tablet does not occur and that the pH of the final reaction residue is slightly acidic (Ferguson, Phillips and Willis, 1976).

When hydrogen is provided from an external source a vented lid is essential, the valves incidently providing a facility for assessing catalyst activity and for testing the jar for leaks. In the BTL jar, however, the needle valves themselves are prone to develop leaks, as is the anaerobic indicator side-arm. Schrader valves, as supplied with the Whitley jar, are very reliable.

Anaerobiosis is achieved somewhat more rapidly when preliminary evacuation of the jar is practised, so that a higher isolation rate of demanding organisms may be expected. Moreover, the cost of disposable gas generators is high compared with that of compressed gas. At the time of writing, the cost of the Gaspak gas generator is about 20p, that of the Whitley Gaskit generator about 15p.

I consider that for laboratories in which experience with strict anaerobes is likely to be limited or sporadic, the Gaskit or Gaspak system is the anaerobic method of choice. For laboratories that deal with anaerobes on a large scale, a method employing evacuation and an external hydrogen–carbon dioxide source is preferred, although it is convenient to have some backup with an internal gas generator system.

Evaluations of these three types of anaerobic jar have been published by Stokes (1958), Williams (1970), Dowell (1972), Collee *et al.* (1972) and Burt and Whitley (1977); (*see also* Laboratory Equipment Test Report, 1958). Some useful comments on the control of anaerobic jars in the clinical laboratory have been published by Jarvis and Bruten (1972).

Modernizing old anaerobic jars

Old vented jars of the McIntosh and Fildes, and Brewer types are easily adapted for use with the Deoxo catalyst, and may be used either with an external hydrogen–carbon dioxide source or with an internal generator (*see* Khairat, 1964a, b; 1965). An excellent conversion kit at reasonable cost is available from Don Whitley Scientific; it consists of two Schrader valves with fitting washers, a 4-g catalyst sachet and spring clip sachet holder.

Almost any sealable containers such as pressure cookers and milk churns (Schaedler, Dubos and Costello, 1965), even biscuit tins and plastic bags, may be adapted for use as anaerobic 'jars', utilizing cold catalysis and internal gas generation. When these adaptions are made, care must be taken to ensure that the quantity of catalyst and the number of gas generators used are consistent with the volume of air in the container; one Gaspak or Gaskit generator is sufficient for about 2000 ml of air.

When a plastic bag is used, care must be taken to ensure that the catalyst capsule is kept well clear of the bag itself, since the capsule gets very hot during catalysis and may melt the plastic. Gaspak manufacture a completely disposable anaerobic system in which the anaerobic container is a clear, triple laminated plastic bag, sealed with a sponge strip and a plastic sealing bar.

OTHER METHODS OF OBTAINING AN OXYGEN-FREE ATMOSPHERE

For the routine isolation and study of anaerobic bacteria from clinical material, cold catalyst anaerobic jars provide quite the most convenient method of providing an anaerobic atmosphere. There are, however, many other effective ways of producing anaerobiosis in a closed container which may be used if conventional anaerobic jars are not available, or which may be more appropriate in particular circumstances. Some of these methods are briefly outlined below.

Alkaline pyrogallol anaerobic jar

Alkaline solutions of pyrogallol have the property of absorbing large amounts of oxygen, and a variety of devices has been invented which make use of this reaction for the production of anaerobic conditions.

The great advantage of this method is that it requires no special equipment. The reagents are readily available and the method is suitable for use with any container that can be sealed. A large, wide-mouthed glass jar with a screw-cap lid and a rubber gasket, or a tin with a tight-fitting lid are suitable containers. It is a wise precaution to ensure an air-tight seal in containers of these types by pressing plasticine around the joint when the jar has been set up. Plastic bags have also been used successfully as sealed containers (Westmacott and Primrose, 1975).

Plates for incubation are placed in the jar together with the anaerobic indicator. Pyrogallol is added to a solution of sodium hydroxide in a large test tube, which is then placed inside the jar and the jar is quickly sealed. For each 100 ml of jar capacity, use 1 g pyrogallol and 10 ml of 2.5M sodium hydroxide.

Though anaerobiosis is fairly rapidly attained by this method, oxidation of the pyrogallate leads to the production of a small amount of carbon monoxide which is inhibitory towards some organisms. Further, excess of sodium hydroxide which is present in the pyrogallate absorbs any carbon dioxide in the jar, which is a disadvantage. This may be overcome by using sodium carbonate instead of sodium hydroxide.

Application of the alkaline pyrogallol method to single cultures

The method may be applied to a single culture tube by moistening the cotton wool plug with a little pyrogallol, adding some sodium carbonate solution, and then sealing the tube with a rubber stopper. Less crude techniques have been described by Lockhart (1953), Pankhurst (1967, 1971) and Sacks (1974).

Pankhurst devised a simple culture tube for the isolation and enumeration of sulphate-reducing anaerobes. The culture apparatus, which is available commercially from Astell Laboratory Scientific Company, consists of two Pyrex containers similar to standard test tubes, one 150 mm long and the other 50 mm long, joined by a horizontal glass tube. The large tube contains 10–20 ml of inoculated medium; the small side-arm tube contains a pledget of absorbent cotton wool to which is added 1 ml of 40% aqueous pyrogallol and 1 ml of a saturated solution of sodium carbonate.

Single plate cultures may be prepared by the method of Mossel *et al.* (1959), in which a dry mixture of pyrogallol, potassium carbonate and diatomaceous earth is used for absorbing the oxygen. Two petri dishes are selected. One of these is poured with medium and streak-inoculated with the organism in the usual way.

To the other dish is attached a filter paper envelope (7 × 4 cm) which contains 2 g of the dry mixed powder. The two plates are fitted together at their rims, the junction made air tight with plasticine, and the apparatus is placed in the incubator. The pyrogallol mixture is activated by the moisture within the closed system, and complete anaerobiosis develops in about 2 h. The dry powder is prepared by mixing together 3 g pyrogallol, 3 g potassium carbonate and 15 g diatomaceous earth. A similar single plate technique was described by Naylor (1963) in which a partitioned polystyrene petri dish, sealed with cellulose tape, is used.

Matthews and Karnauchow (1961) adapted the alkaline pyrogallol method for use with single plate cultures in polythene bags. Here 1 g dry pyrogallol and 2 ml 50% solution of anhydrous sodium carbonate are placed on a Seitz filter pad in a plastic screw cap, and this is placed inside the bag which is then quickly sealed.

Cultivation in atmospheres of indifferent gases

The sealed jar containing the cultures is evacuated with a vacuum pump and refilled with an indifferent gas such as nitrogen, hydrogen or helium (Bridges, Pepper and Chandler, 1952; Futter and Richardson, 1971). In order to replace oxygen completely in this way repeated evacuation and refilling with the inert gas is necessary. Even then complete removal of oxygen is not assured and many strict anaerobes may fail to grow. The method, however, is satisfactory for the culture of the less exacting anaerobes such as *Clostridium histolyticum, Cl. tertium* and *Cl. perfringens* (*see* p. 140), although it suffers from the general disadvantages attendant upon any method that employs extreme evacuation of the jar (*see* p. 7). In studies on the quantitative recovery of clostridia from pure damaged spore suspensions, Futter and Richardson (1971) used four consecutive evacuations of the jar of at least 650 mmHg and refilling with oxygen-free nitrogen. Their recovery rates for *Cl. perfringens, Cl. septicum, Cl. histolyticum* and *Cl. sporogenes* incubated under nitrogen were two- to three-fold those incubated under hydrogen; they attributed this to either nitrogen stimulation or hydrogen toxicity, rather than to an oxidation-reduction effect.

Use of aerobic organisms to absorb oxygen

The use of aerobic organisms to absorb oxygen was originally described by Fortner (1928). Organisms such as coliform bacilli, *Pseudomonas aeruginosa* and *Bacillus subtilis* are suitable for the purpose. Two blood agar plates are selected, one of which is heavily inoculated with the aerobic organism, and the other is streaked with the anaerobic culture. The two plates are fitted together at their rims, the junction made air-tight with plasticine, and the apparatus is placed in the incubator immediately. Anaerobiosis takes some time to develop, so that the method usually fails with exacting anaerobes (*see also* Blum, 1964).

This principle for achieving anaerobiosis in a single plate culture was applied by Kneteman (1957) to a plastic film technique in which an aerobic micrococcus and the anaerobe were grown within the medium as a pour plate culture. Oxygen from the air was prevented from diffusing into the medium by covering the surface of the agar with a film of oxygen-impermeable plastic. Subsequently, Shank (1963) modified this plastic film method by using sodium thioglycollate in the medium as the reductant instead of a micrococcus.

Anaerobiosis with steel wool

Steel wool for the removal of oxygen from the atmosphere in sealed containers was developed by Parker (1955). Commercial steel wool is moulded into a loose pad (10 g) to fit a petri dish lid. The pad is dipped in 500 ml of a solution of 0.25–0.5% copper sulphate, which contains 0.25% of a wetting agent such as Lissapol or Tween 80. The solution is acidified with sulphuric acid to a pH of 1.5–2.0. Excess solution is drained from the pad, which is then returned to the petri dish lid. The activated steel wool is placed wherever convenient in the anaerobic jar. Steel wool treated in this way becomes irregularly coated with metallic copper, the couples so formed being subject to rapid oxidation. Rosenblatt and Stewart (1975) adapted this method for use in a sealed bag (*see also* Barton and Winzar 1973), and it may also be used for the transport of specimens under anaerobic conditions from the bedside to the laboratory (Sutter *et al.*, 1972).

PRE-REDUCED ANAEROBICALLY STERILIZED ROLL TUBE TECHNIQUE

Hungate (1950) first developed this technique for the isolation of oxygen intolerant anaerobes from rumen fluid and sewage. Since then the method has undergone numerous modifications and improvements (Hungate, 1966, 1969), and it has been adapted in various ways to suit the particular requirements of other anaerobic studies (Kistner, 1960; Moore, 1966; Smith, 1966; Attebery and Finegold, 1969; Paynter and Hungate, 1968; Holdeman and Moore, 1972; Macy, Snellen and Hungate, 1972; Berg and Nord, 1973). The method has been used mainly as a research tool for the isolation and study of sewage and intestinal bacteria; it has not been extensively used in clinical diagnostic work (McMinn and Crawford, 1970; Killgore *et al.*, 1972, 1973; Rosenblatt, Fallon and Finegold, 1972, 1973, 1974; Dowell, 1972; Slots, 1975). The brief notes which follow are offered only as an introduction to the technique.

In essence, the method involves the use of pre-reduced anaerobically sterilized media which are prepared in stoppered roll tubes. Exposure of bacteria and culture media to oxygen is avoided by displacing the air in the culture tubes with an oxygen-free gas such as carbon dioxide. When the stoppers are removed from the tubes for manipulation of the culture, anaerobiosis in the tubes is maintained by continuous flushing with an oxygen-free gas.

Preparation of oxygen-free gas

The inert gas (carbon dioxide or nitrogen) from a cylinder supply is rendered oxygen-free by passing it through a vertical column of heated copper turnings; the heated copper is periodically reactivated (reduced) by passing hydrogen through the column.

Instead of using a column of hot copper, the inert gas may be rendered oxygen-free with a Model D laboratory Deoxo Purifier. The purifier, which contains Deoxo cold catalyst is inserted in the gas line from a cylinder containing 97% carbon dioxide and 3% hydrogen. The purifier lasts indefinitely and requires no reactivation.

Anaerobic media

Preparation

After the medium has been compounded in the usual way, resazurin (anaerobic indicator) is added and the medium is then boiled in a flask to drive off dissolved oxygen; the pink resazurin is reduced to its colourless form when this has been accomplished. The flask of hot medium is 'gassed out' with the oxygen-free gas supply during cooling, thus ensuring exclusion of air from the flask. A reducing agent, such as cysteine hydrochloride is then added.

Anaerobic dispensing

The cooled medium may be tubed under oxygen-free gas without exposure to air (*see also* PRAS blood culture medium, p. 60). Both the flask of bulk medium and the culture tube are flushed continually with oxygen-free gas. The medium is transferred with a graduated pipette using a fairly long rubber mouth piece. The pipette is inserted into the flask and oxygen-free gas drawn through it before medium is sucked in; with carbon dioxide, the taste of the gas indicates when the air has been flushed out. The required volume of medium is then drawn into the pipette and transferred to the culture tube. The culture tube is closed with a butyl rubber stopper, the gassing needle being withdrawn as the stopper is inserted.

Sterilization

Tubes of medium are sterilized in the autoclave. The rubber closures are firmly wired or clamped in place, or a combined screw-cap and rubber closure may be used. After sterilization the autoclave pressure may be reduced rapidly, but the clamps holding the rubber closures are not removed until the tubes have cooled to a safe temperature. Medium preparation is now complete, and tubes may be stored until required for use as either roll tube dilution or roll tube streak cultures.

Preparation of roll tube streak cultures

The medium is melted and the stoppered tubes of PRAS medium are spun about their long axes in a roll tube apparatus before the medium solidifies. Centrifugal force spreads a thin even layer of the agar medium over the wall of the tube, where it hardens. The tube of medium for inoculation is held vertically on a rotating platform (about 70 rpm), and is inoculated with a charged stainless steel or platinum loop. As the loop is moved upwards over the revolving agar surface a spiral inoculation streak results. This procedure is carried out under the conditions of continuous flushing of the vessels with oxygen-free gas to maintain anaerobiosis, and with preliminary and final flaming of the gas cannulas and necks of tubes to prevent contamination. After the stopper is replaced, the tube is placed in the incubator.

Preparation of roll tube dilution cultures

Dilutions of material for culture are prepared in an anaerobic diluent by the anaerobic dispensing method, and appropriate volumes of these are inoculated into tubes of molten medium (using the same anaerobic technique) to produce shake cultures. The tubes are then spun on the roll tube apparatus until the agar has set, and the tubes incubated.

Inoculation of PRAS tubes may also be carried out by a closed method in which a syringe and needle are used to inoculate the tube through the rubber seal (Hungate, 1969).

Comment

The re-stoppered tube is thus its own anaerobic culture chamber. Culture tubes are incubated without further treatment and they may be examined individually at any time without exposing the cultures to air. That anaerobiosis is maintained in any individual tube is indicated by the resazurin remaining colourless; any trace of oxygen entering the tube rapidly turns the Eh indicator pink. Uninoculated tubes have a relatively long shelf life; they are satisfactory for use until they become oxidized (the resazurin turns pink).

The foregoing notes on the PRAS roll tube technique are intended only as an explanation of the basic concept of the method. Those who wish to learn more about it should consult the publications of Hungate (1950, 1969), Moore (1966), Latham and Sharpe (1971), Eller, Crabill and Bryant (1971), Holdeman and Moore (1972, 1975), Bryant (1972), Macy, Snellen and Hungate (1972). Although some of these publications present detailed instructions with excellent illustrations for the manipulative techniques involved (Holdeman and Moore, 1972; Hungate, 1969), these are best learnt by spending time at a laboratory where the method is in routine use.

A full range of equipment and accessories for this anaerobic cultural technique has been designed by the staff of the Anaerobe Laboratory, Virginia Polytechnic Institute, USA, and is available in the United Kingdom from Bellco Biological (Arnold R. Horwell Ltd, London).

ANAEROBIC CABINETS AND GLOVE BOXES

During the last ten years several groups of workers have used air-tight anaerobic cabinets or glove boxes, in which conventional bacteriological techniques may be practised in an oxygen-free atmosphere (Socransky, Macdonald and Sawyer, 1959; Rosebury and Reynolds, 1964; Drasar, 1967; Lee, Gordon and Dubos, 1968; Drasar and Crowther, 1971; Aranki *et al.*, 1969; Leach, Bullen and Grant, 1971; Aranki and Freter, 1972; Moodie and Woods, 1973; Watt, Collee and Brown, 1974). These techniques have the advantage of the roll tube method in that all the operations of isolating and sub-culturing anaerobes are conducted in the absence of oxygen; in addition, they enable the use of petri dish plate cultures.

An anaerobic cabinet consists essentially of an air-tight chamber constructed of perspex, metal, fibre-glass or flexible vinyl plastic. It is provided with glove ports, and is commonly fitted with an air-lock through which materials are transferred into and out of the chamber. A diversity of methods have been used for the removal of oxygen. In Drasar's cabinet, oxygen was removed by burning a spirit lamp inside the box, and then removing the 'last traces' of oxygen by pumping the cabinet's atmosphere through alkaline pyrogallol or through a column of heated copper. Socransky and his colleagues and Lee and his co-workers rendered their cabinets anaerobic by the ingenious method of inflating a large meteorological balloon inside the cabinet to expel the air, the displaced gas inside the box being replaced with nitrogen mixed with 5% carbon dioxide. The anaerobic glove box described by Aranki and his colleagues is constructed of flexible vinyl plastic which can be almost completely evacuated of air with a vacuum pump, since it is collapsible. The air is then replaced with an atmosphere consisting of nitrogen with 10% hydrogen and 5% carbon dioxide, and this is circulated over a palladium catalyst. In the cabinet designed by Leach and her colleagues, air is displaced by a continuous stream of oxygen-free carbon dioxide, and does not require the use of an air lock, catalyst or a vacuum pumping system.

All anaerobic cabinets are operated with some form of anaerobic indicator, such as resazurin, to ensure that anaerobiosis is achieved and maintained. Some cabinets, for example that of Leach and her co-workers, are intended for use only during manipulation of specimens and cultures, plates being incubated subsequently in anaerobic jars. The vinyl plastic glove box of Aranki and his colleagues is operated continuously, and is used both for manipulative procedures and for incubation of cultures. Consequently the whole glove box must be kept at 37 °C; alternatively a standard incubator is housed inside the anaerobic chamber.

SHAKE CULTURES AND FLUID CULTURES

Exclusion of oxygen from part of the medium is the simplest method of growing anaerobes and is effected by growing the organisms *within* the culture medium, which may be either solid or fluid. In the case of solid media growth is obtained as a shake culture in a tube or bottle, or as a pour-plate culture. This method of culture is favoured by some bacteriologists as an aid to purifying contaminated cultures of anaerobes (Fredette, 1964). Although the deep agar method of culture has a number of advantages, it is inferior to, and less convenient than, surface plating techniques. Beerens and Castel (1958–59) showed that deep cultures had an efficiency of only 26% that of surface cultures in the isolation of anaerobes from pathological material. Stab culture in solid media also allows growth of anaerobes, but this technique finds little practical application.

Fluid media in deep tubes are extensively used in anaerobic bacteriology since most anaerobes grow readily in this type of culture and manipulations with them are easily performed. Before use, media for shake culture and deep fluid culture must be steamed or heated in a bath of boiling water in order to drive off any dissolved oxygen, and then cooled quickly prior to inoculation. When incubated in air there is a tendency for oxygen to diffuse back into shake and fluid cultures, so that anaerobic growth appears some distance below the surface; strict anaerobes grow only in the depths of the medium, and highly exacting species may not grow at all; less oxygen intolerant organisms show growth almost to the surface. The deeper parts of shake cultures remain anaerobic far longer than those of fluid cultures, since convection currents which are always present in fluid media help to distribute oxygen rapidly throughout the medium. For this reason it is a common practice to thicken fluid media by incorporating a small amount of agar (0.05–0.1%) in them. These sloppy agar media are still fluid, but convection currents in them are reduced to a minimum. The growth of anaerobes in such media is greatly enhanced by the addition of reducing substances. The use of petroleum jelly to seal the surface of fluid cultures is not advocated. Petroleum jelly seals were once used to prevent diffusion of atmospheric oxygen into the medium, but they are ineffective, and make manipulation of the culture difficult.

An ingenious agar shake culture technique was developed by Bladel and Greenberg (1965) for the isolation and enumeration of clostridia. The cultures are prepared in transparent pouches fabricated from a laminated plastic film. The pouch cultures are incubated aerobically, and do not require sealing since a narrow column of agar in the neck of the pouch provides a satisfactory oxygen barrier (*see also* Mossel and de Waart, 1968; de Waart, 1971).

Although it is a common practice to incubate shake and fluid cultures under ordinary atmospheric conditions, many strict anaerobes may fail to grow. Consequently, when one is dealing with exacting species or with organisms of unknown identity, as, for example in pathological material, it is a wise precaution to incubate all cultures in the anaerobic jar.

REDUCING AGENTS

Although the addition of reducing agents to fluid and shake cultures of some anaerobes is not necessary, it is essential for the satisfactory culture of the more exacting anaerobes. Consequently, many bacteriologists add reducing substances of one kind or another routinely to all the media (including surface plating media) used for anaerobic work. The following are some of the reducing agents commonly used.

Thioglycollic acid

Thioglycollic acid or its sodium salt may be used in media in the proportion of 0.01–0.2%. Sodium thioglycollate is inhibitory to the growth of some clostridia, and gradually becomes slightly toxic towards a variety of anaerobes during storage (Mossel and Beerens, 1968; Hibbert and Spencer, 1970). Thioglycollate media were first introduced by Brewer (1940 a, b).

Glucose

Glucose is usually incorporated in media in the proportion of 0.5–1.0%. It is a good reducing agent, is non-toxic, and it also serves as an additional nutritional agent for bacterial growth. Shake and pour-plate cultures of saccharolytic anaerobes should not be made in glucose agar media if deep colonies are to be studied. Fermentation of the sugar by gas producing organisms leads to disruption of the medium by gas.

Ascorbic acid

Ascorbic acid may be used in the proportion of 0.1%. Tulloch (1945) advocated the use of ascorbic acid or sodium thioglycollate for the culture of anaerobes in deep pour-plates without other anaerobic precautions. The reducing agent is added to the molten agar medium which is then inoculated and poured into a petri dish to a depth of at least 12 mm. These days, incubation of *all* cultures in the anaerobic jar is preferred. Ascorbic acid is sometimes inhibitory towards some strains of non-sporing anaerobic bacilli.

Sodium sulphide

Bryant and Robinson (1961) recommended the use of sodium sulphide (0.025%) as a reductant for the culture of ruminal anaerobes. Sodium sulphide was superior to cysteine hydrochloride for the growth of ruminococci.

Cysteine

Cysteine is a valuable reducing agent if used in concentrations not exceeding 0.05%. Higher concentrations may inhibit bacterial growth. Hirsch and Grinsted (1954) considered that cysteine was superior to thioglycollate and they included this reductant in their reinforced clostridial medium (*see* p. 41).

Moore (1968) demonstrated the essential nature of L-cysteine for the successful surface culture of *Cl. novyi* type B on solid media, and drew attention to the importance of protecting the cysteine against atmospheric oxidation. In neutral and slightly alkaline solution the oxidation of cysteine proceeds rapidly, so that under conditions prevailing during pouring, setting and drying plating media, loss of cysteine is likely to be appreciable. Moore showed that cysteine in culture media is protected from oxidation by either ascorbic acid, thioglycollate or dithiothreitol, and that luxuriant growth of *Cl. novyi* type B is obtained on appropriate media containing cysteine paired with one of these protective substances. These important observations led to the development of Moore's NP-Medium – (Neopeptone Medium – *see* p. 43), in which cysteine hydrochloride is paired with dithiothreitol, and its use was subsequently extended to the surface culture of *Cl. novyi* types C and D (Walker, Harris and Moore, 1971). Since neither ascorbic acid nor dithiothreitol can replace cysteine in solid media for the culture of these types of *Cl. novyi*, it seems that cysteine is required *per se* for these organisms and not only as a reductant. This may be borne out by the observations of Watt (1971), who found that there was no enhancement of the growth of any of a wide range of anaerobes (other than that of the *Cl. novyi* group) when cultured on cysteine–dithiothreitol media; the organisms studied by Watt, however, are not especially demanding species.

Metallic iron

Metallic iron may be used as a reducing agent in fluid media. Iron wire, filings, nails and steel wool are suitable for this purpose. Ordinary 'tin-tacks' or screws are convenient for use for they are quickly sterilized by flaming before being added to sterile stock media (Spray, 1936; Hayward and Miles 1943).

In their studies of the cultural requirements of the *Cl. novyi* group of organisms, Collee, Rutter and Watt (1971) found that iron filings or powdered iron sprinkled on the surface of agar plates seeded with the organisms favoured the growth of types B, C and D in an anaerobic atmosphere. Under these conditions, they obtained consistent early growth of these exacting species, the surface colonies developing as satellites around particles of iron on the medium. The effect was also obtained when the cultures were prepared as pour plates with the iron spread on the surface, but there was no enhancement of growth when the iron filings were incorporated into the agar. This stimulatory effect of iron appears to be specific for these organisms (Watt, 1971).

Cooked meat medium

Cooked meat medium was introduced by Robertson (1915–16) and is widely used for the culture of anaerobes. The cooked sterile muscle tissue contains reducing substances, particularly glutathione, which permit the growth of many strict anaerobes without the application of other anaerobic methods. The reducing activity of the meat is shown by the pink colour of reduced haemin which develops in its lower layers. In addition to its reducing effect, the meat provides a variety of nutritional substances for bacterial growth. The reducing systems in this medium have been discussed by Lepper and Martin (1929).

APPENDIX

Preparation of Wright's Capsule (Wright, 1943)

Preparation of palladinized asbestos

Dissolve 1 g of palladium chloride in 15 ml distilled water acidified with a few drops of concentrated HC1. Thoroughly soak about 1.5 g of asbestos wool in this solution and then dry it in the incubator. Tease out the impregnated wool and coat it with carbon in a smokey flame, and finally burn off the carbon deposit in a blow lamp. This reduces the palladium chloride and leaves a finely divided deposit of palladium on the asbestos wool fibres. Alternatively, the palladium chloride may be chemically reduced by immersion of the impregnated asbestos in boiling 5% sodium formate solution. The palladinized asbestos is tested by directing a fine jet of hydrogen onto a sample piece; if the preparation is successful the catalyst will glow and ignite the hydrogen. The quantity of catalyst is sufficient for about three anaerobic jars.

Preparation of the capsule

Fold a layer of the catalyst into a piece of fine mesh copper or brass gauze, to form an envelope, the open side of which is then secured with a strip of soft metal to ensure a good seal and to facilitate handling. After completion, capsules should be carefully inspected for integrity of the envelope and for any pieces of catalyst which may project outside it. This is most likely to occur if the mesh of the gauze is too coarse. Any projecting pieces must be removed, otherwise there is a great risk of the glowing catalyst igniting the hydrogen in the jar and causing an explosion.

Use of Wright's capsule

When the anaerobic jar has been loaded and is ready for setting up, the Wright's capsule is heated by bunsen burner until it is red hot, and is then clipped into position on the underside of the lid before sealing the jar. Evacuation of the jar and admission of hydrogen are then conducted in the usual way, except that the hydrogen source is not turned off until the jar has become anaerobic.

The risk of an explosion occurring when this type of capsule is used is greater than with the room temperature catalyst. It should not be used with glass jars, and air should not be removed by flushing with hydrogen. I have no experience of how this capsule might react if hydrogen is supplied from an internal hydrogen generator.

Fildes and McIntosh indicator (Fildes and McIntosh, 1921)

Prepare three stock solutions as follows:

1. 6% glucose in distilled water
2. 6 ml 0.1 M sodium hydroxide diluted to 100 ml with distilled water
3. 3 ml 0.5% aqueous methylene blue diluted to 100 ml with distilled water (0.015%).

Each time the indicator solution is required, mix together equal parts of the three stock solutions in a test tube, and boil until the methylene blue is reduced. Place the tube of colourless indicator immediately in the anaerobic jar and initiate the process of securing anaerobiosis.

If anaerobic conditions are secured and maintained the indicator solution remains colourless. If on the other hand the blue colour returns to the solution, it has been oxidized and indicates that a failure has occurred in the anaerobic equipment.

REFERENCES

Aranki, A. and Freter, R. (1972). 'Use of anaerobic glove boxes for the cultivation of strictly anaerobic bacteria.' *American Journal of clinical Nutrition,* **25**, 1329

Aranki, A., Syed, S. A., Kenney, E. B. and Freter, R. (1969). 'Isolation of anaerobic bacteria from human gingiva and mouse cecum by means of a simplified glove box procedure.' *Applied Microbiology,* **17**, 568

Attebery, H. R. and Finegold, S. M. (1969). 'Combined screw-cap and rubber-stopper closure for Hungate tubes (pre-reduced anaerobically sterilized roll tubes and liquid media).' *Applied Microbiology,* **18**, 558

Baldwin, A. W. F. (1975). 'An improved catalyst sachet for anaerobic jars.' *Medical laboratory Technology,* **32**, 329

Barnes, E. M. and Impey, C. S. (1970). 'The isolation and properties of the predominant anaerobic bacteria in the caeca of chickens and turkeys.' *British Poultry Science,* **11**, 467

Barnes, E. M., Impey, C. S. and Goldberg, H. S. (1966). 'Methods for the characterization of the *Bacteroidaceae.*' In *Identification Methods for Microbiologists.* p. 51. Ed. by B. M. Gibbs and F. A. Skinner. London; Academic Press

Barton, A. P. and Winzar, J. A. (1973). 'Simple economical anaerobiosis.' *Journal of clinical Pathology,* 26, 238

Beerens, H. and Castel, M. M. (1958–59). 'A simple procedure for the surface culture of anaerobic bacteria compared with the technique using deep cultures.' *Annales de l'Institut Pasteur de Lille,* **10**, 183 (French)

Berg, J–O. and Nord, C–E. (1973). 'A method for isolation of anaerobic bacteria from endodontic specimens.' *Scandinavian Journal of dental Research,* **81**, 163

Bladel, B. O. and Greenberg, R. A. (1965). 'Pouch method for the isolation and enumeration of clostridia.' *Applied Microbiology,* **13**, 281

Blum, W. (1964). 'The propagation of anaerobes is easy.' *American Journal of medical Technology,* **30**, 423

Brewer, J. H. (1938–39). 'A modification of the Brown anaerobe jar.' *Journal of Laboratory and clinical Medicine,* **24**, 1190

Brewer, J. H. (1940a). 'A clear liquid medium for the "aerobic" cultivation of anaerobes.' *Journal of Bacteriology,* **39**, 10

Brewer, J. H. (1940b). 'Clinical notes, suggestions and new instruments.' *Journal of the American medical Association,* **115**, 598

Brewer, J. H. and Allgeier, D. L. (1965). 'Disposable hydrogen generator.' *Science,* **147**, 1033

Brewer, J. H. and Allgeier, D. L. (1966). 'Safe self-contained carbon dioxide – hydrogen anaerobic system.' *Applied Microbiology,* **14**, 985

Brewer, J. H., Allgeier, D. L. and McLaughlin, C. B. (1966). 'Improved anaerobic indicator.' *Applied Microbiology,* **14**, 135

Brewer, J. H., Heer, A. A. and McLaughlin, C. B. (1955). 'The use of sodium borohydride for producing hydrogen in an anaerobic jar.' *Applied Microbiology,* **3**, 136

Bridges, A. E., Pepper, R. E. and Chandler, V. L. (1952). 'Apparatus for large scale anaerobic cultures in an atmosphere of helium.' *Journal of Bacteriology,* **64**, 137

Brown, J. H. (1921). 'An improved anaerobe jar.' *Journal of experimental Medicine,* **33**, 677

Brown, J. H. (1922). 'Modification of an improved anaerobe jar.' *Journal of experimental Medicine,* **35**, 467

Bryant, M. P. (1972). 'Commentary on the Hungate technique for culture of anaerobic bacteria.' *American Journal of clinical Nutrition,* **25**, 1324

Bryant, M. P. and Robinson, I. M. (1961). 'An improved non-selective culture medium for ruminal bacteria and its use in determining diurnal variation in numbers of bacteria in the rumen.' *Journal of Dairy Science,* **44**, 1446

Burt, R. and Whitley, D. C. (1977). 'A new anaerobic jar.' In press

Cadogan-Cowper, G. and Wilkinson, R. G. (1974). 'Isolation and identification of *Bacteroides* and *Fusobacterium* from human sources.' *Australian Journal of experimental Biology and medical Science,* **52**, 897

Collee, J. G., Rutter, J. M. and Watt, B. (1971). 'The significantly viable particle: A study of the subculture of an exacting sporing anaerobe.' *Journal of medical Microbiology,* **4**, 271

Collee, J. G. Watt, B., Fowler, E. B. and Brown, R. (1972). 'An evaluation of the Gaspak system in the culture of anaerobic bacteria.' *Journal of applied Bacteriology,* **35**, 71

Dehority, B. A. (1971). 'Carbon dioxide requirement of various species of rumen bacteria.' *Journal of Bacteriology,* **105**, 70

Department of Health and Social Security (1972). *Safety in Pathology Laboratories.* London; HMSO

Dowell, V. R. (1972). 'Comparison of techniques for isolation and identification of anaerobic bacteria.' *American Journal of clinical Nutrition,* **25**, 1335

Dowell, V. R. and Hawkins, T. M. (1968). *Laboratory Methods in Anaerobic Bacteriology.* Public Health Service Publication No. 1803; Washington; US Government Printing Office

Drasar, B. S. (1967). 'Cultivation of anaerobic intestinal bacteria.' *Journal of Pathology and Bacteriology,* **94**, 417

Drasar, B. S. and Crowther, J. S. (1971). 'The cultivation of human intestinal bacteria.' In *Isolation of Anaerobes.* p. 93, Ed. by D. A. Shapton and R. G. Board. London; Academic Press

Drollette, D. D. (1969). 'Anaerobic indicators: A review.' *American Journal of medical Technology,* **35**, 758

Eller, C., Crabill, M. R. and Bryant, M. P. (1971). 'Anaerobic roll tube media for nonselective enumeration and isolation of bacteria in human feces.' *Applied Microbiology,* **22**, 522

Evans, J. M., Carlquist, P. R. and Brewer, J. H. (1948). 'A modification of the Brewer anaerobic jar.' *American Journal of clinical Pathology,* **18**, 745

Ferguson, I. R., Phillips, K. D. and Tearle, P. V. (1975). 'An evaluation of the carbon dioxide component in the GasPak anaerobic system.' *Journal of applied Bacteriology,* **39**, 167

Ferguson, I. R., Phillips, K. D. and Willis, A. T. (1976). 'An evaluation of GasKit–a disposable hydrogen–carbon dioxide generator – for the culture of anaerobic bacteria.' *Journal of applied Bacteriology,* **41**, 433

Fildes, P. and McIntosh, J. (1921). 'An improved form of McIntosh and Fildes' anaerobic jar.' *British Journal of experimental Pathology,* **2**, 153

Fortner, J. (1928). 'A simple plate culture method for strict anaerobes (anaerobic bacilli – filterable anaerobic bacteria –*Spirochaeta pallidum*).' *Zentralblatt fur Bakteriologie, Parasitenkunde, Infecktionskrankheiten und Hygiene* (Abteilung I), **108**, 155 (German)

Fredette, V. (1964). 'The role of the anaerobic bacteria, with particular reference to the virulence of *Clostridium perfringens.*' *Review of Canadian Biology,* **23**, 85

Futter, B. V. and Richardson, G. (1971). 'Anaerobic jars in the quantitative recovery of clostridia.' In *Isolation of Anaerobes*. p. 81. Ed. by D. A. Shapton and R. G. Board. London; Academic Press

Hayward, N. J. and Miles, A. A. (1943). 'Iron media for cultivation of anaerobic bacteria in air.' *Lancet,* **1**, 645

Heller, C. L. (1954). 'A simple method for producing anaerobiosis.' *Journal of applied bacteriology,* **17**, 202

Hibbert, H. R. and Spencer, R. (1970). 'An investigation of the inhibitory properties of sodium thioglycollate in media for the recovery of clostridial spores.' *Journal of Hygiene, Cambridge,* **68**, 131

Hirsch, A. and Grinsted, E. (1954). 'Methods for the growth and enumeration of anaerobic spore-formers from cheese, with observations on the effect of Nisin.' *Journal of Dairy Research,* **21**, 101

Holdeman, L. V. and Moore, W. E. C. (1972). 'Roll-tube technique for anaerobic bacteria.' *American Journal of clinical Nutrition,* **25**, 1314

Holdeman, L. V. and Moore, W. E. C. (1975). *Anaerobe Laboratory Manual.* Blacksburg; Virginia Polytechnic Institute and State University

Holland, D., Barker, A. N. and Wolf, J. (1970). 'The effect of carbon dioxide on spore germination in some clostridia.' *Journal of applied Bacteriology,* **33**, 274

Hungate, R. E. (1950). 'The anaerobic mesophilic cellulolytic bacteria.' *Bacteriological Reviews,* **14**, 1

Hungate, R. E. (1966). *The Rumen and its Microbes.* New York; Academic Press

Hungate, R. E. (1969). 'A roll tube method for cultivation of strict anaerobes.' In *Methods in Microbiology.* Vol. 3B, p. 117. Ed. by J. R. Norris and D. W. Ribbons. London; Academic Press

Jarvis, J. D. and Bruten, D. M. (1972). 'Observations in the use of anaerobic jars and the methods of their control.' *Medical laboratory Technology,* **29**, 325

Khairat, O. (1940). 'The effect of a CO_2 atmosphere on blood cultures.' *Journal of Pathology and Bacteriology,* **50**, 491

Khairat, O. (1964a). 'Routine incubation in an 80 plate CO_2 container.' *Canadian Journal of Microbiology,* **10**, 499

Khairat, O. (1964b). 'Modernizing Brewer and other anaerobic jars.' *Journal of Bacteriology,* **87**, 963

Khairat, O. (1965). 'A modern anaerobic jar accommodating sixty-six plates.' *American Journal of clinical Pathology,* **43**, 596

Killgore, G. E., Starr, S. E., Del Bene, V. E., Whaley, D. N. and Dowell, V. R. (1972). 'Comparison of three anaerobic systems for the isolation of anaerobic bacteria from clinical specimens.' *Abstracts of the 72nd Annual Meetings of the American Society of Microbiology,* p. 94

Killgore, G. E., Starr, S. E., Del Bene, V. E., Whaley, D. N. and Dowell, V. R. (1973). 'Comparison of three anaerobic systems for the isolation of anaerobic bacteria from clinical specimens.' *American Journal of clinical Pathology,* **59**, 552

Kistner, A. (1960). 'An improved method for viable counts of bacteria of the ovine rumen which ferment carbohydrates.' *Journal of general Microbiology,* **23**, 565

Kneteman, A. (1957). 'A method for the cultivation of anaerobic spore forming bacteria.' *Journal of applied Bacteriology,* **20**, 101

Laboratory Equipment Test Report (1958). 'A new anaerobic jar.' *Laboratory Practice,* **7**, 145

Laidlaw, P. P. (1915). 'Some simple anaerobic methods.' *British medical Journal,* **1**, 497

Latham, M. J. and Sharpe, M. E. (1971). 'The isolation of anaerobic organisms from the bovine rumen.' In *Isolation of Anaerobes.* p. 133. Ed. by D. A. Shapton and R. G. Board. London; Academic Press

Leach, P. A., Bullen, J. J. and Grant, I. D. (1971). 'Anaerobic CO_2 cabinet for the cultivation of strict anaerobes.' *Applied Microbiology,* **22**, 824

Lee, A., Gordon, J. and Dubos, R. (1968). 'Enumeration of the oxygen sensitive bacteria usually present in the intestine of healthy mice.' *Nature,* **220**, 1137

Lepper, E. and Martin, C. J. (1929). 'The chemical mechanisms exploited in the use of meat media for the cultivation of anaerobes.' *British Journal of experimental Pathology,* **10**, 327

Lockhart, W. R. (1953). 'A single tube method for anaerobic incubation of bacterial cultures.' *Science,* **118,** 144

McIntosh, J. and Fildes, P. (1916). 'A new apparatus for the isolation and cultivation of anaerobic micro-organisms.' *Lancet,* **1**, 768

McMinn, M. T. and Crawford, J. J. (1970). 'Recovery of anaerobic micro-organisms from clinical specimens in pre-reduced media versus recovery by routine clinical laboratory methods.' *Applied Microbiology,* **19**, 207

Macy, J. M. Snellen, J. E. and Hungate, R. E. (1972). 'Use of syringe methods for anaerobiosis.' *American Journal of clinical Nutrition,* **25**, 1318

Matthews, A. D. and Karnauchow, P. N. (1961). 'A simple technique for the cultivation of anaerobes.' *Canadian medical Association Journal,* **84**, 793

Moodie, H. L. and Woods, D. R. (1973). 'Isolation of obligate anaerobic faecal bacteria using an anaerobic glove cabinet.' *South African medical Journal,* **47**, 1739

Moore, W. B. (1968). 'Solidified media suitable for the cultivation of *Clostridium novyi* type B.' *Journal of general Microbiology,* **53,** 415

Moore, W. E. C. (1966). 'Techniques for routine culture of fastidious anaerobes.' *International Journal of systematic Bacteriology,* **16,** 173

Mossel, D. A. A. and Beerens, H. (1968). 'Studies on the inhibitory properties of sodium thioglycollate on the germination of wet spores of clostridia.' *Journal of Hygiene, Cambridge,* **66,** 269

Mossel, D. A. A., Golstein Brouwers, G. W. M.v. and de Bruin, A. S. (1959). 'A simplified method for the isolation and study of obligate anaerobes.' *Journal of Pathology and Bacteriology,* **78,** 290

Mossel, D. A. A. and Waart, J. de (1968). 'The enumeration of clostridia in foods and feeds.' *Annales de l'Institut Pasteur,* **19**, 13

Naylor, P. G. D. (1963). 'A novel polystyrene dish for the production of anaerobiosis.' *Journal of applied Bacteriology,* **26**, 219

Pankhurst, E. S. (1967). 'A simple culture tube for anaerobic bacteria.' *Laboratory Practice,* **16**, 58

Pankhurst, E. S. (1971). 'The isolation and enumeration of sulphate-reducing bacteria.' In *Isolation of Anaerobes* p. 219. Ed. by D. A. Shapton and R. G. Board. London; Academic Press

Parker, C. A. (1955). 'Anaerobiosis with iron wool.' *Australian Journal of experimental Biology and medical Science,* **33**, 33

Paynter, M. J. B. and Hungate, R. E. (1968). 'Characterization of *Methanobacterium mobilis* sp. n, isolated from the bovine rumen.' *Journal of Bacteriology,* **95,** 1943

Peach, S. and Hayek, L. (1974). 'The isolation of anaerobic bacteria from wound swabs.' *Journal of clinical Pathology,* **27,** 578

Pittman, M. (1946). 'A study of the fluid thioglycollate medium for the sterility test.' *Journal of Bacteriology,* **51,** 19

Roberts, T. A. and Hobbs, G. (1968). 'Low temperature growth characteristics of clostridia.' *Journal of applied Bacteriology*, **31**, 75

Robertson, M. (1915–16). 'Notes upon certain anaerobes isolated from wounds.' *Journal of Pathology and Bacteriology*, **20**, 327

Rosebury, T. and Reynolds, J. B. (1964). 'Continuous anaerobiosis for cultivation of spirochetes.' *Proceedings of the Society of Experimental Biology and Medicine*, **117**, 813

Rosenblatt, J. E., Fallon, A. M. and Finegold, S. M. (1972). 'Recovery of anaerobes from clinical specimens.' *Abstracts of the 72nd Annual Meeting of The American Society of Microbiology*, p. 94

Rosenblatt, J. E., Fallon, A. and Finegold, S. M. (1973). 'Comparison of methods for isolation of anaerobic bacteria from clinical specimens.' *Applied Microbiology*, **25**, 77

Rosenblatt, J. E., Fallon, A. and Finegold, S. M. (1974). 'Comparison of methods for isolation of anaerobic bacteria.' In *Anaerobic Bacteria, Role in Disease*. p. 21. Ed. by A. Balows, R. M. DeHaan, V. R. Dowell, and L. B. Guze. Springfield; Thomas

Rosenblatt, J. E. and Stewart, P. R. (1975). 'Anaerobic bag culture method.' *Journal of clinical Microbiology*, **1**, 527

Sacks, L. E. (1974). 'An anaerobic screw-cap culture tube.' *Journal of applied Bacteriology*, **37**, 451

Schaedler, R. W., Dubos, R. and Costello, R. (1965). 'The development of the bacterial flora in the gastrointestinal tract of mice.' *Journal of experimental Medicine*, **122**, 59

Shank, J. L. (1963). 'Applications of the plastic film technique in the isolation and study of anaerobic bacteria.' *Journal of Bacteriology*, **86** 95

Slots, J. (1975). 'Comparison of five growth media and two anaerobic techniques for isolating bacteria from dental plaque.' *Scandinavian Journal of dental Research*, **83**, 274

Smillie, W. G. (1917). 'New anaerobic methods.' *Journal of experimental Medicine*, **26**, 59

Smith, L. DS. (1955). *Introduction to the Pathogenic Anaerobes*. Chicago; University of Chicago Press

Smith L. DS. and Holdeman, L. V. (1968). *The Pathogenic Anaerobic Bacteria*. Illinois; Charles C. Thomas

Smith, P. H. (1966). 'The microbial ecology of sludge methanogenesis.' *Developmental industrial Microbiology*, **7**, 156

Socransky, S., MacDonald, J. B. and Sawyer, S. (1959). 'The cultivation of *Treponema microdentium* as surface colonies.' *Archives of oral Biology*, **1**, 171

Spaulding, E. H., Vargo, V., Michaelson, T. C. and Swenson, R. M. (1974). 'A comparison of two procedures for isolating anaerobic bacteria from clinical specimens.' In *Anaerobic Bacteria, Role in Disease*. p. 37. Ed. by A. Balows, R. M. DeHaan, V. R. Dowell, and L. B. Guze. Springfield; Thomas

Spears, R. W. and Freter, R. (1967). 'Improved isolation of anaerobic bacteria from the mouse cecum by maintaining continuous strict anaerobiosis.' *Proceedings of the Society of Experimental Biology and Medicine*, **124**, 903

Spray, R. S. (1936). 'Semisolid media for cultivation and identification of the sporulating anaerobes.' *Journal of Bacteriology*, **32**, 135

Stalons, D. R., Thornsberry, C. and Dowell, V. R. (1974). 'Effect of culture medium and carbon dioxide concentration on growth of anaerobic bacteria commonly encountered in clinical specimens.' *Applied Microbiology*, **27**, 1098

Starr, S. E. (1974). 'Comparison of isolation techniques for anaerobic bacteria.' In *Anaerobic Bacteria, Role in Disease.* p. 47. Ed. by A. Balows, R. M. DeHaan, V. R. Dowell, and L. B. Guze. Springfield; Thomas

Stokes, E. J. (1958). 'Anaerobes in routine diagnostic cultures.' *Lancet,* **1,** 668

Stokes, E. J. (1968). *Clinical Bacteriology.* 3rd edn, London; Arnold

Sutter, V. L., Attebery, H. R., Rosenblatt, J. E., Bricknell, K. S. and Finegold, S. M. (1972). *Anaerobic Bacteriology.* California; University of California

Tulloch, W. J. (1945). 'Notes on the examination of exudates for pathogenic anaerobes.' *Journal of Pathology and Bacteriology,* **57**, 67

Ulrich, J. A., and Larsen, A. M. (1948). 'A single solution indicator for anaerobiosis.' *Journal of Bacteriology,* **56,** 373

Waart, J. de (1971). 'The enumeration of clostridia in food.' *Archiv fur Lebensmittelhygiene,* **22,** 149

Walker, P. D., Harris, E. and Moore, W. B. (1971). 'The isolation of clostridia from animal tissues.' In *Isolation of Anaerobes.* p. 25. Ed. by D. A. Shapton and R. G. Board. London; Academic Press.

Warren, A. (1973). 'Alligator clips on anaerobic jars.' *Gazette of the Institute of Medical Laboratory Technology,* **17**, 277

Watt, B. (1971). 'The recovery of clinically important anaerobes on solid media.' *Journal of medical Microbiology,* **5,** 211

Watt, B. (1973). 'The influence of carbon dioxide on the growth of obligate and facultative anaerobes on solid media.' *Journal of medical Microbiology,* **6,** 307

Watt, B., Collee, J. G. and Brown, R. (1974). 'The isolation of strict anaerobes: The use of an anaerobic cabinet compared with a conventional procedure.' *Journal of medical Microbiology,* **7,** 315

Watt, B., Hoare, M. V. and Brown, R. (1973). 'Some variables affecting the recovery of anaerobic bacteria: A quantitative study.' *Journal of general Microbiology,* **77**, 447

Westmacott, D. and Primrose, S. B. (1975). 'An anaerobic bag for photoheterotrophic growth of some *Rhodospirillaceae* in petri dishes.' *Journal of applied Bacteriology,* **38**, 205

Williams, G. A. (1970). 'Gaspak system.' *Medi-Miscellanea,* **1**, No. 13, 2

Willis, A. T. (1964). *Anaerobic Bacteriology in Clinical Medicine.* 2nd. edn. London; Butterworths

Wilson, G. S. and Miles, A. A. (1964). *Topley and Wilson's Principles of Bacteriology and Immunity.* 5th edn. London; Arnold

Wright, B. M. (1943). 'Improved catalyst for McIntosh and Fildes' anaerobic jar.' *Army Pathology Laboratory Service, Current Notes* No. 9, 3

2

Media

For convenient reference, this chapter describes briefly the preparation and uses of the most important media that are of value in the routine study of anaerobes.

The use of pre-reduced anaerobically sterilized (PRAS) media is not generally necessary in the routine laboratory, since almost all of the anaerobes that are pathogenic for man can be isolated and identified by conventional anaerobic techniques: PRAS medium, however, is recommended for anaerobic blood culture (*see* p. 59). Those who wish to know more about PRAS media and methods should consult the appropriate end-references in Chapter 1.

Excellent sections on media are presented in the manuals of anaerobic bacteriology published by Sutter *et al.* (1972), Dowell and Hawkins (1974) and Holdeman and Moore (1975).

INOCULATION OF MEDIA

Plate cultures

Plate cultures are inoculated in the usual way. When dealing with very strict, or otherwise demanding anaerobes, it is desirable to use freshly prepared plates, since during storage the medium takes up oxygen from the atmosphere in sufficient amounts to prevent the growth of these, even though complete anaerobiosis has apparently been obtained in the jar. Alternatively, but less satisfactorily, it may be convenient to store a set of uninoculated plates under anaerobic conditions, the whole jar being kept in the refrigerator. After plates have been inoculated, they should be placed under anaerobic conditions as quickly as possible. Similarly, surface cultures of such organisms must be subcultured to an anerobic environment promptly after removal from the jar.

For the preparation of half-antitoxin plates, the medium is first thoroughly dried (but not dehydrated) in the incubator. Three or four drops of antitoxin are then pipetted on the plate and spread over half its surface. A convenient spreader consists of a pulled pasteur

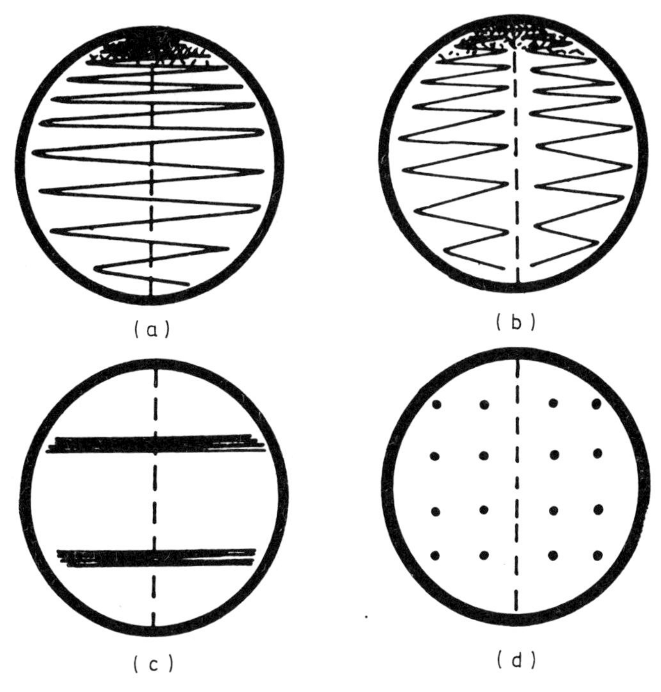

Figure 2.1. Diagram showing four methods of inoculating half-antitoxin plates. In each case antitoxin is spread on the left-hand half of the plate. (a) The whole plate is streaked at right-angles to the antitoxin area. (b) The primary inoculum is spread out separately on the two halves of the plate. (c) Separate heavy streaks of inoculum are applied at right-angles to the antitoxin area. (d) Corresponding areas on each side of the plate are spot-inoculated

pipette, the distal end of which has been bent in the form of a hockey stick. The spreader must, of course, be flamed and cooled before use. After antitoxin treatment, the plate is allowed to stand for a few minutes to allow the serum to be absorbed into the medium.

The method of inoculating half-antitoxin plates varies according to the requirements. Four conventional patterns are shown in *Figure 2.1*. Methods (a) and (b) provide separate colonies, and are therefore suitable for mixed cultures, while methods (c) and (d) are most usefully employed for pure culture studies. Replicate spotting (method (d)), though economical of medium, is the least satisfactory.

Fluid media

Fluid media are inoculated after dissolved oxygen has been driven off by steaming or heating in a bath of boiling water. When inoculating with a pasteur pipette, the inoculum should be pipetted gently into the depths of the medium, care being taken not to introduce any air bubbles. Very thorough flaming of the pipette before use is most important; the use of unsterile pipettes is one of the commonest causes of culture contamination. Growth develops rapidly in fluid media that are inoculated while still warm, preferably at 37 °C, and are incubated in a water bath. Some strict anaerobes, such as some species of *Bacteroides* and *Fusobacterium*, and *Cl. novyi* type D will not grow, however, unless cultures are incubated in the anaerobic jar. By using an inoculum of higher specific gravity than the medium, very early basal growth of most clostridia is obtained, and this can be detected much sooner than diffuse growth (Willis, 1957). The specific gravity of the inoculum is easily adjusted by the addition of a small quantity of 10% sterile (air-free) glucose solution.

PREPARATION OF MEDIA

The general methods of preparation of media for the culture of anaerobes are the same as for other bacteriological media. The nutrient basis for most of the media used in my laboratory is a modification of the VL (Viande–Levure) broth medium of Beerens and Fievez (1971). Other useful basic media are Reinforced Clostridial Medium (Hirsch and Grinsted, 1954), Columbia Agar (Ellner *et al.*, 1966), and Tryptose Blood Agar Base (Difco). Basic broth media may be solidified by the addition of New Zealand agar; a concentration of 1.5% agar provides a medium with a satisfactory degree of solidity for plate cultures, 2% or more is used for concentrated agar media (*see* p. 48), and 0.5% or less for sloppy agar. Soft agar plating media, containing 0.75% New Zealand agar, may be used to advantage when it is necessary to 'maximize' the colonial size of anaerobes (Bullen, Willis and Williams, 1973). An agar concentration of 0.65% in agar shake cultures similarly allows the development of large deep colonies and facilitates their enumeration (Eller, Rogers and Wynne, 1967).

To these basic media, whether fluid or solid, may be added any required enrichment, selective or indicator substance that is compatible with bacterial growth.

Most routine laboratories will find it convenient to select one or two basic media (Reinforced Clostridial Medium and Columbia Agar are available commercially), from which most other media can then be prepared.

Agar plate media are poured in 90 mm plastic petri dishes. It is preferable to use vented dishes, otherwise an aerobic atmosphere may become trapped within individual dishes when a vacuum is applied to the anaerobic jar (*see* p. 8).

Although it is fashionable to dispense many bacteriological fluid media in screw-capped bottles, this is generally less appropriate for anaerobic work. Because it is impracticable for most clinical laboratories to use freshly prepared media each time anaerobic culture is required, stock fluid media is usually used, but it must be heated in a boiling water bath (or in steam) in order to drive off any dissolved oxygen. Subsequent rapid cooling in a cold water bath and immediate inoculation then ensures that cultures are placed under anaerobic conditions before significant amounts of atmospheric oxygen again dissolve in the medium. These essential procedures of heating and cooling are technically awkward with small screw-capped bottles (universals – 28 ml volume, and bijoux – 5 ml volume), and are associated with a high contamination rate of cultures and a high breakage rate of bottles. For these reasons, most fluid media for anaerobic work are best dispensed in 150 × 18 mm capped test tubes. Working cultures of anaerobes in cooked meat medium are conveniently held in similar test tubes, either capped or plugged with non-absorbent cotton wool. Stock cultures in cooked meat medium for short term holding, and for transmission by post should be contained in 28 ml screw-capped bottles.

SOME MEDIUM ADDITIVES

Selective agents

Selective agents, which are considered later in this chapter, may be added to solid plating media or to fluid enrichment media.

Glucose

Glucose, added to almost any medium, promotes more rapid growth of most anaerobic species. It must be remembered, however, that the growth of glucose-fermenting organisms is accompanied by the accumulation of acid metabolites in the medium, and this may cause 'self-sterilization' of the culture in a matter of hours. Clearly, broth base media containing glucose must not be used for fermentation tests, and it is best to omit glucose from deep shake and pour plate cultures, which may otherwise be disrupted by gas formation. Glucose must also be excluded from sensitivity test media since some antibiotics are inactivated by acid conditions.

pH Indicator solutions

pH indicator solutions are always added to the culture media *after* incubation, since most of these dyes are reduced under anaerobic conditions to colourless leuco-forms. In my laboratory, bromthymol blue is used for fermentation tests, and phenol red for urease activity.

There is generally no need to determine acid production by pH meter.

Reducing agents

Reducing agents, which are considered briefly in Chapter 1, may be added to virtually any compatible medium. Glucose is an excellent reductant, but has certain other limitations already discussed. The reducing agent most frequently used in my laboratory is cysteine hydrochloride; this may be added up to a maximum final concentration of 0.05% to almost any solid or fluid medium. It is the active reductant in cysteine hydrochloride–dithiothreitol solution (Moore, 1968; Collee, Rutter and Watt, 1971):

CYSTEINE–DITHIOTHREITOL SOLUTION

Cysteine hydrochloride	1 g
Dithiothreitol	0.1 g
Distilled water	10.0 ml

This solution is prepared immediately before use, and sterilized by filtration. It is added to any sterile, cooled medium in the proportion of 1 ml per 100 ml medium.

Resazurin

Resazurin is widely used in pre-reduced anaerobically sterilized (PRAS) media as an indicator of anaerobiosis. Under anaerobic conditions resazurin is colourless, but it rapidly turns pink on exposure to traces of oxygen. For this reason it is a useful addition to anaerobic blood culture media in the proportion of 1 mg resazurin per litre medium (*see* p. 60).

Sodium bicarbonate

Sodium bicarbonate may be added to any medium that is used for the culture of organisms whose growth is encouraged by carbon dioxide. There is no doubt that the growth of many anaerobes is enhanced by the presence of carbon dioxide in the atmosphere (*see* p. 8). Bacteria, however, may not assimilate carbon dioxide directly from the atmosphere, but as bicarbonate ions (HCO_3^-) formed by solution of carbon dioxide in the medium. Under these circumstances, it is clearly desirable to provide available bicarbonate ions in the medium which is then equilibrated in an appropriate atmosphere of carbon dioxide. For most solid and fluid media at about pH 7.2, and incubated in an atmosphere containing 10% carbon dioxide, the addition of 1 g $NaHCO_3$ per litre of medium is appropriate. The $NaHCO_3$, prepared as a sterile solution by filtration, is added to the cooled sterilized base medium just prior to dispensing. Alternatively, complete media for special purposes, e.g. blood culture, may be fully prepared with bicarbonate added, and held in stock in an atmosphere containg 10% carbon dioxide.

Bile

Bile (ox gall) stimulates the growth of some anaerobes, notably *Bacteroides fragilis*, but it is inhibitory towards that of others, especially some fusobacteria. The addition of 20% bile (2% dehydrated ox gall, Oxoid) to basic fluid and solid media used for the primary culture of pathological material, may be expected to enhance the growth of *B. fragilis*.

Vitamin K and haemin

Vitamin K and haemin are growth factors required for the successful culture of some strains of *Bacteroides melaninogenicus*. Consequently, these substances must be added to media on which *B. melaninogenicus* is to be grown, especially when pure culture studies are undertaken (*see* p. 178).

STOCK HAEMIN SOLUTION

Haemin	50 mg
1 M NaOH	1 ml
Distilled water	100 ml

Dissolve the haemin in the NaOH, add the water and sterilize by autoclaving at 115 °C for 15 min.

STOCK MENADIONE SOLUTION

Menadione	100 mg
Ethanol (95%)	20 ml

Dissolve and sterilize the solution by filtration.

VITAMIN K–HAEMIN SOLUTION

Add 1 ml of stock menadione solution to 100 ml of stock haemin solution.

This sterile solution is added to sterile base media in the proportion of 1%.

COOKED MEAT MEDIUM (Robertson, 1915–16)

To 1 litre boiling distilled water add 500 g minced fresh bullock's heart, horse meat or beef, and simmer for 1 h. Drain off the liquid and partly dry the cooked minced meat, first by pressing between layers of cloth or filter paper, and then by standing in the air, spread out on a cloth. Place the mince in 150 mm test tubes or in 28 ml screw-capped bottles so that there is about 25 mm meat in each. Cover the meat with 10–12 ml nutrient broth base (pH 7.4) and autoclave at 121 °C for 10 min. The liquid filtered from the meat may be used to prepare the nutrient broth. Minced liver broth is prepared in the same way.

This medium is suitable for the culture of all anaerobes. It is useful for enrichment culture, is excellent for the production of toxins by most clostridia, and is ideal for the preservation of many anaerobes as stock cultures. Clostridia remain viable in it for many months. The viability of stock cultures of non-sporing anaerobes in cooked meat medium is improved by the addition of 10% sterile glycerol.

Glucose cooked meat medium

Glucose cooked meat medium, prepared as above, but containing 0.2% glucose, gives more luxuriant growth of many anaerobes.

VL (VIANDE–LEVURE) BROTH

(Modified from Beerens and Fievez, 1971)

Tryptone (Oxoid)	10 g
NaC1	5 g
Meat extract (Lab-Lemco, Oxoid)	2 g
Yeast extract (Oxoid)	5 g
Cysteine HC1	0.4 g
New Zealand agar	0.6 g
Distilled water	1000 ml

Heat to dissolve, and adjust the pH to 7.2–7.4. Dispense and sterilize by autoclaving at 115 °C for 10 min.

This is a most useful basal broth medium from which an almost complete range of fluid and solid media for the identification of anaerobes can be prepared. In my laboratory it forms the basis of media for conventional fermentation reactions, for various agar plate media and for gas liquid chromatographic analysis of volatile products of metabolism.

VL glucose broth

This is VL broth prepared as above, but containing 0.2% glucose. It gives a more luxuriant growth of many anaerobic species.

REINFORCED CLOSTRIDIAL MEDIUM

(Hirsch and Grinstead, 1954: Gibbs and Hirsch, 1956)

Yeast extract (Oxoid)	3 g
Meat extract (Lab-Lemco, Oxoid)	10 g
Peptone (Oxoid)	10 g
Soluble starch	1 g
Glucose	5 g
Cysteine HC1	0.5 g
NaC1	5 g
Sodium acetate	3 g
Distilled water	1000 ml

Steam to dissolve and filter through paper pulp. Adjust to pH 7.4, dispense and sterilize by autoclaving at 121 °C for 15 min.

This broth medium is useful for enrichment culture of anaerobes. With the addition of 1%–1.5% New Zealand agar, it may be used directly as a plating medium, or may be used as the nutrient agar base for media such as fresh blood agar, milk agar and egg yolk agar.

Reinforced clostridial medium is available commercially in the dehydrated form.

COLUMBIA AGAR BASE

(Ellner *et al.*, 1966)

This is an excellent peptone agar base medium from which a variety of agar plate media may be prepared. It is improved by the addition of 0.05% cysteine hydrochloride at the time of preparation, prior to autoclaving. It is available in the dehyrated form from Oxoid Ltd.

FRESH BLOOD AGAR

This is a nutrient agar base to which sterile blood (commonly horse blood) is added. Typtose Blood Agar (Difco) and columbia Agar Base (Oxoid) are suitable base media, the anaerobic performance of which is improved by the addition of 0.05% cysteine hydrochloride before sterilization. VL broth, VL glucose broth and reinforced clostridial medium are also appropriate base media.

Prepare and sterilize 100 ml of the agar base medium. Cool to 50 °C in a water bath. Add 7 ml sterile defibrinated horse blood and mix gently to avoid bubbles. Pour into petri dishes. If layer plates are used, the nutrient agar layer should contain cysteine hydrochloride.

Sterile sheep, human or other blood may be used instead of horse blood. It should be remembered, however, that the pattern of haemolysis produced by different anaerobes is partly dependent on the type of erythrocytes used as the substrate. For example, *Clostridium chauvoei* blood agar medium of Batty and Walker (1966) employs sheep blood because horse erythrocytes are not attacked by *Cl. chauvoei* haemolysin.

Fresh blood agar is widely used for direct plating of pathological material, and for observation of colonial morphology and haemolytic activity. The growth of many anaerobic species is greatly improved by the presence of blood in the medium. The addition of 0.5% (final concentration) of $CaCl_2$ to horse blood agar encourages the development of 'target' haemolysis among strains of *Cl. perfringens*.

Cl. chauvoei blood agar
(After Batty and Walker, 1966)

VL broth base	94 ml
Liver extract (Oxoid)	3 g
Glucose	1 g
New Zealand agar	1.6 g
Defibrinated sheep blood	5 ml

Mix together the broth base, liver extract, glucose and agar, and sterilize by autoclaving at 115 °C for 10 min. Cool to 50 °C, add the sheep blood, mix gently and pour plates.

Cl. chauvoei grows poorly on ordinary blood agar, but good growth is obtained by supplementing the medium with liver extract. Sheep erythrocytes are susceptible to *Cl. chauvoei* haemolysin.

Cl. novyi (types B–D) blood agar
(NP medium of Moore, 1968)

Formula of basal medium

Neopeptone (Difco)	1.0 g
Yeast extract (Difco)	0.5 g
Proteolyzed liver (Pabryn Labs)	0.5 g
Glucose	1.0 g
New Zealand agar	2.0 g
Salts solution*	0.5 ml
Distilled water	100.0 ml

*Salts solution contains (g/1) $MgSO_4.7H_2O$, 40; $MnSO_4.4H_2O$, 2; $FeCl_3$ (anhydrous), 0.4; concentrated HCl, 0.5.

Preparation of basal medium

Dissolve the neopeptone, yeast extract, liver extract and glucose in 50 ml of the water, add the salts solution and adjust to pH 7.6–7.8. Mix this with a hot solution of agar, prepared from the agar and the remaining 50 ml water. Dispense in 18 ml volumes in 28 ml screw-capped bottles and sterilize by autoclaving at 115 °C for 10 min. Store the basal medium at 4 °C.

Preparation of the reducing solution

Cysteine hydrochloride	120 mg
Dithiothreitol	120 mg
Glutamine	60 mg
Distilled water	10 ml

Dissolve the constituents in the water, adjust to pH 7.6–7.8 and sterilize by filtration. This solution must be freshly prepared before use.

Preparation of the complete medium

Melt an 18 ml volume of the basal medium and cool to 50 °C. Add to it 2 ml defibrinated horse blood and 0.15 ml reducing solution, mix gently and pour plates immediately.

Comment

This medium facilitates luxuriant growth of fastidious clostridia such as *Cl. novyi* types B, C and D, and *Cl. botulinum* types C and D (Moore, 1968; Walker, Harris and Moore, 1971). The success of this medium for the culture of these fastidious organisms is largely attributable to the presence of cysteine *plus* dithiothreitol (Collee, Rutter and Watt, 1971). Under the usual conditions of medium preparation, pouring and inoculating, oxidation of cysteine in the medium proceeds rapidly, so that the cysteine is likely to be largely inactivated by the time cultures are set up. Dithiothreitol protects the free thiol group of cysteine from oxidation, thus preserving its reducing activity.

Although Moore's medium allows optimal growth of these fastidious organisms, they may be grown successfully on defibrinated horse blood agar prepared from a modified VL broth base. VL broth is prepared with the glucose and agar added, but without cysteine hydrochloride. Freshly prepared cysteine hydrochloride–dithiothreitol reducing solution (1 ml per 100 ml VL base) is added along with the horse blood immediately prior to pouring the plates.

HEATED BLOOD (CHOCOLATE) AGAR

To molten sterile nutrient agar base at about 65 °C add 10% of sterile defibrinated horse blood. After mixing, place the medium in a water bath at 75 °C for about 10 min, until it develops a rich chocolate-brown colour. Pour plates immediately. If the agar is too hot when the blood is added, denaturation occurs too rapidly and a coarse, curdy precipitate results. The same thing happens in the poured plates if glass dishes are still very hot from sterilization. Heated blood agar is a generally useful medium for the culture of most anaerobes. Cultures of anaerobes may produce partial clearing of the medium due to proteolytic activity, or slight bleaching due to hydrogen peroxide production (*see* p. 77).

BENZIDINE HEATED BLOOD AGAR
(Gordon and McLeod, 1940)

Proceed as for heated blood agar. Immediately before pouring the plates, add to the medium a solution of benzidine in the proportion of 1 ml to 11 ml of medium. The benzidine solution is as follows:

Benzidine	0.25 g
1M HC1	0.3 ml
Distilled water	50 ml

The solution is sterilized by boiling. Some difficulty may be experienced in producing a satisfactory medium owing to the low solubility of benzidine.

This medium is used to detect hydrogen peroxide formation by cultures of anaerobes exposed to air, and has been used as an indicator medium for *Cl. novyi* (*see* p. 77). Because benzidine is a carcinogenic substance, this medium should not be used for routine purposes.

EGG YOLK AGAR AND HUMAN SERUM AGAR

Human serum agar
(Hayward, 1941, 1943)

To sterile molten VL broth base containing 1.2% New Zealand agar, cooled to 50–55 °C, add 20% sterile human serum. Mix and pour plates immediately.

This medium may be used as a half-antitoxin plate, and shows diffuse opalescence with lecithinase C producing organisms. Except in the case of *Cl. novyi* type A, bacterial lipolysis is not produced on this medium (*see* p. 85).

Egg yolk agar
(McClung and Toabe, 1947)

Prepare an egg yolk emulsion by mixing egg yolk with an equal volume (about 20 ml per egg yolk) of sterile normal saline solution. The egg yolk is separated from the white by the usual culinary techniques; no special aseptic precautions are necessary. Add the egg yolk emulsion to sterile molten VL agar base (at 50–55 °C) in the proportion of 10%. Pour plates immediately.

This medium is used in the same way as human serum agar, and provides the same information about clostridia. In addition, however, all lipolytic organisms produce a restricted opacity and pearly layer on egg yolk media. Lecithinase reactions on egg yolk agar are much stronger and are more consistent than on human serum agar.

Lactose egg yolk milk agar
(Willis and Hobbs, 1959)

Prepare a basal medium by mixing together 100 ml VL broth, 1.2 g New Zealand agar, 1.2 g lactose and 0.3 ml of a 1% solution of neutral red. Autoclave the mixture at 121 °C for 15 min. Cool this base medium to 50–55 °C, and add to it 3.5 ml egg yolk emulsion (*see* above) and 15 ml sterile skimmed milk (*see below*). Mix gently and pour plates immediately.

In addition to serving as a half-antitoxin egg yolk plate, this medium also shows bacterial lipolysis, lactose fermentation and proteolytic effects (*see* Willis and Hobbs, 1959).

Selective egg yolk media
(Willis and Hobbs, 1959)

By adding neomycin sulphate (Upjohn) in the proportion of 50–100 μg/ml to the above egg yolk media, they are rendered partially selective for anaerobes. An aqueous solution of neomycin sulphate is added to the cooled basal medium just before pouring the plates. Alternatively, a few drops of a 1% aqueous solution of neomycin sulphate may be spread over the surface of a neomycin-free plate medium before inoculation (*see* Hobbs, 1960).

MILK MEDIUM AND LITMUS MILK

Ordinary fresh milk is centrifuged in order to separate the cream; the milk is pipetted off, distributed in 10 ml amounts in test tubes, and autoclaved. The preparation of litmus milk is similar except that 2.5% of an alcoholic solution of litmus is added to the milk before it is tubed off. These media may be used in testing for lactose fermentation, clotting and digestion of milk. Although they are little used in modern diagnostic work, anaerobic cultures in fluid milk do demonstrate some important principles.

MILK AGAR

(Reed and Orr, 1941)

Melt the required amount of VL agar base which contains 1.25% of agar. Cool to 55 °C and add sterile skimmed milk to a concentration of 25%. Pour plates immediately.

Skimmed milk is prepared from fresh cow's milk by centrifugation in order to separate the cream. The milk is pipetted off, distributed in 20 ml volumes in screw-capped bottles, and autoclaved at 121 °C for 10 min.

This medium is used for the detection of proteinase activity and for the demonstration of rennin-like activity of the culture.

CASEIN AGAR

Casein agar is an alternative to milk agar. It is prepared as described for milk agar, except that a sterile aqueous solution of pure soluble casein (BDH) is used instead of sterile skimmed milk. To each 100 ml VL agar base add 5 ml of 8% casein solution that has been sterilized by autoclaving at 115 °C for 10 min. Proteolytic activity only is demonstrated on this medium.

GELATIN AGAR

(Frazier, 1926)

Prepare the required amount of VL agar base and add to it 0.4% gelatin. Sterilize by autoclaving at 115 °C for 10 min, and pour plates. This medium is used for the detection of gelatinase activity.

DEOXYRIBONUCLEASE AGAR (DNAse AGAR)

Prepare the required amount of VL agar base and add to it 0.2% of thymus gland sodium deoxyribonucleate (BDH). Autoclave at 115 °C for 10 min and pour plates.

This medium is used for the detection of deoxyribonuclease activity.

TRIBUTYRIN AGAR

Melt the required amount of VL agar base and add to it 0.2% glycerol tributyrate (BDH). Emulsify the hot mixture with a Silverson mixer for about 3 min. Sterilize by autoclaving at 115 °C for 10 min, and pour plates.

This medium is used for the detection of lipolytic activity

CONCENTRATED (FIRM) AGAR

Concentrated agar media (Miles, 1943) are used to obtain discrete colonies of organisms that otherwise tend to spread over the surface of solid media, obscuring and contaminating colonies of other organisms. The isolation of anaerobes from wounds is often complicated by the presence of *Proteus* species, and by swarming anaerobic species such as *Cl. tetani*. All organisms grow well on concentrated agar (2% or more New Zealand agar in any suitable medium) and swarming does not occur. The only disadvantage is that the colonial characteristics of most micro-organisms are rendered atypical, and colonies are always smaller than an ordinary agar media.

Swarming growth of *Cl. tetani* and *Cl. septicum* can be specifically prevented by treatment of ordinary agar media with appropriate antiserum before inoculation (*see* pp. 115 and 150).

MEDIA FOR SPORULATION OF CL. PERFRINGENS

Ellner's Medium

(Ellner, 1956)

The formula of the medium is as follows:

Peptone (Evans)	1.0 g
Yeast extract (Difco)	0.3 g
Starch	0.3 g

$MgSO_4$	0.01 g
KH_2PO_4	0.15 g
$Na_2HPO_4.7H_2O$	5.0 g
Distilled water	100.0 ml

Dissolve the ingredients in the water, and adjust to pH 7.8. Dispense, and sterilize by autoclaving at 121 °C for 15 min.

Medium of Duncan and Strong (1968)

The formula of the medium is as follows:

Proteose peptone	1.5 g
Yeast extract (Difco)	0.4 g
Starch	0.4 g
Sodium thioglycollate	0.1 g
$Na_2HPO_4.7H_2O$	1.0 g
Distilled water	100.0 ml

Dissolve the ingredients in the water, and adjust to pH 7.2. Dispense, and sterilize by autoclaving at 121 °C for 15 min.

These media are used to promote sporulation of *Cl. perfringens*. Before use tubes of the medium are steamed to drive off dissolved oxygen, and after rapid cooling the medium is inoculated with about 0.5 ml of a heavy suspension of organisms obtained from an actively growing broth culture. Very little growth occurs in Ellner's medium. Strains that sporulate usually do so abundantly. In both of these media it seems likely that the starch plays an important part in promoting sporulation (Labbe and Duncan, 1975; Labbe, Somers and Duncan, 1976).

FERMENTATION MEDIA

Tube method

Sterile solutions of the various fermentable substances are added to tubes of sterile VL broth base to a concentration of 1%. The different fermentable substances are identified by the use of coloured test tube caps. A few drops of indicator, for example, 0.4% solution of bromthymol blue (yellow at pH 6.0, blue at pH 7.6) are added to the cultures *after* incubation. This is necessary, since most indicators are reduced under anaerobic conditions to colourless leuco-forms.

Aesculin hydrolysis is determined in VL broth containing 0.5% aesculin. A few drops of 1% ferric ammonium citrate are added to the culture after incubation; development of a black colour indicates hydrolysis.

Agar plate method
(Phillips, 1976)

Spread 0.5 ml of a sterile 20% solution of the fermentable substrate over the surface of a dried horse blood agar plate. Allow the solution to be absorbed, re-drying if necessary. Spot inoculate the test cultures onto the plate (each plate conveniently accommodates 6 cultures) and incubate anaerobically for 48 h; a control horse blood agar plate should always be included.

Cultures that have hydrolysed the fermentable substance are surrounded by zones of discoloration of the medium. To confirm fermentation, remove a plug of the agar medium from the area of growth, and expose it to a few drops of bromthymol blue indicator solution on a glass microscope slide. A short length of drinking straw makes a convenient 'hollow tyne' for 'biopsy' of the medium.

To check for aesculin hydrolysis, flood the plate with ferric ammonium citrate solution. The medium surrounding aesculin-positive cultures turns brown-black.

BILE BROTH

This is VL broth containing 1% glucose and 20% (v/v) ox gall (Oxoid), dispensed in capped test tubes. It is used to determine if growth is inhibited or stimulated by bile.

HYDROGEN SULPHIDE PRODUCTION

A lead acetate strip is wedged in the top of a culture tube of VL broth, VL glucose broth or cooked meat medium.

PREPARATION OF LEAD ACETATE STRIPS

Cut white filter paper strips about 5 × 50 mm in size, and soak them in a saturated solution of lead acetate. Sterilize the strips in cotton wool plugged test tubes and dry in a hot air oven at 120 °C.

INDOLE PRODUCTION

Indole is tested for in cultures growing in Trypticase Nitrate Broth (Indole Nitrate Medium, BBL). On the addition of Kovac's reagent to an aliquot of the culture, a dark red colour in the amyl alcohol surface layer constitutes a positive indole test.

KOVAC'S REAGENT

Paradimethylaminobenzaldehyde	5 g
Amyl alcohol	75 ml
Hydrochloric acid (concentrated)	25 ml

TEST

Add 0.2 ml Kovac's reagent to a 3 ml volume of the culture and shake. Allow to stand for 10 min and read.

NITRATE REDUCTION

Nitrate reduction is tested for in cultures growing in Trypticase Nitrate Broth (Indole Nitrate Medium, BBL). On the addition of sulphanilic acid and Cleve's acid (Crosby, 1967) to an aliquot of the culture, a pink colour developing in the culture indicates the presence of nitrite.

REAGENT A

Sulphanilic acid (Analar)	0.5 g
Glacial acetic acid (Analar)	30.0 ml
Distilled water	120.0 ml

REAGENT B

1-napthylamine-7 sulphonic acid (Cleve's acid)	0.2 g
Glacial acetic acid (Analar)	30.0 ml
Distilled water	120.0 ml

Add the water to the Cleve's acid and warm in a water bath. Filter, cool and add the acetic acid.

TEST

Add 2 ml reagent A followed by 2 ml reagent B to a 5 ml volume of the culture. Mix, allow to stand for 10 min and read.

Since absence of nitrite in the presence of good growth may indicate reduction of nitrate beyond the nitrite stage, a negative result must be followed by a test for nitrate. Nitrate is tested for by adding a little zinc dust to the tube to which the nitrite reagents were previously added. If nitrate is present it will be reduced to nitrite with the development of the characteristic pink colour within a few minutes.

CATALASE PRODUCTION

Catalase production may be tested for on any ordinary plate culture that does not contain fresh blood in the medium. Cultures on heated blood agar or on egg yolk agar are appropriate.

Expose the culture to the air for 30 min, then pipette a little 3% hydrogen peroxide onto the bacterial growth. Evolution of gas bubbles indicates catalase activity.

The blue slide catalase test of Thomas (1963, 1974) is a useful alternative. Flame and cool a coverslip, and smear its centre with a colony picked from a catalase-negative medium. Invert the coverslip onto a drop of blue peroxide (equal volumes of 20 volume hydrogen peroxide and of Loeffler's methylene blue solution) on a glass slide. Clear bubbles form around catalase-positive colonies within 30 seconds. A drop of suitable fluid culture may be used instead of a picked colony.

UREA MEDIUM

To 100 ml of sterile VL broth base add aseptically 5 ml of 40% urea solution (Oxoid), mix well and distribute in sterile capped test tubes. On the addition of phenol red (yellow at pH 6.8, red at pH 8.4) to the culture, a pink colour indicates hydrolysis of urea.

THE API MICROSYSTEM FOR ANAEROBES

Following the successful introduction of their multitest microsystem for the biochemical characterization of the *Enterobacteriaceae,* Analylab

Products Inc. (USA) have developed a similar system for use with obligate anaerobes (the API 20 Anaerobe System). In essence, the system consists of 20 biochemical tests presented in individual micro-containers pressed into a single strip of plastic of size 220 mm × 30 mm. The dehydrated test substrates and indicators are simultaneously reconstituted and inoculated by the addition of a suspension of the test organism in a nutrient suspension medium. Owing to their length, the complete units can be accommodated in a Gaspak jar only by placing the jar horizontally; the complete units are too long to fit into the BTL and standard Whitley jars.

The 20 biochemical tests at present included on the strip are indole, urease, glucose, mannitol, lactose, sucrose, maltose, salicin, xylose, arabinose, gelatin, aesculin, glycerol, cellobiose, mannose, melezitose, raffinose, sorbitol, rhamnose and trehalose.

Published evaulations show that the microsystem gives reliable results except for indole and gelatin which are too insensitive (Starr *et al.*, 1973; Moore, Sutter and Finegold, 1975; Nord, Dahlback and Wadstrom, 1975).

Although the API system allows performance of biochemical tests in those laboratories that do not have adequate medium facilities, it will be appreciated that for the identification of most clinically important anaerobes the API strip must be supplemented with other tests such as egg yolk reactions, antibiotic sensitivity patterns and gas liquid chromatography. Moreover, the strips are relatively expensive, and, like all multi-test methods, the strip is inflexible–a whole strip of 20 tests must be sacrificed when only a few reactions are required.

In the United Kingdom the API 20 Anaerobe System is available from API Laboratory Products Ltd, Philpot House, Rayleigh, Essex.

SELECTIVE AGENTS

In order to facilitate the separation of anaerobic species from mixtures with one another, and from mixtures with facultative anaerobes, various substances may be incorporated into fluid or solid media. Some of these selective agents are more generally useful than others, but none of them is ideal. The main difficulty with selective agents is that they are often partially inhibitory towards the growth of the organisms that they are used to select. This is of little consequence if they are used exclusively as an aid to the purification of anaerobic cultures, where the aim is a purely qualitative one.

Selective agents may also be usefully employed in parallel with routine culture media for the isolation of anaerobes from pathological material, but *they must never replace the use of routine media*. Like enrichment techniques, primary selective anaerobic culture almost always gives a grossly distorted picture of the bacterial flora, from which equally distorted clinical conclusions are likely to be drawn.

Gentian violet

Gentian violet, in a concentration of 1 in 10^5 in broth media, was suggested by Hall (1919, 1920) as a means of eliminating aerobic spore-forming bacilli from anaerobic cultures. For a similar purpose, Spray (1936) recommended the use of crystal violet. While gentian violet is effective up to a point, the growth of many anaerobes is inhibited by the dye, which, in addition, has little inhibitory effect on the growth of Gram-negative facultative organisms. Slanetz and Rettger (1933) recommended the use of 1 in 10^4 gentian violet in plate cultures for the isolation of fusobacteria, while Omata and Disraely (1956) found that a mixture of crystal violet (1 in 10^5) and streptomycin sulphate (10 μg/ml) in plate cultures facilitated isolation of these organisms.

Sodium azide

Sodium azide was used by Johansson (1953) as a selective agent for clostridia. Using 0.02% of the salt in egg yolk agar plates, he found that the growth of aerobic spore-forming bacilli was completely inhibited, while that of clostridia was unaffected. Wetzler and his colleagues (1956b) used a selective blood agar medium for clostridia, which contained sodium azide (0.02%) and chloral hydrate (2 ppm). Although it was selective, and prevented swarming anaerobic growth, it caused atypical colonies, retarded haemolysis and induced phase variation.

A combination of sodium azide with brilliant green or bile was introduced by Beerens and Tahon-Castel (1965) and Beerens and Fievez (1971). Sodium azide bile agar is selective for some species of bacteroides, while sodium azide brilliant green agar is used for the isolation of bacteroides and fusobacteria (*see below*, and *Table 2.1*).

Ellner and O'Donnell (1971) recommended a combination of sodium azide (0.02%) with neomycin sulphate (150 μg/ml) for the selective culture of histotoxic clostridia on agar plate media.

Table 2.1 Selective agents for anaerobic bacteria

Agents	*Amounts/100 ml of medium*	*Organisms selected*	*Reference*
Crystal violet Streptomycin	1.0 mg 1.0 mg	*Fusobacterium*	Omata and Disraely (1956)
Sodium azide	20.0 mg	*Clostridium*	Johansson (1953)
Sodium azide Bile (ox gall)	20.0 mg 1.7 g	Some *Bacteroides*	Beerens and Fievez (1971)
Sodium azide Brilliant green	30.0 mg 1.8 mg	*Bacteroides* and *Fusobacterium*	Beerens and Fievez (1971)
Sorbic acid	0.12 g	*Clostridium*	Emard and Vaughn (1952)
Sorbic acid Polymyxin B	0.12 g 2.0 mg	*Clostridium*	Wetzler, Marshall and Cardella (1956a)
Phenylethyl alcohol	0.25 g	*Clostridium*, *Bacteroides*, *Fusobacterium*	Dowell, Hill and Altemeier (1964)
Neomycin	10.0 mg	*Clostridium*, anaerobic cocci, *Bacteroides*	Lowbury and Lilly (1955) Willis and Hobbs (1959)
Kanamycin	7.5 mg	Gram positive non-sporing anaerobic bacilli	Finefold, Sugihara and Sutter (1971)
Kanamycin Vancomycin	10.0 mg 750 μg	*Bacteroides* and *Fusobacterium*	Finegold, Sugihara and Sutter (1971)
Neomycin Vancomycin	10.0 mg 750 μg	*Fusobacterium* and *Veillonella*	Finegold, Sugihara and Sutter (1971)
Oleandomycin Polymyxin Sulphadiazine	50 μg 1000 units 10.0 mg	*Cl. perfringens*	Handford (1974)

Sorbic acid

Sorbic acid (0.12%) in broth and solid media was used by Emard and Vaughn (1952) for the isolation of clostridia and lactobacilli. This agent was more selective for clostridia in fluid media when polymyxin B sulphate (20 μg/ml) was added (Wetzler, Marshall and Cardella, 1956a).

Phenylethyl alcohol

Lilley and Brewer (1953) noted that anaerobes could be isolated from mixtures with Gram-negative facultative organisms such as *Proteus* species by culturing mixtures on media containing phenylethyl alcohol. Dowell, Hill and Altemeier (1964) extended this general observation to a variety of anaerobic species, and showed that 0.25% of the alcohol incorporated in fluid or solid media was a useful selective agent for many non-clostridial anaerobes.

Paracresol

Hafiz and Oakley (personal communication) (*see* Hafiz and his colleagues, 1975) introduced paracresol as an agent for the selective isolation of *Cl. difficile*. Paracresol (0.2%) is added to an appropriate enrichment broth, such as reinforced clostridial medium.

Antibiotics

Antibiotics have been used by a number of workers as selective agents for anaerobes. Polymyxin B sulphate was first used for the isolation of clostridia by Hirsch and Grinsted (1954). Neomycin sulphate was introduced by Lowbury and Lilly (1955) for the isolation of *Cl. perfringens* type A, and it was later used for the isolation of most of the common clostridia (Willis and Hobbs, 1959). Finegold, Siewert and Hewitt (1957) used a mixture of neomycin (200 μg/ml) and vancomycin (7.5 μg/ml) in fresh blood agar for the isolation of bacteroides species. Streptomycin sulphate has been used for the isolation of fusobacterial species (Omata and Disraely, 1956) and clostridia (Willis, 1957).

A range of antibiotic selective media for the isolation of anaerobes of human origin, many of which were developed by Finegold and his colleagues, have been usefully reviewed by Finegold, Sugihara and Sutter (1971) (*see below* and *Table 2.1*).

Note that although selective agents are commonly incorporated into solid media, they may also be used very effectively in fluid enrichment cultures.

SELECTIVE MEDIA

Phenylethyl alcohol blood agar

(Dowell and his colleagues, 1964)

This is ordinary horse blood agar that contains 0.25% of β-phenylethyl alcohol. The alcohol is added to the nutrient agar base before sterilization. This medium is useful for the isolation of a variety of anaerobic species, especially the non-sporing anaerobes and poorly heat resistant clostridia.

Sodium azide bile blood agar

(After Beerens and Fievez, 1971)

VL glucose broth	100.0 ml
Ox gall (Oxoid)	1.7 g
Sodium azide	0.02 g
New Zealand agar	1.2 g
Defibrinated horse blood	10.0 ml

To 100 ml VL glucose broth, add the ox gall, sodium azide and agar. Sterilize by autoclaving at 115 °C for 10 min. Cool to 50 °C, add the horse blood and pour plates.

This is a selective medium for some species of bacteroides.

Sodium azide brilliant green blood agar

(After Beerens and Fievez, 1971)

VL glucose broth	100.0 ml
Sodium azide	0.03 g
Brilliant green	0.0018 g
New Zealand agar	1.2 g
Defibrinated horse blood	10.0 ml

To 100 ml VL glucose broth add the sodium azide, brilliant green and agar. Sterilize by autoclaving at 115 °C for 10 min. Cool to 50 °C, add the horse blood, mix gently and pour the plates.

It is convenient to prepare the sodium azide and brilliant green as a stock solution:

Sodium azide	0.3 g
Brilliant green	0.018 g
Distilled water	100.0 ml

Add 10 ml of this solution to the VL glucose broth base (with water content adjusted) before sterilization.

This medium is selective for bacteroides and fusobacteria.

Neomycin blood agar

This is ordinary horse blood agar that contains 100 μg/ml of neomycin (0.5 g neomycin sulphate ≡ 0.35 g neomycin). Neomycin is heat-stable and may be added to the agar base medium before sterilization. This medium is selective for clostridia and anaerobic cocci. For the isolation of clostridia, egg yolk agar containing neomycin is often of value.

Kanamycin blood agar
(Finegold, Sugihara and Sutter, 1971)

This is ordinary horse blood agar that contains 75 μg/ml of kanamycin (1 g kanamycin sulphate ≡ 1 g kanamycin base). Like neomycin, kanamycin is heat-stable, and may be added to the agar base medium before sterilization. This medium is selective for Gram-positive non-sporing anaerobic bacilli.

Kanamycin vancomycin blood agar
(Finegold, Sugihara and Sutter, 1971)

This is ordinary horse blood agar that contains 100 μg/ml of kanamycin and 7.5 μg/ml of vancomycin. The kanamycin is added to the base medium before sterilization. Vancomycin hydrochloride (intravenous) is dissolved in distilled water and is added with the blood to the cooled agar base. This medium is selective for bacteroides and fusobacteria.

Neomycin vancomycin blood agar
(Finegold, Sugihara and Sutter, 1971)

This is ordinary horse blood agar that contains 100 μg/ml of neomycin and 7.5 μg/ml of vancomycin. The neomycin is added to the base medium before sterilization; vancomycin is added as a sterile solution along with the blood. This medium is selective for fusobacteria and veillonellae.

MEDIA FOR ANAEROBIC BLOOD CULTURE

The chief requirement of a blood culture medium for anaerobic bacteria is that it should support early and rapid growth of the most exacting species during incubation under ordinary atmospheric conditions. For the investigation of bacteraemic states many laboratories adopt the 'shot-gun' technique of blood culture, in which standard *sets* of blood culture media are inoculated. The aim of this technique is to cover the whole spectrum of possible offending organisms from *Brucella abortus* and *Pseudomonas aeruginosa* through the more commonly encountered facultative species such as streptococci, staphylococci, *Escherichia coli* and *Neisseria meningitidis*, to the exacting anaerobes such as the fusobacteria (*see* Stokes, 1974). The present text is concerned primarily with anaerobic blood culture.

There is undoubtedly a small group of patients in whom multiple cultural methods of the sort advocated by Stokes (1974) are necessary. Much more commonly, however, the clinical history, symptoms and signs provide a firm guide to the nature of the bacteraemia, as indeed is the case with most anaerobic bacteraemias. It is the duty of the clinical microbiologist to inform himself of the clinical condition of all patients from whom blood cultures are made, for it is only with this knowledge that he can initiate discriminating studies in the laboratory, and offer balanced advice to his clinical colleagues on empirical antimicrobial therapy.

With the upsurge of interest in non-clostridial anaerobic infections, and the recognition that they are prone to progress to a bacteraemic state, a good deal of attention has been paid recently to the effectiveness of different anaerobic blood culture media (Sonnenwirth, 1973; Bartlett, 1973; Finegold, 1973; Washington, 1973; Rosner, 1973; Bartlett, Ellner and Washington, 1974; Forgan-Smith and Darrell, 1974; Gantz *et al.*, 1974; Shanson, 1974; Stokes, 1974; Shanson and Barnicoat, 1975; Washington, 1975). With the findings of these various studies in mind, and on the basis of experience in my laboratory, the following considerations are of importance when choosing a blood culture medium for anaerobic bacteria.

The medium should be easy to prepare, preferably from a commercially available base, and it must have a long shelf storage life. Clearly, it must support the growth of exacting anaerobic species, preferably without resorting to incubation in the anaerobic jar. It is an advantage (but not essential) if the medium is clear, so that delayed bacterial growth may be detected as the blood cells settle. The medium should contain an Eh indicator, such as resazurin, which must remain colourless throughout its shelf life and its subsequent use as a culture; clearly, the addition of blood to the medium should not neutralize its reduction effect (*see* Zwarun, 1975). It is a great advantage if the medium contains antibiotic inhibitors, especially antagonists for penicillin, sulphonamide and the aminoglycosides. Whether or not polyanethol sulphonate should be added to anaerobic blood culture media remains a matter of debate. Although there is no doubt that it facilitates the isolation of many bacteria, including many anaerobic species, the substance is inhibitory towards some anaerobic cocci (Hoare, 1939; Kocka, Arthur and Searcy, 1974; Shanson, 1974).

Thiol Broth (Difco)

This commercially available dehydrated medium fulfils most of the requirements of an anaerobic blood culture base medium outlined above (Shanson and Barnicoat, 1975; Szawatkowski, 1976). For the successful culture of most facultative organisms and obligate anaerobes, it is sufficient to prepare the medium according to the manufacturer's instructions. The performance of the medium for the isolation of anaerobes may be marginally improved by preparing it as a PRAS medium.

Preparation of thiol broth as a PRAS medium

Prepare the required amount of thiol broth medium (according to the manufacturer's instructions) in a large conical flask of suitable size. Add resazurin (0.001 g per litre medium) and glucose (10 g per litre medium), and heat the medium to boiling until the resazurin has decolorized. Fit an anaerobic dispensing apparatus (Willingham and Oppenheimer, 1964) into the mouth of the flask (*Figure 2.2*), turn off the heat, and slowly bubble oxygen-free carbon dioxide through the medium until ambient temperature is reached. Dispense the prereduced medium in 100 ml volumes into screw-capped blood culture bottles as follows.

Thoroughly flush the empty blood culture bottle from the bottom with oxygen-free carbon dioxide escaping from the apparatus through the delivery tube (D); the rubber washer (W) serves as a valve for the escaping gas. Dispense the medium into the flushed bottle by pinching tube A closed; the subsequent slow release of tube A discontinues the dispensing and recommences the gas flushing. Rapidly firmly cap the

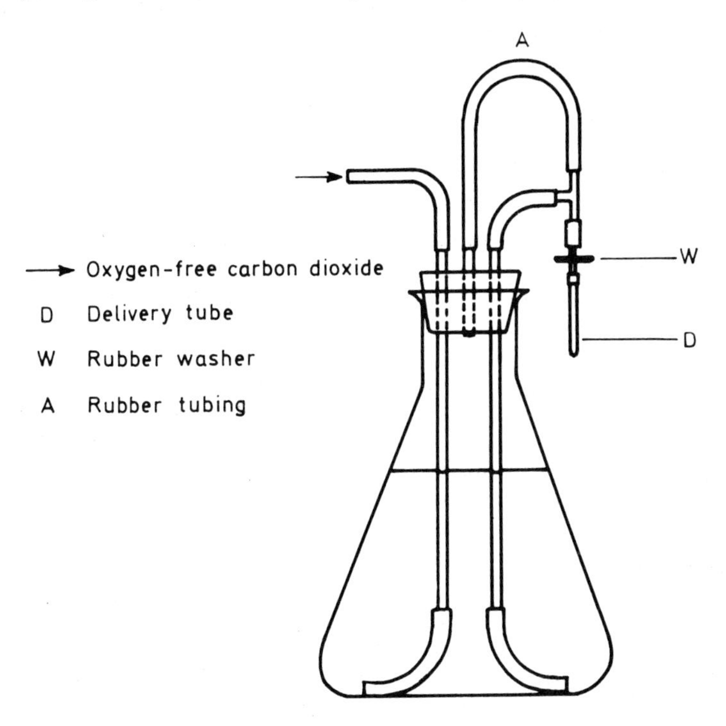

Figure 2.2. Apparatus for dispensing anaerobic medium (Willington and Oppenheimer, 1964)

medium bottle after its slow removal from the delivery tube (D). If anaerobic dispensing has been successful, the medium in each blood culture bottle remains colourless; in cases of failure the medium turns pink.

Sterilize the prereduced bottles of medium by autoclaving at 121 °C for 15 min, and allow to cool. To each 100 ml bottle of medium add aseptically with a sterile syringe and needle 1 ml of sterile vitamin K–haemin solution (*see* p. 39), and 1 ml of a sterile 10% sodium bicarbonate solution.

This PRAS thiol broth medium should remain colourless. Bottles of medium that become pink during storage should be discarded.

STOCK CULTURES OF ANAEROBES

Freeze-drying of anaerobic cultures is the most satisfactory method of preserving them (Lapage and Redway, 1974). However, the preparation of freeze-dried material requires special and expensive apparatus, which is not available in most laboratories.

Cooked meat medium

Preservation of clostridia in cooked meat medium is simple and reliable, and has been used successfully in my laboratory for many years. Inoculated screw-capped bottles of cooked meat medium are incubated until obvious growth has developed (usually 24 h). Screw caps on bottles are tightened, and the cultures are stored at room temperature in the dark.

For non-clostridial anaerobes, cooked meat medium cultures in screw-capped bottles are incubated for 3 days. Sterile glycerol is then added to each culture to give a final concentration of glycerol of 10%. Caps are tightened and the cultures stored at −20 °C.

Brain medium

Professor L. DS. Smith (personal communication) devised a complex and unlikely storage medium for anaerobes. Its preparation is arduous, but its performance excellent.

1. STERILE BEEF BRAIN STOCK

Chop 250 g beef brain and place in a conical flask. Add enough distilled water to cover. Autoclave at 115 °C for 15 min and store at 4 °C.

2. PEPTONE LIVER BROTH

Mince 500 g fresh beef liver, add 1 litre distilled water and infuse at 4 °C overnight. Heat to boiling to coagulate the protein, cool and clarify by filtration through 'super cell'. To the filtrate add 1 g peptone (Evans), and adjust to pH 7.6–7.8.

3. THE COMPLETE MEDIUM

To the sterile beef brain stock add 1 egg yolk and homogenize in a blender. Mis the homogenate with the liver broth in the proportion 1 to 3, and dispense into bijoux bottles, three-quarters filling them. To each bottle add 1 or 2 small pieces of oyster shell and a wisp of steel wool. Sterilize by autoclaving at 121 °C for 15 min.

4. STORAGE OF CULTURES

A heavy suspension of the culture for storage is inoculated into a bottle of the brain holding medium, which is then stored at–20 °C.

Rejuvenation of stock cultures is effected by subculturing from the holding medium to agar plate media and to cooked meat medium.

REFERENCES

Bartlett, R. C. (1973). 'Contemporary blood culture practices.' In *Bacteremia, Laboratory and Clinical Aspects*. p. 15. Ed. by A. C. Sonnenwirth. Springfield; Thomas

Bartlett, R. C., Ellner, P. D. and Washington, J. A. (1974). 'Blood cultures.' *American Society for Microbiology*, Cumitech 1

Batty, I. and Walker, P. D. (1966). 'Colonial morphology and fluorescent labelled antibody staining in the identification of species of the genus *Clostridium*.' In *Identification Methods for Microbiologists*, Society for Applied Bacteriology Technical Series No. 1. p. 81. Ed. by B. M. Gibbs and F. A. Skinner. London; Academic Press

Beerens, H. and Fievez, L. (1971). 'Isolation of *Bacteroides fragilis* and *Sphaerophorus-Fusiformis* groups.' In *Isolation of Anaerobes*; Society for Applied Bacteriology Technical Series No. 3. p. 109. Ed. by D. A. Shapton and R. G. Board. London; Academic Press

Beerens, H. and Tahon-Castel, M. (1965). *Infections humaines à bactéries anaérobies non-toxigénes*. Brussels; Presses Academiques Europeenes

Bullen, C. L., Willis, A. T. and Williams, K. (1973). 'The significance of bifidobacteria in the intestinal tract of infants.' In *Actinomycetales: Characteristics and Practical Importance*. p. 311. Ed. by G. Sykes and F. A. Skinner. London; Academic Press

Collee, J. G., Rutter, J. M. and Watt, B. (1971). 'The significantly viable particle: A study of the subculture of an exacting sporing anaerobe.' *Journal of medical Microbiology*, **4**, 271

Crosby, N. T. (1967). 'The determination of nitrite in water using Cleve's acid, 1-naphthylamine-7-sulphonic acid.' *Proceedings of the Society for Water Treatment and Examination*, **16**, 51

Dowell, V. R. and Hawkins, T. M. (1974). *Laboratory Methods in Anaerobic Bacteriology*. Public Health Service Publication No. 1803: Washington; US Government Printing Office

Dowell, V. R., Hill, E. O. and Altemeier, W. A. (1964). 'Use of phenethyl alcohol in media for isolation of anaerobic bacteria.' *Journal of Bacteriology,* **88,** 1811

Duncan, C. L. and Strong, D. H. (1968). 'Improved medium for sporulation of *Clostridium perfringens*.' *Applied Microbiology,* **16,** 82

Eller, C., Rogers, L. and Wynne, E. S. (1967). 'Agar concentration in counting *Clostridium* colonies.' *Applied Microbiology,* **15,** 55

Ellner, P. D. (1956). 'A medium promoting rapid quantitative sporulation in *Clostridium perfringens*.' *Journal of Bacteriology,* **71,** 495

Ellner, P. D. and O'Donnell, E. D. (1971). 'A selective differential medium for histotoxic clostridia.' *American Journal of clinical Pathology,* **56,** 197

Ellner, P. D., Stoessel, C. J., Drakeford, E. and Vasi, F. (1966). 'A new culture medium for medical bacteriology.' *American Journal of clinical Pathology,* **45,** 502

Emard, L. O. and Vaughn, R. H. (1952). 'Selectivity of sorbic acid media for the catalase-negative lactic acid bacteria and clostridia.' *Journal of Bacteriology,* **63,** 487

Finegold, S. M. (1973). 'Early detection of bacteremia.' In *Bacteremia; Laboratory and Clinical Aspects*. p. 36. Ed. by A. C. Sonnenwirth. Springfield; Thomas

Finegold, S. M., Siewert, L. A. and Hewitt, W. L. (1957). 'Simple selective media for *Bacteroides* and other anaerobes.' *Bacteriological Proceedings,* **57,** 59

Finegold, S. M., Sugihara, P. T. and Sutter, V. L. (1971). 'Use of selective media for isolation of anaerobes from humans.' In *Isolation of Anaerobes,* Society for Applied Bacteriology Technical Series No. 5. p. 99. Ed. by D. A. Shapton and R. G. Board. London; Academic Press

Forgan-Smith, W. R. and Darrell, J. H. (1974). 'A comparison of media used *in vitro* to isolate non-sporing Gram-negative anaerobes from blood.' *Journal of clinical Pathology,* **27,** 280

Frazier, W. C. (1926). 'A method for the detection of changes in gelatin due to bacteria.' *Journal of infectious Diseases,* **39,** 302

Gantz, N. M., Medeiros, A. A., Swain, J. L. and O'Brien, T. F. (1974). 'Vacuum blood-culture bottles inhibiting growth of candida and fostering growth of bacteroides.' *Lancet,* **2,** 1174

Gibbs, B. M. and Hirsch, A. (1956). 'Spore formation by *Clostridium* species in an artificial medium.' *Journal of applied Bacteriology,* **19,** 129

Gordon, J. and McLeod, J. W. (1940). 'A simple and rapid method of distinguishing *Cl. novyi* (*B. oedematiens*) from other bacteria associated with gas gangrene.' *Journal of Pathology and Bacteriology,* **50,** 167

Hafiz, S., McEntegart, M. G., Morton, R. S. and Waitkins, S. A. (1975). '*Clostridium difficile* in the urogenital tract of males and females.' *Lancet,* **1,** 420

Hall, I. C. (1919). 'Selective elimination of hay bacillus from cultures of obligative anaerobes.' *Journal of the American medical Association,* **72,** 274

Hall, I. C. (1920). 'Practical methods in the purification of obligate anaerobes.' *Journal of infectious Diseases,* **27,** 576

Handford, P. M. (1974). 'A new medium for the detection and enumeration of *Clostridium perfringens* in foods.' *Journal of applied Bacteriology,* **37,** 559

Hayward, N. J. (1941). 'Rapid identification of *Cl. welchii* by the Nagler reaction.' *British medical Journal,* **1,** 811, 916

Hayward, N. J. (1943). 'The rapid identification of *Cl. welchii* by Nagler tests in plate cultures.' *Journal of Pathology and Bacteriology*, **55**, 285

Hirsch, A. and Grinsted, E. (1954). 'Methods for the growth and enumeration of anaerobic spore-formers from cheese, with observations on the effect of nisin.' *Journal of Dairy Research*, **21**, 101

Hoare, E. D. (1939). 'The stability of "Liquoid" for use in blood culture media, with particular reference to anaerobic streptococci.' *Journal of Pathology and Bacteriology*, **48**, 573

Hobbs, B. C. (1960). 'Sources of food-poisoning bacteria and their control.' In *Recent Advances in Clinical Pathology*, Series III, p. 91. Ed. by S. C. Dyke. London; Churchill

Holdeman, L. V. and Moore, W. E. C. (1975). *Anaerobe Laboratory Manual*. Blacksburg; Virginia Polytechnic Institute and State University

Johansson, K. R. (1953). 'A modified egg-yolk medium for detecting lecithinase-producing anaerobes in faeces.' *Journal of Bacteriology*, **65**, 225

Kocka, F. E., Arthur, E. J. and Searcy, R. L. (1974). 'Comparative effects of two sulfated polyanions used in blood culture on anaerobic cocci.' *American Journal of clinical Pathology*, **61**, 25

Labbe, R. G. and Duncan, C. L. (1975). 'Influence of carbohydrates on growth and sporulation of *Clostridium perfringens* type A.' *Applied Microbiology*, **29**, 345

Labbe, R., Somers, E. and Duncan, C. (1976). 'Influence of starch source on sporulation and enterotoxin production by *Clostridium perfringens* type A.' *Applied and environmental Microbiology*, **31**, 455

Lapage, S. P. and Redway, K. F. (1974). *Preservation of Bacteria with Notes on Other Micro-organisms.* Public Health Laboratory Service Monograph Series No. 7; London; HMSO

Lilley, B. D. and Brewer, J. H. (1953). 'The selective antibacterial action of phenylethyl alcohol.' *Journal of the American Pharmaceutical Association*, **42**, 6

Lowbury, E. J. L. and Lilley, H. A. (1955). 'A selective plate medium for *Cl. welchii.' Journal of Pathology and Bacteriology*, **70**, 105

McClung, L. S. and Toabe, R. (1947). 'Egg-yolk plate reactions for presumptive diagnosis of *Clostridium sporogenes* and certain species of gangrene and botulinum groups.' *Journal of Bacteriology*, **53**, 139

Miles, A. A. (1943). 'Inhibition of swarming in plate culture.' *Army Pathology Laboratory Service, Current Notes*, No. 9, 3

Moore, H. B., Sutter, V. L. and Finegold, S. M. (1975). 'Comparison of three procedures for biochemical testing of anaerobic bacteria.' *Journal of Clinical Microbiology*, **1**, 15

Moore. W. B. (1968). 'Solidified media suitable for the cultivation of *Clostridium novyi* type B.' *Journal of general Microbiology*, **53**, 415

Nord, C–E., Dahlback, A. and Wadstrom, T. (1975). 'Evaluation of a test kit for identification of anaerobic bacteria.' *Medical Microbiology and Immunology*, **161**, 239

Omata, R. R. and Disraely, M. N. (1956). 'A selective medium for oral fusobacteria.' *Journal of Bacteriology*, **72**, 677

Phillips, K. D. (1976). 'A simple and sensitive technique for determining the fermentation reactions of non-sporing anaerobes.' *Journal of applied Bacteriology*, **41**, 325

Reed, G. B. and Orr, J. H. (1941). 'Rapid identification of gas gangrene anaerobes.' *War Medicine, Chicago*, **1**, 493

Robertson, M. (1915–16). 'Notes upon certain anaerobes isolated from wounds.' *Journal of Pathology and Bacteriology*, **20**, 327

Rosner, R. (1973). 'A quantitative evaluation of three different blood culture systems.' In *Bacteremia; Laboratory and Clinical Aspects*. p. 61. Ed. by A. C. Sonnenwirth. Springfield; Thomas

Shanson, D. C. (1974). 'An experimental assessment of different anaerobic blood culture methods.' *Journal of clinical Pathology*, **27**, 273

Shanson, D. C. and Barnicoat, M. (1975). 'An experimental comparison of Thiol broth with Brewer's thioglycollate for anaerobic blood cultures.' *Journal of clinical Pathology*, **28**, 407

Slanetz, L. W. and Rettger, L. F. (1933). 'A systematic study of the fusiform bacteria.' *Journal of Bacteriology*, **26**, 599

Sonnenwirth, A. C. (1973). 'Bacteremia–extent of the problem.' In *Bacteremia; Laboratory and Clinical Aspects*. p. 3. Ed. by A. C. Sonnenwirth. Springfield; Thomas

Spray, R. S. (1936). 'Semisolid media for cultivation and identification of the sporulating anaerobes.' *Journal of Bacteriology*, **32**, 135

Starr, S. E., Thompson, F. S., Dowell, V. R. and Balows, A. (1973). 'Micromethod system for identification of anaerobic bacteria.' *Applied Microbiology*, **25**, 713

Stokes, E. J. (1974). 'Blood culture technique.' *Association of Clinical Pathologists: Broadsheet 81*

Sutter, V. L., Attebery, H. R., Rosenblatt, J. E., Bricknell, K. S. and Finegold, S. M. (1972). *Anaerobic bacteriology:* California; University of California

Szawatkowski, M. V. (1976). 'A comparison of three readily available types of anaerobic blood culture media.' *Medical Laboratory Sciences*, **33**, 5

Thomas, M. (1963). 'A blue peroxide slide catalase test.' *Monthly Bulletin of the Ministry of Health and the Public Health Laboratory Service*, **22**, 124

Thomas, M. E. M. (1974). 'Isolation and identification of lactobacilli from the genital tract.' In *Laboratory Methods 1*. p. 30. Ed. by A. T. Willis and C. H. Collins. Public Health Laboratory Service Monograph Series No. 5: London; HMSO

Walker, P. D., Harris, E. and Moore, W. B. (1971). 'The isolation of clostridia from animal tissues.' In *Isolation of Anaerobes*. Society of Applied Bacteriology Technical Series No. 5. p. 25. Ed. by D. A. Shapton and R. G. Board. London; Academic Press

Washington, J. A. (1973). 'Bacteremia due to anaerobic, unusual and fastidious bacteria.' In *Bacteremia; Laboratory and Clinical Aspects*. p. 47. Ed. by A. C. Sonnenwirth. Springfield; Thomas

Washington, J. A. (1975). 'Blood cultures. Principles and techniques.' *Mayo Clinic Proceedings*, **50**, 91

Wetzler, T. F., Marshall, J. D. and Cardella, M. A. (1956a). 'Rapid isolation of clostridia by selective inhibition of aerobic flora. II. A systematic method as applied to surveys of clostridia in Korea.' *American Journal of clinical Pathology*, **26**, 418

Wetzler, T. F., Marshall, J. D. and Cardella, M. A. (1956b). 'Rapid isolation of clostridia by selective inhibition of aerobic flora. I. Use of sorbic acid and polymyxin B sulphate in a liquid medium.' *American Journal of clinical Pathology*, **26**, 345

Willingham, C. A. and Oppenheimer, C. H. (1964). 'Modified device for anaerobic dispensing of reduced media.' *Journal of Bacteriology*, **88**, 541

Willis, A. T. (1957). 'A rapid method for the purification of some clostridia from mixtures with other organisms, especially the aerobic spore-formers.' *Journal of Pathology and Bacteriology,* **74**, 113

Willis, A. T. and Hobbs, G. (1959). 'Some new media for the isolation and identification of clostridia.' *Journal of Pathology and Bacteriology,* **77**, 511

Zwarun, A. A. (1975). 'Measurement of redox potential changes in anaerobic culture media caused by addition of blood.' *Journal of Laboratory and clinical Medicine,* **85**, 174

3

Examination and Identification of Anaerobes

Before applying the various methods of identifying an anaerobic organism it is most important to ensure that the culture in question is pure. It must be free from both aerobic and anaerobic contaminants. Methods of purifying cultures of anaerobes that are mixed with other organisms are discussed in Chapter 4. It is pertinent to mention here, however, that whenever subcultures are made from a supposedly pure culture of an anaerobe, an aerobic control plate must always be inoculated, and this plate should remain sterile if aerobic contaminants are absent. Some aerotolerant organisms, however, notably *Clostridium histolyticum, Cl. tertium* and *Cl. carnis* do produce small surface colonies under aerobic conditions. Some organisms which grow only anaerobically on primary isolation may become aerotolerant after two or three subcultures; it seems likely that many of these are, in fact, carbon dioxide-dependent rather than anaerobic. Organisms of these types must be borne in mind when the aerobic control plate is examined.

In order to ensure that only one anaerobic species is present, smears from the culture are examined microscopically and the culture is plated on suitable media, such as horse blood agar and heated blood agar. If the original culture is pure then usually only one type of colony will develop on these plates. An individual strain may show slight colonial variation on different media, especially when different concentrations of agar are used.

The confusion which may be caused by studying impure cultures is thoroughly understood. At the same time, however, it must be remembered that individual organisms in pure culture are likely to occur in two or more phases from time to time, showing differences in morphology and in cultural and biochemical characters. In such cases care must be taken not to confuse contamination with phase variation; and similarly, two or more phases of the same species must not be regarded as different organisms (Frobisher, 1933). Pure cultures of anaerobes are not always easy to maintain; it is always as well to be prepared for the appearance of contaminants, and sudden changes and variations in the properties of a culture should lead one to suspect their presence.

When it is certain that the organism is in pure culture, then the process of recording its characteristics can begin. Since the chief aim of the clinical bacteriologist is to identify any particular strain, in the following pages the main emphasis is laid on tests and methods of examination that yield the maximum amount of clinically useful information in the shortest possible time, and more academic considerations have been avoided. Some of the tests mentioned are not subsequently referred to, either because they are the least useful in differentiating one anaerobe from another, or because the information given by the particular test is more conveniently obtained by another. Useful approaches to the identification of anaerobes have been published by Marymont, Holdeman and Moore (1972), Ellner, Granato and May (1973), Sutter and Finegold (1973), Moore and Holdeman (1972) and Duerden *et al.* (1976).

MORPHOLOGY AND STAINING

The shape, size, arrangement, Gram-reaction, presence or absence of spores, and presence or absence of branching determine in which genus a particular anaerobe belongs. *Table 3.1* which summarizes the main morphological characters of some anaerobic genera provides a general plan for the separation of genera but should not be regarded as complete or definitive. Rather, it shows where the genera might fall in relation to one another in an ideal taxonomic situation. In practice, individual isolates may be much more elusive, and it is important to recognize that, except in its broadest sense, morphology is not taxonomic.

The following notes illustrate the sorts of difficulties that are imposed by a morphological classification of the anaerobes.

The presence of vacuoles in some strains of *Bacteroides* may be mistaken for spores, it being well recognized that many clostridia lose their Gram-reaction in cultures more than a few hours old. Some clostridia, on the other hand, may fail to produce discernable spores and might, therefore, be regarded as *Eubacterium* or *Propionibacterium,*

Table 3.1 Some characteristics of the main anaerobic genera

Genus	*Gram reaction*	*Shape*	*Arrangement*	*Other*
Actinomyces	+	Long filamentous bacilli or mycelia	True branching	*A. israelii* the main human pathogen
Bacteroides	–	Bacillary with rounded ends; may be cocco-bacillary or pleomorphic	None typical	*B. fragilis* and *B. melaninogenicus* the common pathogens
Bifidobacterium	+	Bacillary; clubbed or bifurcated ends	None typical	Not pathogenic
Clostridium	+	Bacillary	None typical	Histotoxic or neuro-toxic; spore-formers
Eubacterium	+	Bacillary; often pleomorphic	Pairs and short chains	
Fusobacterium	–	Bacillary; often with tapered ends; often pleomorphic	None typical	
Lactobacillus	+	Bacillary; often pleomorphic	Chain formation common	Not pathogenic
Peptococcus	+	Coccal	Singly, pairs, short chains and clumps	
Peptostreptococcus	+	Coccal	Singly, pairs and in chains	
Propionibacterium	+	Pleomorphic, club-shaped bacilli	None typical	
Veillonella	–	Coccal	Pairs, short chains and irregular clumps	

or even as *Bacteroides* in older cultures. In the same way strains of *Bifidobacterium* that do not show the classic bifid form may be confused with *Clostridium* or *Eubacterium,* and branching forms of *Bifidobacterium* may be identified as *Actinomyces.* Coccoid or coccobacillary forms of *Eubacterium* may be identified as *Peptostreptococcus,* while the absence of chain formation in some strains of peptostreptococci may make it impossible to differentiate them morphologically from strains of *Peptococcus.* The fact that some strains of *Veillonella* retain the Gram stain in young cultures may lead to their identification as *Peptostreptococcus* or *Peptococcus.* And it is a not uncommon experience to encounter strains of anaerobes whose shape and Gram-reaction are so indeterminate that they could fall into a number of the groups covered by the Gram-reaction and by coccal and bacillary morphology.

Having thus destroyed all faith in the microscopy of anaerobes, it must be said at once that very many anaerobic isolates do fit neatly into the morphological scheme of separation; indeed, it is on the basis of the *typical* that such a scheme is devised. In common with many aerobic organisms the typical morphology of anaerobes is often best developed in pathological material.

Gram reaction

The conventional methods of Gram-staining are quite satisfactory for anaerobes, but it should be noted that over-decolorization of Gram-positive anaerobes is more likely to occur if acetone instead of alcohol is used as the decolorizing agent. Gram-positive anaerobes not infrequently stain Gram-negatively, especially in old cultures. This is especially the case with strict anaerobes such as *Cl. tetani, Cl. novyi* and the peptostreptococci, and is probably due to the toxic effects of oxygen on these organisms. In the case of the clostridia, the presence of spores will exclude Gram-negative anaerobic bacilli, but Gram-negative peptostreptococci may be confused with veillonellae. This variability in the Gram-reaction is utilized by Prévot (1957, 1966) in his classification of the clostridia, in which he divides them into two orders, four families and nine genera. The author considers this undesirable because it needlessly complicates the nomenclature and because it places too much emphasis on minor differences.

Spores

In Gram-stained preparations a spore shows as a circular or oval unstained area in the body of the bacillus. The size, shape and position of the spore should be noted, for sporulating cultures of some clostridia, for example *Cl. tetani*, present a characteristic microscopic appearance on which a presumptive diagnosis may be made. Free spores may occur singly, or more commonly in groups, and show in Gram-stained films as circular or oval unstained bodies, each delineated by a faintly Gram-negative 'wall'. Spores may be stained by any of the common methods; the modified Ziehl–Neelsen spore stain is satisfactory (*see below*). Generally, however, little is to be gained by staining the spores of clostridia since they are readily seen in the Gram-stained preparations. Though some taxonomic systems attach importance to the size and position of the spore, these features are usually far too variable to be of any diagnostic value. A complete absence of spores in culture should also be noted, for with some clostridia, notably *Cl. perfringens* type A, this is a characteristic feature. It is worth noting here that the appearance of the spore can often be used to differentiate between Gram-positive aerobic and anaerobic spore-forming bacilli. The former produce spores within the bacillary body usually without distorting its shape, while clostridial spores commonly distend it.

When spores cannot be detected microscopically, their presence may sometimes be inferred by demonstrating resistance of the culture to heat (pasteurization) or to treatment with alcohol (*see* p. 126).

Modified Ziehl–Neelsen spore stain

Prepare a film and fix it with heat. Stain with hot carbol fuchsin (Ziehl–Neelsen) for 3 min. Wash with tap water, then flood the film with 30% aqueous ferric chloride solution for 2 min. Wash with tap water and decolorize with 5% sodium sulphite solution. Wash and counterstain with 1% malachite green for 2 min. Wash, drain and blot dry.

Spores are stained red; vegetative material and cells are green.

Spore stain of Schaeffer and Fulton (1933)

Prepare a film and fix it with heat. Flood the slide with hot 5% aqueous malachite green for 1 min. Wash with tap water, then counterstain with 0.5% aqueous safranin for 30 seconds. Wash, drain and blot dry.

Spores are stained green; vegetative cells are red.

Capsule staining

Capsule staining (*see below*) does not find much application, since only two of the commonly occurring anaerobes are capsulated, namely

Cl. perfringens and *Cl. butyricum*. Butler (1945) drew attention to the correlation between capsule formation by *Cl. perfringens* and the severity of wound infections caused by that organism. Capsule formation may be demonstrated in pathological material and in cooked meat broth cultures to which has been added a little blood or serum. In the direct examination of pathological material from some *Cl. perfringens* infections, capsule staining may be of considerable importance (*see* p. 314).

Capsule stain of Howie and Kirkpatrick (1934)

STAINING SOLUTION

10% water-soluble eosin or erythrosin in distilled water	4 parts
Serum (human, rabbit or sheep, heated to 56 °C)	1 part
One crystal of thymol	

Store the mixture at room temperature for several days. Centrifuge, and retain the supernatant fluid stain, which keeps at room temperature for many months.

METHOD

On a slide mix one drop of culture suspension or pathological exudate with one drop of a 1 in 5 dilution of Ziehl–Neelsen's carbol fuchsin. Allow to stain for 30 seconds. Add one drop of capsule stain and allow to stand for 1 min. Spread the film with the edge of another slide and allow to dry. There is a homogeneous red background against which the unstained capsules show prominently; bacterial bodies stain red.

Muir's capsule stain

Solution 1
Carbol fuchsin (Ziehl–Neelsen).

Solution 2

Saturated aqueous solution of mercuric chloride	2 volumes
20% aqueous solution of tannic acid	2 volumes
Saturated aqueous solution of potassium alum	5 volumes

Solution 3
Methylene blue.

METHOD

Prepare a thin film and fix by heat. Flood the slide with hot strong carbol fuchsin for 1 min. Wash rapidly with tap water and then with ethanol. Treat the smear with solution 2 (Muir's mordant) for 30 seconds and wash with water. Decolorize with absolute ethanol for 30 seconds – the film should now have a pale red appearance. Apply the methylene blue counterstain for 1 min, wash with tap water, drain and blot dry. Organisms appear red with blue capsules.

Flagella staining

Flagella staining is not recommended for the routine identification of anaerobes, since the techniques are difficult and time consuming and the information gained is not highly definitive. With the important exceptions of *Cl. perfringens* and *Cl. barati* most of the clostridia are motile and flagellated. Many bacteroides and fusobacteria are peritrichous.

Flagella staining is discussed by Holdeman and Moore (1975), and Cowan (1974).

Motility

Absence of motility is of no significance in identifying any organism; for not only do non-motile variants of motile species occur, but also it is common amongst the anaerobes for motility to be lost due to death of the organism or to interference with its flagellar mechanism. Hence the routine examination of anaerobic cultures for motility is not an essential procedure. As in other fields of bacteriology, a positive motility test is of much greater significance than a negative one.

Since most anaerobes are adversely affected by the presence of free oxygen, a result of which may be loss of motility, it is inadvisable to look for motility using hanging-drop or wet-mount coverslip preparations. The best method is to use a capillary tube filled with an actively growing fluid culture (4–10 h) and sealed at both ends (Stanbridge and Preston, 1969). Such a capillary tube is easily prepared from a pasteur pipette. The preparation is examined under the high-power objective with reduced illumination. Three or four preparations from cultures of different ages should be examined before a negative result for motility is recorded.

An alternative to this direct method of observing motility is the use of stab cultures in semi-solid agar (0.5%) (Tittsler and Sandholzer, 1936). Motile organisms show a diffuse zone of growth spreading from the line of inoculation, while the growth of non-motile species is confined to the area of inoculation. Since the test is slow (up to 6 days' incubation may be required), and since interpretation of the result is often difficult, this method is not recommended for routine use.

COLONIAL APPEARANCE

Any medium that supports good surface growth of anaerobes is satisfactory for the study of colonial morphology. Media used for this purpose should contain not more than 1.5% New Zealand agar as higher concentrations appreciably alter some of the colonial characters Fresh horse blood agar and heated blood agar are suitable media. For the full development of characteristic colonies some organisms require longer periods of incubation than others.

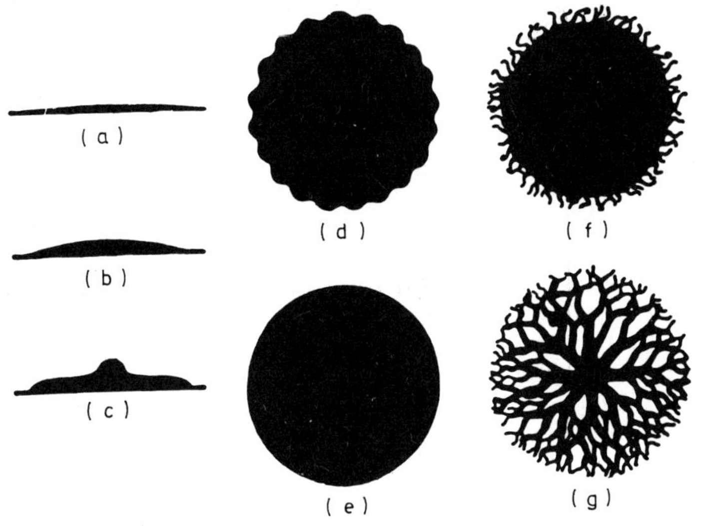

Figure 3.1. The main types of anaerobe colonies; (a) flat, (b) convex, (c) umbonate, (d) crenated, (e) entire, (f) fimbriate, and (g) rhizoid

Most anaerobic organisms produce semitranslucent or clear surface growths, important exceptions being the *Actinomyces* and some members of the *Fusobacterium* and *Bacteroides* groups. A description of the types of colonies produced by the different anaerobic bacteria is to be found in Chapters 4 and 5.

Figure 3.1 illustrates the main types of colonies that may be encountered. Swarming growth, which is not easily illustrated, is characteristic of some anaerobes, *Cl. tetani* and *Cl. septicum* being notable examples. The size and shape of colonies are not static features; with continued incubation convex colonies may become umbonate, circular colonies may become irregular, clear or semitranslucent colonies may become opaque or pigmented. Colonial appearances may also vary on different media, being affected by such factors as the concentration of the agar, and the amount of moisture present on the surface of the medium.

The clostridia characteristically produce colonies of irregular shape, some species forming frankly rhizoidal colonies, for example, *Cl. sporogenes*. Others, such as *Cl. tetani*, rarely produce discrete colonies, except in young culture, their characteristic habit of growth being a swarming film. *Cl. perfringens* is one of the few commonly occurring clostridia which produces a circular, entire colony.

The non-sporing anaerobes form colonies that are almost always discrete, and are commonly circular and entire. They are usually rather smaller than those of the clostridia, and some, such as those of *P. avidum* may appear as mere pin-points. Pitting of the agar surface is characteristically produced by colonies of *B. corrodens* and *A. israelii*.

CULTURAL CHARACTERISTICS

Fresh horse blood agar

Most of the anaerobes that are pathogenic for man will grow well on fresh horse blood agar and on heated blood agar (*see below*) in the anaerobic jar, although the addition to the medium of reductants or enrichments may be helpful or essential for satisfactory growth of some species (*see* Chapter 2). Like some aerobic species, some anaerobes produce partial or complete haemolysis on horse blood agar; of these complete haemolysis is the commoner. Though many anaerobes are haemolytic, different strains of a single species may vary considerably in this respect; for example, some strains of *Cl. perfringens* are non-haemolytic on horse blood agar, others produce extensive zones of haemolysis and yet others produce only a limited partial haemolysis. Thus haemolysis, considered alone, is not of much diagnostic value. It is worth noting that the haemolysins elaborated by individual anaerobic species are not equally active against the erythrocytes of all animals. For example, the haemolysis produced by *Cl. perfringens* on horse blood agar is due to its θ toxin, which is relatively inactive against red cells of the mouse, while *Cl. perfringens* α-toxin haemolyses the erythrocytes of the mouse but not those of the horse (*see* p. 131).

Another change that is produced in horse blood agar by some strains of anaerobic cocci is browning of the medium, which develops as a zone immediately surrounding the colony.

Bacteroides melaninogenicus is an unusual organism in that blackening of its colonies occurs with prolonged incubation. This pigmentation, which takes place on blood-containing media, commences in the centre of the colony and gradually extends to its periphery, through shades of brown, to a shiny black. The responsible pigment is haematin (Schwabacher and Lucas, 1947). Colonies of some strains of peptococci and peptostreptococci may be dark, or even black, but this is not a constant feature. Colonies of many strains of *B. melaninogenicus* (and possibly those of some species of *Veillonella*) show brick-red fluorescence in ultraviolet light, a property which is not shared with any other commonly encountered anaerobe. Fluorescence is often exhibited by young cultures, long before pigmentation has developed. The fluorescent substance, and the black pigment appear to be distinct.

Many anaerobic organisms growing on horse blood agar produce no change in the medium.

Heated blood (chocolate) agar

The most important and the commonest change on this medium is partial clearing due to proteolytic activity of the anaerobic culture. Examined by reflected light, proteolysis in this medium appears as a darkening of the medium, whereas by transmitted light it shows as partial clearing. Proteolytic anaerobes are confined almost exclusively to the genus *Clostridium;* the only common proteolytic non-clostridial organism is *Bacteroides melaninogenicus* subspecies *asaccharolyticus.*

Bleaching of heated blood agar, a change commonly associated in the mind with *Streptococcus pneumoniae* and *Str. viridans,* is also produced by some anaerobes, and is particularly well shown by some strains of *Cl. novyi*. Heated blood agar plate cultures of this organism show no evidence of bleaching when they are first removed from their anaerobic environment. The greenish discoloration in the medium develops around the colonies after they have been exposed to the air for about half an hour. This occurs most rapidly at 37 °C, and is due to the production of hydrogen peroxide. As a specific indicator of hydrogen peroxide one may incorporate benzidine (Gordon and McLeod, 1940; Hayward, 1942) or *m*-aminoacetanilide hydrochloride (Gordon, 1954) in heated blood agar or fresh blood agar media. On exposure to free oxygen the colonies of hydrogen peroxide-producing organisms are blackened. Organisms which have been reported to produce this effect are *Cl. novyi, Cl. botulinum, Cl. multifermentans, Cl. cochlearium* and some anaerobic streptococci and bacteroides. *M*-aminoacetanilide hydrochloride is inferior to benzidine, which is unfortunate since the latter is carcinogenic.

Cultures of many anaerobic organisms on heated blood agar show a zone of reddening of the medium about the colonies when plates are first removed from the anaerobic jar. This discoloration, which was first observed in anaerobic cultures by McLeod (1928) and Taylor (1929), is due to the strong reducing effect on the medium by the growing organisms, and probably results from the formation of haemochromogen. Although some anaerobes are more active than others (the reddening is most marked in cultures of *Cl. perfringens*), all anaerobic organisms produce it to some extent, whereas facultative anaerobes are quite inactive. The reaction is thus of value in identifying colonies of anaerobes when they are mixed with those of facultative organisms. Since the reddening is essentially a reduction phenomenon, it fades rapidly on exposure to air. A good colour illustration of this effect is provided by Gordon and McLeod (1940).

Gelatinase media

Depending on whether or not an organism hydrolyses gelatin, it can be said that the organism is at least slightly proteolytic or not proteolytic at all. All strongly proteolytic anaerobes digest gelatin, but there are many gelatinase-producing anaerobes which do not break down more complex proteins. Gelatinase activity may be demonstrated by growing the organism in nutrient gelatin or glucose gelatin media (Willis and Hobbs, 1959). After incubation it is necessary to cool the culture at 4 °C for half an hour, since gelatin is fluid at incubator temperature (37 °C). A positive result is recorded if the medium remains fluid after cooling. It is important to include a negative control tube when performing this test since with repeated melting and solidification gelatin ultimately fails to solidify.

Quite the most satisfactory method of determining gelatinase activity is that described by Frazier (1926), in which the test organisms are cultured on 0.4% gelatin agar. After satisfactory growth has developed, cultures are flooded with acid mercuric chloride solution (15% mercuric chloride in 1M hydrochloric acid), which rapidly denatures (and renders opaque) any unhydrolysed gelatin. Spot inocula of gelatinase-producing cultures are surrounded by a wide zone of clearing in the medium. A standard 9 cm diameter petri dish conveniently accommodates nine test cultures.

Inspissated serum

Loeffler's medium is used for the determination of proteolytic properties of anaerobes. It has already been noted that gelatin is attacked by both strongly and weakly proteolytic organisms. Liquefaction of inspissated serum, however, is produced only by strongly proteolytic anaerobes, almost all of which are clostridia. When an organism is designated 'proteolytic' it is generally taken to mean that it will attack complex proteins such as heated serum. In Loeffler slope cultures of proteolytic anaerobes there is first a surface erosion of the medium, followed later by its liquefaction or disintegration, or both, accompanied by a putrefactive odour. Three or four days' incubation is usually required before liquefaction is evident, and much longer may be required.

Milk agar and casein agar

Milk agar was devised by Reed and Orr (1941) and serves the same purpose as inspissated serum. Plate cultures of proteolytic anaerobes produce zones of clearing beneath and around areas of growth, due to digestion of the casein. Clear-cut reactions develop in 24–48 h, and consequently this medium is considered superior to inspissation serum.

Cultures of lactose-fermenting anaerobes, such as *Cl. perfringens*, commonly produce a diffuse opacity on milk agar, due to acid precipitation of the casein, while some clostridia, notably *Cl. tetani* and *Cl. novyi* type A produce a diffuse opacity due to the activity of a rennin-like enzyme (Willis and Williams, 1970; Williams, 1971). Although these opacity reactions are quite distinct from proteolytic changes, they may be found confusing. In this event, casein agar, which shows only proteolytic activity, should be used.

Litmus milk

The various changes which may be produced by anaerobes growing in litmus milk are: acid, gas, rapid or slow clotting, and digestion. Some of the alterations may occur simultaneously, or a sequence of changes may be observed. Acid production is due to fermentation of the lactose in the milk, and is indicated by a change in the colour of the litmus from blue to red. Acid formation is usually accompanied by coagulation of the casein. The formation of gas may be due either to fermentation of the lactose or to proteolytic activity, but gas is produced much more vigorously and in greater quantity by lactose-fermenting anaerobes than by proteolytic species. Highly saccharolytic organisms, such as *Cl. perfringens* and *Cl. butyricum*, attack the lactose so vigorously that a 'stormy clot' results – acid production causes the casein to coagulate and the clot thus formed is then disrupted by gas bubbles. The disrupted curd is surrounded by the clear whey. Contrary to popular belief, this reaction is not limited to strains of *Cl. perfringens*, and reliance should not be placed on it as a diagnostic feature of this organism. The high degree of acidity attained by a reaction of this type is sufficient to prevent further bacterial activity.

In addition to coagulation by acid production, curding is also associated with the growth of some non-lactose-fermenting organisms, due to the fact that these organisms secrete a rennin-like enzyme capable of hydrolysing casein to soluble paracasein, which then reacts with the soluble calcium salts present in the milk to form a precipitate of calcium paracaseinate. As in the case of acid coagulation, the curd is surrounded by the clear supernatant whey. A rennin curd once formed is usually digested, since the organisms that coagulate milk in this way are commonly proteolytic. After complete peptonization of the casein has occurred a clear purplish solution remains.

From these results it will be clear that while gross changes in milk such as stormy clot or clot followed by digestion are useful diagnostic features, finer alterations, such as slight gas production due to proteolysis or to slow lactose fermentation, are less likely to be observed and are therefore of less value. While the culture of anaerobes in tubed milk medium demonstrates some important principles, the chief information that is given, namely lactose fermentation and proteolytic activity, can be more rapidly and more accurately determined by the use of other cultural methods.

Human serum agar and egg yolk agar media

Seiffert (1939) and Nagler (1939) showed independently that when *Cl. perfringens* is grown in media containing human serum a dense opalescence is produced; this reaction – the Nagler reaction – is given by culture filtrates of *Cl. perfringens* as well as by the growing organisms, and is inhibited by *Cl. perfringens* antitoxin (Nagler, 1939). Later, Macfarlane, Oakley and Anderson (1941) reported that similar but stronger opalescence was produced in filtered egg yolk emulsion (the lecithovitellin or LV reaction), and showed that the reactions are due to the α-toxin of *Cl. perfringens*. Hayward (1941, 1943), using human serum agar plates, showed that *Cl. perfringens, Cl. novyi* and members of the *Cl. bifermentans–sordellii* group produced a dense diffuse opalescence on this medium; the opacity due to *Cl. perfringens* and *Cl. novyi* was inhibited by specific antitoxic sera, while that due to *Cl. bifermentans* was inhibited by *Cl. perfringens* antitoxin.

These opalescent changes are due in many cases to the production by the organisms of specific lecithinase (*see below*). As Miles and Miles (1947) showed, the lecithinases of *Cl. perfringens* and *Cl. bifermentans* are antigenically related, but distinct. Hayward favoured the use of human serum instead of egg yolk in her lecithinase plates, no doubt because it was at that time easier to obtain; different samples, however, varied considerably in sensitivity. Subsequent work in this field (Nagler, 1944, 1945; McClung, Heidenreich and Toabe, 1945; McClung and Toabe, 1947; Oakley, Warrack and Clarke, 1947; Fievez, 1963) has shown that various anaerobes are able to produce opalescence in egg yolk media (*Table 3.2*), and that this is sometimes due to specific lecithinase secreted by the organisms. Some anaerobes produce, in addition to an opalescence, a pearly layer or iridescent film, which

Table 3.2 Reactions of various anaerobes on egg yolk media (from Willis and Hobbs, 1958)

Organism	*Opalescence* *Diffuse (lecithinase C)*	*Restricted (lipase)*	*Pearly layer (lipase)*
Cl. perfringens A–E	+	—	—
Cl. barati	+	—	—
Cl. bifermentans – sordellii	+	—	—
Cl. botulinum A–F	—	+	+
Cl. sporogenes	—	+	+
Cl. novyi			
Type A	+	+[1]	+
Type B	+	—	—
Type C	—	—	—
Type D	+	—	—
F. necrophorum	—	+	+

[1] Note that the restricted lipase opacity is obscured by the lecithinase C opacity

Table 3.3 Reactions of clostridia on half-antitoxin[1] lactose egg yolk milk agar (modified from Willis and Hobbs, 1959)

Organism	Lactose egg yolk milk agar				
	Diffuse lecithinase C opacity — *Produced*	*Inhibited*	*Restricted opacity and pearly layer (lipase)*	*Lactose fermentation*	*Proteolysis*
Cl. perfringens A–E	+	+	—	+	—
Cl. barati	+	—	—	+	—
Cl. bifermentans	+	+	—	—	+
Cl. sordellii	+	+	—	—	+
Cl. botulinum A	—	—	+	—	+
B	—	—	+	—	+ or —
C	—	—	+	—	—
D	—	—	+	—	—
E	—	—	+	—	—
F	—	—	+	—	+ or —

B	+	—	—	—	—
C	—		—	—	—
D	+	—	—	—	—
Cl. histolyticum	—		—	—	+
Cl. septicum	—		—	+	—
Cl. chauvoei	—		—	+	—
Cl. tetani	—		—	—	—
Cl. cochlearium	—		—	—	—
Cl. cadaveris	—		—	—	—
Cl. sphenoides	—		—	+	—
Cl. tertium	—		—	+	—
Cl. butyricum	—		—	+	—
Cl. fallax	—		—	+	—

[1] Mixture of *Cl. perfringens* type A and *Cl. novyi* type A antitoxic sera

[2] Note that the restricted lipase opacity is obscured on the non-antitoxin half of the plate by the lecithinase C opacity

Table 3.4 A scheme for the identification of some clostridia using lactose egg yolk milk half-antitoxin[1] plates[2]

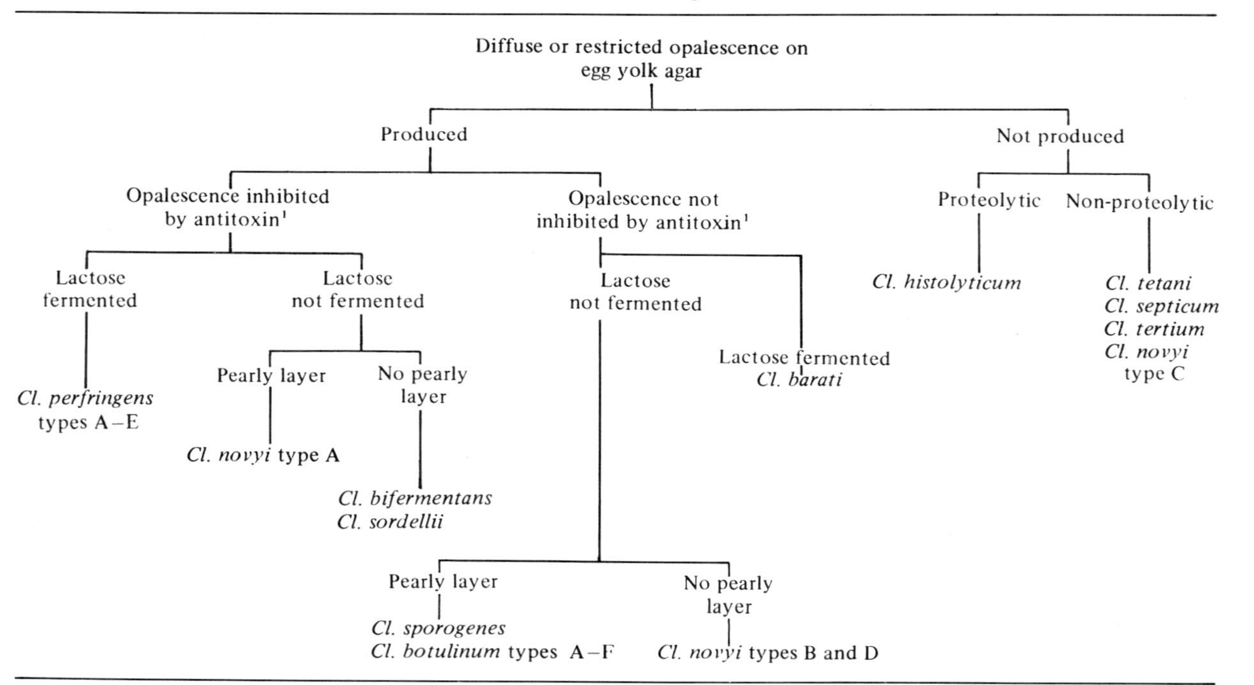

[1] A mixture of *Cl. perfringens* type A and *Cl. novyi* type A antisera

[2] Instead of using the complex egg yolk medium, may use separate tests:

covers the colonies and in some cases extends beyond their edge on to the surface of the medium (Nagler, 1945). This is due to lipolysis (Willis, 1960a, b; 1962) (*see below*). Egg yolk media are thus of considerable value in identifying opalescence-producing anaerobes and, used as half antitoxin plates, in distinguishing between them. It should be noted that while lecithinase C *and* lipase activity may be demonstrated on egg yolk media, lecithinase C activity *only* is developed on human serum agar. This is so because most samples of human serum do not contain any appreciable amount of free fat.

Willis and Hobbs (1957, 1958, 1959) developed a lactose egg yolk milk agar medium. It is a complex medium which not only provides more information about the organisms growing on it, but it also enables a presumptive diagnosis of some anaerobes to be made from a single plate culture. The egg yolk content of the medium enables lecithinase C- and lipase-producing organisms to produce their characteristic opalescence and pearly layer effects. The presence of lactose and an indicator (neutral red) in the medium provides a means of identifying lactose-fermenting organisms, which is of particular value in distinguishing between cultures of *Cl. perfringens* on the one hand, and *Cl. bifermentans* and *Cl. sordellii* on the other. Lactose-fermenting anaerobes produced diffuse red haloes in the medium around areas of growth, the colonies remaining uncoloured until they are exposed to the air. In the immediate vicinity of growth the indicator is reduced, so that organisms which tend to spread over the plate produce extensive reduction of the dye, and evidence of lactose fermentation may not be present. The presence of milk in the medium enables proteolytic activity to be determined, since it is attacked by proteolytic species and the result is the development of zones of clearing around areas of growth.

Lactose egg yolk milk agar, used as a half-antitoxin plate with a mixture of *Cl. perfringens* type A and *Cl. novyi* type A antitoxic sera as the antitoxin, not only provides a great deal of information about clostridial species in general, but is also differentially diagnostic for *Cl. perfringens, Cl. bifermentans–sordellii, Cl. novyi* type A, *Cl. histolyticum* and non-proteolytic strains of *Cl. botulinum*. The reactions given on this medium by the commoner species of clostridia are shown in *Tables 3.3* and *3.4*.

Lecithinase C and lipase activity

It has been noted above that lecithinase C activity and lipolysis are conveniently determined on media containing egg yolk. The reactions produced by these types of enzyme on egg yolk agar are quite distinct (Willis, 1962).

LECITHINASE C ACTIVITY

Lecithinase C attacks the egg yolk lecithin at the glycerol–phosphorylcholine bond (*Figure 3.2*), splitting it into phosphorylcholine and an

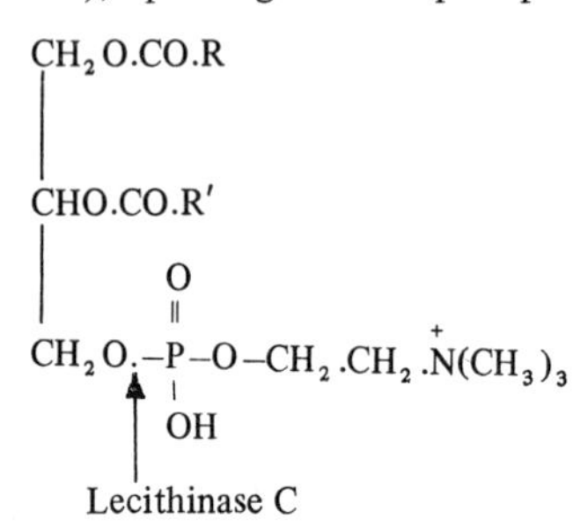

Figure 3.2. The lecithin molecule showing the point at which it is attacked by lecithinase C

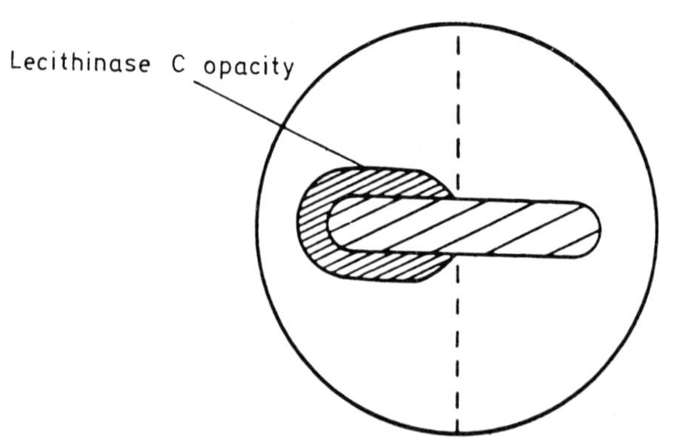

Figure 3.3. Cl. perfringens *on half-antitoxin egg yolk agar.* Cl. perfringens *antitoxin on the right-hand half of the plate inhibits the lecithinase C opalescence*

insoluble diglyceride; this is what happens when a lecithinase C-producing organism is grown on an egg yolk agar plate (Willis and Gowland, 1962). The soluble lecithinase diffuses away from the surface colonies and splits the naturally occurring lecithin in the egg yolk, the visible result of which is the development of an opaque halo in the medium about the zones of growth. It has been noted that the bacterial lecithinases C are inhibited by appropriate antitoxic sera and that use of this is made in the half-antitoxin plate (*Figure 3.3*). Anaerobes which show lecithinase C activity on egg yolk agar are *Cl. perfringens, Cl. bifermentans, Cl. sordellii, Cl. novyi* types A, B and D, and *Cl. barati.* For the demonstration of lecithinase C activity in the absence of lipolysis, organisms may be grown on a nutrient agar base medium containing pure egg lecithin. On lecithin agar (Willis, 1960b) lecithinase C reactions are as described above, and are inhibited by appropriate antitoxic sera (*see also* Ellner and O'Donnell, 1971).

LIPASE ACTIVITY

The lipases produced by anaerobes break down the naturally occurring free fats present in egg yolk (but not compound fats such as lecithin) with the production of glycerol and free fatty acids. The fatty acids are, in the main, insoluble substances, and as some of them have melting points above 37 °C they appear in the egg yolk agar medium as zones of opacity. These zones of anaerobic lipase opacity are usually restricted, so that they are limited to that part of the medium immediately beneath the colonies. Lipase opacity is thus easily distinguished from the diffuse lecithinase C type of egg yolk reaction. In addition to producing a restricted opacity in egg yolk agar, the lipolytic anaerobes also produce a surface pearly layer which covers the colonies and often extends for a short distance over the surface of the medium (*Figure 3.4*). The pearly layer, so called because of its oily mother-of-pearl sheen, is a thin

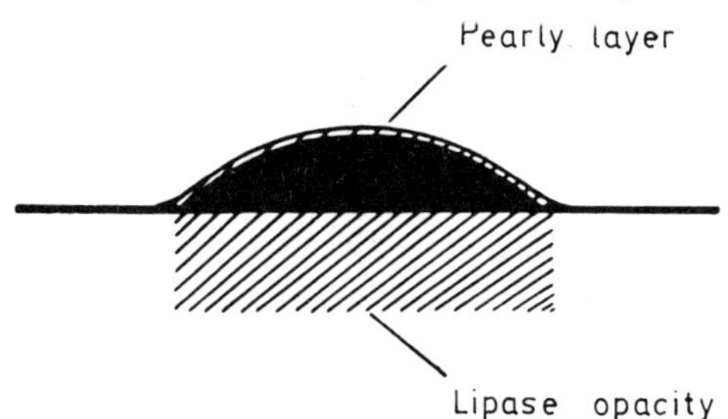

Figure 3.4. Schematic elevation of a lipolytic colony on egg yolk agar showing restricted opacity and pearly layer

film composed of fatty acids which can be readily washed off the colonies and medium. These reactions require 48 h incubation for their full development, unlike the lecithinase C effect which requires only 24 h. Though these lipases are antigenic, high titre anti-lipase sera are not available so that half-antitoxin plates similar to those used with anti-lecithinase sera are of no value. Anaerobes that show lipolytic activity on egg yolk agar are *Cl. sporogenes*, all types of *Cl. botulinum* (the lipolytic activity of *Cl. botulinum* type G has not been recorded), *Cl. novyi* type A and *F. necrophorum.*

Other fat media may also be used to demonstrate bacterial lipolysis. Ox tallow (Eijkman, 1901), incorporated in a nutrient agar plate has been used in the past, but is less satisfactory than media such as cow's cream agar, trioleine agar and tributyrin agar (Willis, 1960a, b). It should be noted that lipolysis on tributyrin agar does not result in an opacity and a pearly layer, but in zones of clearing about areas of growth. This is because the breakdown products of glycerol tributyrate (glycerol and butyric acid) are both water soluble substances. Bacterial lipolysis may be rendered more obvious by the addition of night blue, victoria blue or aniline blue S/S to the medium (Jones and Richards, 1952; Holt, 1971), or by treating plate cultures after incubation with saturated aqueous copper sulphate solution (Berry, 1933; Willis, 1960a, b). The copper sulphate reacts in the cold with fatty acids to form greenish-blue insoluble copper soaps. Since fats are not 'stained' in this way, the method serves also to distinguish between lecithinase and lipase opacities in egg yolk agar media (*see also* Willis and Gowland, 1962).

Sierra (1957) made use of the water soluble 'Tweens' for detection of bacterial lipolysis, and Sebald and Prévot (1960) and Gonzalez and Sierra (1961) adapted the method to detect lipolytic activity of anaerobic bacteria. The test depends on the production of zones of opacity in the medium due to the formation of precipitates of calcium salts of fatty acids. Although the 'Tween' test appears to be more sensitive than other methods for the detection of lipolysis (Gonzalez and Sierra recorded positive results for strains of *Cl. perfringens, Cl. sordellii* and *Cl. tetani*, which are negative by other methods), the reaction is slow to develop. Thus, the shortest incubation period for the development of a positive result was 4 days for *Cl. botulinum* type A, while *Cl. sporogenes* and *Cl. novyi* require 10 days' incubation. These organisms give positive results on other fat media in 48 h.

Antitoxin nutrient agar

Antitoxin nutrient agar for the identification of toxigenic clostridia was developed by Petrie and Steabben (1943). It involves the addition of specific antitoxin (8 units/ml) to nutrient agar which is used as the plating medium. Colonies of species against which the antitoxin was prepared are surrounded by concentric rings of precipitate. This method has been successfully applied to cultures of *Cl. perfringens, Cl. novyi*, *Cl. septicum* and *Cl. tetani*. Its chief disadvantage is that the precipitation reaction on which identification depends is often slow to develop.

Fermentation media

Many anaerobes do not grow satisfactorily in the ordinary peptone water sugar media used in the identification of enteric bacilli, and it is necessary to use an enriched basal medium for tube fermentation tests. VL broth base is generally satisfactory for this purpose. If the fluid medium fails to support growth of a particular organism, the horse blood agar plate fermentation method of Phillips (1976) is almost always successful.

For the clostridia the most important fermentable substances are glucose, maltose, lactose and sucrose, in that order. Useful additions to this basic set of fermentable substances are mannitol and salicin. Strongly saccharolytic species such as *Cl. perfringens* and *Cl. butyricum* ferment all four basic sugars, while weakly saccharolytic organisms may ferment glucose or glucose and maltose only. Among the clostridia, organisms that attack glucose usually also ferment maltose; but organisms which do not ferment glucose do not attack any of the basic sugars, and constitute the group of non-saccharolytic clostridia. Another general rule which is often of value in the identification of clostridia is that lactose-fermenting anaerobes are rarely strongly proteolytic.

For the Gram-negative non-sporing anaerobic bacilli, the most important fermentable substances are again glucose, maltose, lactose and sucrose, together with starch and aesculin; xylose and arabinose may be added to this basic set. For the sub-grouping of *B. fragilis*, mannitol, trehalose and rhamnose are required.

For the Gram-positive non-sporing anaerobic bacilli, the fermentable substances required are glucose, maltose, lactose, sucrose, mannitol, salicin, xylose and arabinose.

For the anaeronic cocci, the most useful fermentable substances are glucose, maltose, lactose and sucrose. It is useful to note that the *Veillonella* (Gram-negative cocci) are non-saccharolytic.

Indole production

Indole is tested for in the usual way by adding Kovac's reagent to a trypticase nitrate broth culture of the anaerobe. Amongst the clostridia, *Cl. bifermentans, Cl. sordellii, Cl. tetani* and *Cl. cadaveris* are the only important indole-positive species. Among the non-sporing anaerobes, indole is produced by *B. melaninogenicus, F. necrophorum,* and by some strains of *B. fragilis* and *P. acnes.* Some peptostreptococci are also indole producers.

Nitrate reduction

Some organisms are able to reduce nitrates to nitrites, ammonia or free nitrogen. In order to test this ability, organisms are cultured in trypticase nitrate broth. The reduction of nitrate to nitrite is subsequently tested for by adding to the culture a few drops each of a sulphanilic acid solution and Cleve's acid (*see* Chapter 2); a distinct pink colour develops in the presence of nitrite. The test for nitrate reduction is rarely necessary for the identification of the clostridia, as it is not particularly definitive. It is of more value in the study of the non-sporing anaerobes; thus, among the anaerobic cocci the veillonellae reduce nitrate, but the peptococci and peptostreptococci do not. The test is also of value in species identification among the *Propionibacterium, Eubacterium* and *Bacteroides.*

Hydrogen sulphide production

Hydrogen sulphide is detected in bacterial cultures by observing the blackening which it produces in the presence of salts of lead, iron or bismuth. For this purpose the metallic salt may be incorporated in the medium or an indicator strip of filter paper impregnated with the salt may be used. The strip technique is favoured, since the growth of some organisms is inhibited by the presence of these metallic salts in the medium. In carrying out the test the organism is inoculated into any suitable medium which contains available sulphur compounds. After inoculation the indicator strip is wedged in the top of the tube so that about 1 cm of the strip projects below the closure. During incubation the culture is examined regularly to observe whether or not blackening of the test strip has occurred. The test is not of very great value in routine diagnostic work since many of the commonly occurring anaerobes produce hydrogen sulphide.

Cooked meat broth

Practically all anaerobes grow well in cooked meat broth, especially laboratory-prepared medium; some of the commercially available dehydrated cooked meat media are unsatisfactory. The growth of saccharolytic species is often associated with the development of a pink colour in the meat particles probably due to the low Eh attained, with consequent reduction of the haematin. Proteolytic species attack and digest the meat particles, with the formation of a dark red or grey-black 'sludge' in the bottom of the tube, and an accompanying putrid odour. Gas is commonly produced in cooked meat medium by both saccharolytic and proteolytic anaerobes. Some organisms produce a butyrous scum on the surface of the broth, which probably indicates lipolytic activity. Cooked meat medium cultures of some proteolytic species, notably *Cl. histolyticum*, develop a deposit of small rounded masses of white tyrosine crystals, which increases with age, and gives the culture a 'snowed-up' appearance. This is probably attributable to the fact that these species do not deaminate tyrosine to any appreciable extent (Bessey and King, 1934).

While cooked meat medium thus provides evidence of both saccharolytic and proteolytic activity these changes are more conveniently determined by other methods.

ANTIBIOTIC DISC RESISTANCE

Finegold and his colleagues (Finegold, Harada and Miller, 1967; Finegold and Miller, 1968; Sutter and Finegold, 1971), showed that the determination of sensitivity or resistance to a number of antimicrobial agents was a valuable aid to the rapid presumptive identification of Gram-negative anaerobic bacilli. The drugs and their concentrations found most useful are colistin (10 μg), erythromycin (60 μg), kanamycin (1 mg), neomycin (1 mg), penicillin (2 units) and rifampicin (15 μg). Although, as might be expected, some strains of *Bacteroides* and *Fusobacterium* do not conform to the expected patterns of resistance or sensitivity, the frequency of aberrant results is no greater than is experienced with other more conventional tests. Resistance tests, which must be clearly distinguished from sensitivity tests performed for therapeutic purposes, are carried out on pure plate cultures in the usual way, using the appropriate single discs (*see also* Peach, 1975; Duerden *et al.*, 1976).

Single discs of gentamicin (10 μg) and metronidazole (5 μg), used together on primary anaerobic plate cultures of pathological material, and on anaerobic plate subcultures of blood cultures, facilitate early differentiation of the growth of anaerobes from that of facultative and carbon dioxide-dependent organisms. Since all anaerobes are sensitive to metronidazole and all other organisms are universally resistant to it, a zone of inhibition about the metronidazole disc indicates the presence of anaerobes. Since the converse is generally true of gentamicin, a zone of inhibition about this disc is indicative of facultative growth.

GROWTH INHIBITION AND STIMULATION

The effects of various inhibitors (other than antibiotics) and of growth stimulants have been usefully applied to the recognition of some anaerobes. The effect of bile on the growth of different Gram-negative anaerobes was first reported upon by Beerens and Castel (1960) and subsequently by Shimada, Sutter and Finegold (1970); its stimulant effect upon the growth of *B. fragilis* and *F. mortiferum* is now widely used in the recognition of these organisms (*see for example* Barnes, Impey and Goldberg, 1966; Holdeman and Moore, 1975). Definitive use may also be made of the inhibitory effect of bile of the growth of some species such as *B. melaninogenicus* and *B. oralis.* Other growth stimulants include haemin, which is notably effective with *B. melaninogenicus,* and Tween 80, which is markedly stimulatory to the growth of peptococci, peptostreptococci and propionibacteria.

Various dyes such as gentian violet, crystal violet, brilliant green, victoria blue and ethyl violet, are among the inhibitors that have received the most attention (Slanetz and Rettger, 1933; Omata and Disraely, 1956; Beerens and Tahon-Castel, 1965; Suzuki, Ushijima and Ichinose, 1966; Duerden *et al.*, 1976). Generally speaking, dyes have found their widest application in the preparation of selective media, and are less frequently used as discriminants upon pure cultures.

ANTIBIOTIC SUSCEPTIBILITY TESTING OF ANAEROBES

Antibiotic susceptibility testing of anaerobes differs in no fundamental way from the testing of facultative bacteria, a detailed consideration of which is beyond the scope of the present work (*see* Garrod *et al.*, 1973).

Because the susceptibilities of different anaerobic species to various antimicrobial agents are moderately constant and predictable, some workers consider that their routine testing is unnecessary, and may be conveniently replaced by a system of periodic monitoring (Martin, Gardner and Washington, 1972; Kwok *et al.*, 1975). It is considered that a working policy of this sort must be operated under close supervision and must be supported by a very thorough understanding of the pathogenetics of anaerobic bacterial infections. Moreover, reliance on this approach is likely to be misplaced if significant isolates are not 'speciated', and if the use of drugs such as tetracycline, erythromycin and co-trimoxazole is contemplated; the action of these drugs against different strains of the same anaerobic species is likely to be erratic. For those laboratories that do not perform routine susceptibility testing of anaerobes, it is important to note that testing is always relevant in serious infections such as generalized peritonitis, bacteraemia, endocarditis and brain abscess.

Of the various methods available for determining the sensitivity of anaerobes to antimicrobial agents (Wilkins and Appleman, 1976), disc diffusion with regression line analysis has been widely recommended (Sapico *et al.*, 1972; Wilkins *et al.*, 1972; Sutter, Kwok and Finegold, 1972; Dornbusch, Nord and Olsson, 1975; Kwok *et al.*, 1975), but minimum inhibitory concentration (MIC) determinations alone have also been used (Stalons and Thornsberry, 1975; Rotilie *et al.*, 1975). The methods of testing recommended by Holdeman and Moore (1975) are a Kirby–Bauer type disc diffusion method, and an elegant broth-disc method developed by Wilkins and Thiel (1973). In the broth-disc method the approximate blood level concentration of an antibacterial agent is put into a tube of broth medium using commercial antibiotic discs as the carrier of the drug. Good growth of the test organism at this single concentration indicates resistance to clinically useful levels of the drug (*see also* Blazevic, 1975).

Laboratories that are called upon to determine the sensitivity of anaerobes to antibiotics are likely simply to apply the routine method that they use for testing facultative anaerobes; and this will very commonly be a simple disc diffusion method interpreted by comparison with a control. This is entirely adequate provided that the usual basic rules of disc sensitivity testing and of anaerobic culture are adhered to. Care must be taken, however with some additional factors that are known to influence the results of anaerobic sensitivity determination. Although the addition of 10% carbon dioxide to the anaerobic atmosphere is essential for the satisfactory growth of many anaerobes, the fall in pH of the medium due to the presence of the gas is known to partially inactivate drugs such as erythromycin, lincomycin and spectinomycin (Ingham *et al.*, 1970; Rosenblatt and Schoenknecht, 1972; Phillips and Warren, 1975). It seems likely, also, that the presence of even small amounts of fermentable substances in the sensitivity test medium will lead to similar anomalous results among pH-sensitive antibiotics (Watt and Brown, 1975).

Although it is conventional and convenient to use the Oxford staphylococcus as the control organism, there is some uncertainty as to the validity of using an aerobe for the control of anaerobic sensitivities. Indeed, in the case of metronidazole testing, facultatively anaerobic control organisms are entirely inappropriate. For this reason it is customary in my laboratory to make comparisons against a standard laboratory strain of *B. fragilis* or *Cl. sporogenes*.

ANIMAL INOCULATION AND PROTECTION TESTS

Ideally, especially for the pathogenic clostridia, final identification of a species is obtained by inoculating two animals (guinea pigs or mice), one of which has been protected with specific antitoxin. In practice, however, animal inoculation is often unnecessary, as most of the common clostridia are easily recognized by other means. One important exception is *Cl. botulinum*, some strains of which are very similar in their cultural reactions to *Cl. sporogenes*. Final diagnosis can then be made only on the results of animal tests.

Since the pathogenic clostridia are either histotoxic or neurotoxic, some amount of toxin is usually present in the material (commonly a fluid culture) used for the inoculation of laboratory animals. If large amounts of toxin are present, disease may ensue very quickly and be rapidly fatal without the typical sequence of symptoms developing.

All pathogenic anaerobes require a nidus of local necrosis at the site of their inoculation into the tissues in order that multiplication and toxin production shall occur. It is therefore a common practice to add 2% of calcium chloride to the culture for inoculation, the salt acting as a necrotizing agent. Since calcium chloride is toxic to micro-organisms the salt is added to the culture immediately before inoculation.

Another important use of laboratory animals in anaerobic work is the demonstration of pathogenicity or toxigenicity of a known organism. *Cl. tetani*, for example, can easily be identified by its cultural and morphological characteristics, but its toxigenicity can be established only by animal inoculation and protection tests. Absence of toxigenicity in such cases does not contradict a diagnosis of *Cl. tetani*, since non-toxigenic strains of this organism are not uncommon.

In addition to these procedures, the typing of some clostridia, such as *Cl. perfringens* and *Cl. novyi*, is partly dependent on the inoculation of animals with culture filtrates. Here, the aim is to identify specific exotoxins, the presence of which can be recognized only by *in vivo* methods (*see* Oakley, 1967; Willis, 1969). Similarly, the use of animals is essential for the titration of many toxin preparations and for the standardization of toxoids and antitoxins.

Animal inoculation methods are of little value in the identification of non-clostridial anaerobes.

SEROLOGICAL AND TOXICOLOGICAL ASPECTS OF ANAEROBES

The serology of the pathogenic anaerobes was reviewed by Smith (1955). With the exception of immunofluorescence, the various serological reactions, such as somatic and flagellar agglutination, precipitation and complement fixation, do not play an important part in the routine recognition or subdivision of these organisms. Serological methods are, however, of great value in research studies of the anaerobes, especially in relation to taxonomy. Significant studies of this sort are at present concerned, for example, with the *Actinomyces* (*see below*).

The introduction of fluorescent labelled antibodies by Coons, Creech and Jones (1941) made available a new approach to the serology of these organisms, and has proved to be of diagnostic value in the differentiation of some important clostridia. Of great importance in this respect is the pioneer work of Batty and Walker (1963a, b; 1965; 1966; 1967) (*see also* Batty, 1967), who developed the fluorescent antibody technique for the detection, identification and differentiation of clostridial species. The successful application of the immunofluorescent staining method to bacteria depends on the fact that all strains of the particular species share at least one common O antigen that is not possessed by any other species.

CL. SEPTICUM AND CL. CHAUVOEI

All strains of *Cl. septicum* are divisible into two groups on the basis of the O antigen, and neither of these two groups cross reacts with *Cl. chauvoei*; all strains of *Cl. chauvoei* share a common O antigen (Moussa, 1958, 1959). Strains of both species are identifiable in a single fluorescent antibody (FA) stained smear using *Cl. septicum* antiserum conjugated with fluorescein (green fluorescence) and *Cl. chauvoei* antiserum conjugated with rhodamine (orange fluorescence).

CL. NOVYI

All types of *Cl. novyi* fluoresce with a *Cl. novyi* type A antiserum conjugated with fluorescein. Attempts to produce type-specific antisera have not been successful.

CL. BOTULINUM

Three fluorescent antibody groups of *Cl. botulinum* are distinguished. One group comprising types A, B and F, another comprising types C and D, and a single group consisting of type E only. Organisms within these groups fluoresce with fluorescein-labelled antisera prepared respectively against *Cl. botulinum* types A, C and E.

CL. TETANI

All strains of *Cl. tetani* share a common O antigen and may be specifically stained with fluorescent-labelled antiserum.

Preliminary work as led Dowell (1974) and Stauffer *et al.* (1975) to suggest that it may be possible to develop polyvalent FA conjugates for the commonly encountered non-sporing anaerobic bacilli. At present the matter remains experimental, except among species of *Actinomyces* and *Arachnia* for which FA staining is definitive (Snyder, Bullock and Parker, 1967; Baboolal, 1968; Brock and Georg, 1969; Holmberg and Forsum, 1973).

Conjugated antisera for the FA technique are available commercially from Wellcome Reagents as follows: *Cl. septicum* (fluorescein), *Cl. chauvoei* (fluorescein), *Cl. chauvoei* (rhodamine), *Cl. novyi* (fluorescein), and *Cl. botulinum* types A, C and E (fluorescein).

The toxicology of the pathogenic anaerobes is dealt with under the appropriate species headings in later chapters. It is sufficient to note here that soluble bacterial antigens can be used, not only for the recognition of some anaerobic species but also in the division of species into types or similar small groups (Oakley, 1955, 1962).

GAS LIQUID CHROMATOGRAPHY

In the past, workers encountered considerable difficulty in establishing the specific identities of a variety of non-sporing anaerobic bacteria. Not unreasonably, emphasis was placed on the habitat, microscopic morphology and conventional biochemical reactions of these organisms for their classification, taxonomic determinants which were (and still are) well suited to the classification of many other groups of bacteria, including the clostridia. Generally speaking, attempts to classify the non-sporing anaerobes by these criteria met with failure, so that once a consistent set of characteristics had been obtained for any particular strain, it was often impossible to assign a specific name to the organism. Thus, Eggerth and Gagnon (1933) isolated 118 strains of fusiform rods from human faeces, and divided them into 18 species, only two of which had been described previously. New species, often represented by only one or two strains were frequently recorded, but were not encountered by later workers. The confusion regarding the Gram-negative anaerobic bacilli was emphasized by Dack (1940) who listed 22 synonyms of *F. necrophorum.*

In their notes on the 36 species (and 6 variants) of anaerobic cocci listed in the seventh edition of *Bergey's Manual*, Breed, Murray and Smith (1957) admitted that some species, possibly many of them, may be identical with one another. Such uncertainty was due in part to the rarity with which individual strains could be identified with described species, and to the characteristic variation which these organisms showed in their cultural reactions when tested by conventional methods. Referring to this problem, Smith (1955) commented that 'primarily, the trouble lies in the fact that variability in cultural characteristics is the rule rather than the exception, and, given a sufficient number of isolates, unbroken chains of organisms leading from one species to another may be arranged'.

It was suggested by Beerens *et al.* (1962) that these problems might be resolved by the application of other biochemical tests, in particular the detection of the types of volatile fatty acids produced from fermentation of glucose. This suggestion followed from an earlier observation of Guillaumie *et al.* (1956) that among the metabolic end products of bacterial growth various combinations of formic, acetic, propionic and butyric acids were characteristic of the different genera of the *Bacteroidaceae*. In the third edition of his manual, Prévot (1957) made frequent reference to the products of bacterial metabolism although he did not use them as cardinal definitive features.

At that time the techniques for the extraction and identification of fatty acids from cultures were laborious and time-consuming (Charles and Barrett, 1963), and this type of analysis was widely regarded in clinical microbiological circles as essentially of academic or of limited taxonomic interest. In 1952, James and Martin developed the technique of gas chromatography; they used a moving gas phase to separate fatty acids on a column containing silicone oil supported on kieselguhr. This experiment initiated a surge of interest in gas chromatography which led to the development of the highly sophisticated gas chromatography systems that are available today.

American workers were quick to explore the application of this new technique to the analysis of volatile fatty acids produced by bacterial metabolism, and it soon became evident that the concept of Beerens and his colleagues was of fundamental importance in the characterization of the non-sporing anaerobic bacteria. Largely due to the careful and detailed studies of Professor W. E. C. Moore and his colleagues at the Virginia Polytechnic Institute Anaerobe Laboratory, much valuable information is now available concerning the relationships of metabolic products to the taxonomy of the anaerobic bacteria (*see* Moore, 1970; Bricknell, Sutter and Finegold 1975; Moore, Cato and Holdeman, 1966, 1969; Holdeman and Moore, 1975; Moore and Holdeman 1972).

At present it is fair to say that in the routine clinical laboratory detailed specific identification of most non-sporing anaerobic isolates is not mandatory. For those who wish to 'speciate' their isolates, however, chromatographic analysis of volatile metabolic products is essential.

Although this may be accomplished by paper chromatographic methods (Slifkin and Hercher, 1974), gas liquid chromatography is much more convenient. In addition to its proven value in the specific identification of pure cultures of anaerobes, gas liquid chromatography also finds a place in the examination of mixed cultures (Mitruka *et al.*, 1973). Of special interest are the studies of Gorbach *et al.*, (1974, 1976) and Phillips, Tearle and Willis, (1976) which show that direct gas liquid chromatographic analysis of clinical specimens (notably pus) enables anaerobic infections to be recognized within a few minutes of their collection.

Principle of gas liquid chromatography

Gas liquid chromatography (partition gas chromatography) is the process by which a mixture is separated by solution partition between two immiscible phases – a 'liquid stationary phase' distributed on a solid inert support material in a column, and a 'moving gas phase'. The separating principle depends on the difference in the partition coefficients between the liquid and gas phases of the constituents of the mixture. Those constituents which favour the gas phase have low partition coefficients and move quickly through the column, while those that favour the liquid phase have high partition coefficients and are delayed in their transit through the column.

In practice, a little of the liquid mixture for analysis is inoculated onto the column which is held at a temperature that ensures vaporization of the volatile components of the mixture; clearly, the stationary phase in the column must be non-volatile at this temperature. A steady flow of carrier gas (commonly nitrogen) carries the gas phase of each volatile component through the column until each is eluted. The volatile components pass through the column at different rates, so that they are eluted from the end of the column at different times. A detector at the end of the column records the emergence of the various volatile constituents of the original mixture, which is thus analysed.

The detector most suited to microbiological analyses is the flame ionization detector (FID). In this system the detector is used to measure an ionization current in a hydrogen flame. When an eluted substance is burned in the flame there is an increase in the ionization current; this signal is fed to an amplifier and then to a strip chart recorder. The recorder plots the attenuated detector output against a range of time scales.

A recommended specification is the Pye Unicam Series 104 chromatograph, the design concept of which allows for continuing updating of the instrument as requirements expand. A range of add-on modules is available which convert the basic Series 104 chromatograph to varying levels of automation. To the basic instrument in my laboratory has been added the Auto-jector S4, which enables analysis of up to 100 samples to be carried out automatically.

In the Series 104 Chromatograph the column packing used for the analysis of volatile acids and methyl derivatives of non-volatile acids in polyethylene glycol adipate (10%) (liquid phase) adsorbed onto Celite (mesh size 100–120) (support). For the analysis of alcohols, the column packing used is polyethylene glycol 400 (10%) (liquid phase) adsorbed onto Celite (mesh size 100–120) (support).

Analysis of acid and alcohol products of bacterial metabolism

Bacterial cultures

Cultures for gas liquid chromatographic analysis are grown in VL glucose broth in an anaerobic atmosphere containing 10% carbon dioxide. Analysis may be carried out once good growth has developed (24–48 h for many species), although some organisms, notably proteolytic species, require 5 days' incubation for maximum yields of butyric acid.

Analysis of volatile fatty acids and alcohols

Acidify the culture by adding 0.2 ml of 50% H_2SO_4 to each 12 ml of culture. This ensures that fatty acids are present in the free form, and are therefore soluble in ether.

Pipette 2 ml of the acidified culture into a centrifuge tube, add 1 ml of ethyl ether, stopper the tube and mix by inversion. Lightly centrifuge to separate the ether and aqueous phases, and pipette off the ether layer for analysis.

Inject 1 μl of the ether extract into the column.

Analysis of methyl derivatives of non-volatile fatty acids

Non-volatile fatty acids (pyruvic, lactic, fumaric and succinic acids) are converted to volatile methyl derivatives for chromatographic analysis as follows.

Place 1 ml of the acidified culture into a small test tube. Add 0.4 ml of 50% H_2SO_4 and 2 ml of methanol. Stopper, and heat at 55 °C in a water bath for 30 min. Add 0.5 ml of chloroform and 1 ml of distilled water, and mix by inversion. If necessary, separate the chloroform and aqueous phases by light centrifugation, and withdraw the chloroform (lower) phase for analysis.

Inject 1 μl of the chloroform extract into the column.

Standard solutions

Each time chromatographic analyses are carried out it is necessary to prepare 'control' chromatograms from mixtures of standard solutions of the different substances being sought.

VOLATILE ACID STANDARD STOCK SOLUTIONS

Nine standard stock solutions are prepared as follows:

1. Acetic acid	5.7 ml
2. Propionic acid	7.5 ml
3. Isobutyric acid	9.2 ml
4. Butyric acid	9.1 ml
5. Isovaleric acid	12.7 ml
6. Valeric acid	12.5 ml
7. Isocaproic acid	12.6 ml
8. Caproic acid	12.6 ml
9. Heptanoic acid	12.6 ml

Make up each of these nine standard solutions to 100 ml with distilled water, and keep them well stoppered.

Working mixed standard solution. Mix together 1 ml of each stock standard, and make up the volume to 100 ml with distilled water. Keep this working standard well stoppered.

Analysis of working mixed standard. Proceed as described under analysis of volatile fatty acids and alcohols in cultures (p. 99).

ALCOHOL STANDARD STOCK SOLUTIONS

Six standard stock solutions are prepared as follows:

1. Ethanol	10.0 ml
2. Propanol	3.5 ml
3. Isobutanol	0.5 ml
4. Butanol	1.0 ml
5. Isopentanol	0.5 ml
6. Pentanol	0.5 ml

Make up each of these six standard stock solutions to 100 ml with distilled water, and keep them well stoppered.

Working mixed standard solution. Mix together 1 ml of each stock standard, and make up the volume to 100 ml with distilled water. Keep this working standard well stoppered.

Analysis of working mixed standard. Proceed as described under analysis of volatile fatty acids and alcohols in culture (p. 99).

NON-VOLATILE ACID STANDARD STOCK SOLUTIONS

Eight standard stock solutions are prepared as follows:

1. Pyruvic acid	6.8 ml
2. Lactic acid	8.4 ml
3. Oxalacetic acid	6.0 g
4. Oxalic acid	6.0 g
5. Methyl malonic acid	6.0 g
6. Malonic acid	5.0 g
7. Fumaric acid	6.0 g
8. Succinic acid	6.0 g

Make up each of these standard solutions to 100 ml with distilled water, and keep them well stoppered.

Working mixed standard solution. Mix together 1 ml of each stock standard, and make up the volume to 100 ml with distilled water. Keep this working standard well stoppered.

Analysis of working mixed standard. Acidify 12 ml of the working mixed standard solution by adding 0.2 ml of 50 per cent H_2SO_4. Then proceeds as directed under analysis of methyl derivatives (p. 100).

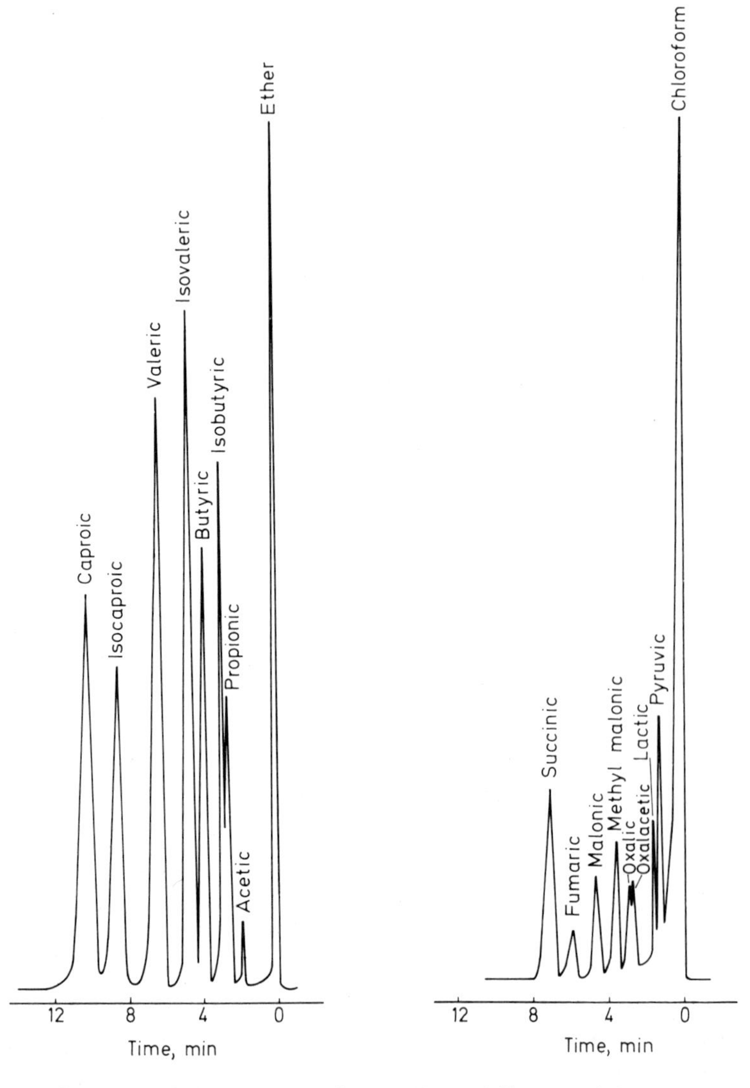

Figure 3.5a. Typical chromatogram of volatile acids

Figure 3.5b. Typical chromatogram of methylated acids

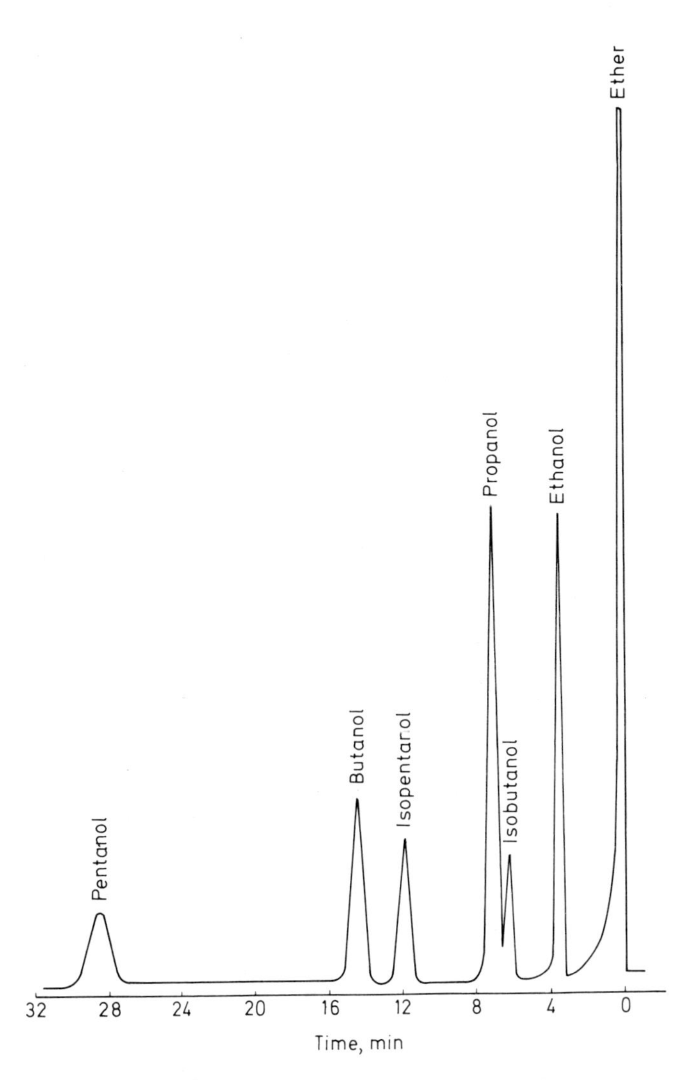

Figure 3.6. Typical chromatogram of alcohols

Medium control

Since different batches of uninoculated culture medium may contain varying small amounts of acids, especially acetic, lactic and succinic acids, corrections for these must be made when chromatograms of cultures are interpreted. For this reason a 'control' analysis of the uninoculated medium is performed with each new batch of culture medium.

Interpretation of chromatograms

The presence and approximate amounts of acids and alcohols in a culture extract are determined by comparing the culture chromatograms (duly corrected by the chromatograms of the uninoculated medium) with the chromatograms of the working standard solutions (*Figures 3.5* and *3.6*). Identification of the culture is then accomplished by comparing its chromatograms against those of known species.

Chromatographic patterns of many of the clinically important anaerobes are summarized in later chapters. Chromatographic patterns of a much wider range of anaerobes have been published by Holdeman and Moore (1975).

REFERENCES

Baboolal, R. (1968). 'Identification of filamentous micro-organisms of the human dental plaque by immuno-fluorescence.' *Caries Research*, **2**, 273

Barnes, E. M., Impey, C. S. and Goldberg, H. S. (1966). 'Methods for the characterization of the *Bacteroidaceae*.' In *Identification Methods for Microbiologists*. p. 51. Ed. by B. M. Gibbs and F. A. Skinner. London; Academic Press

Batty, I. (1967). 'Fluorescent-labelled antibodies as tools in the detection, identification and differentiation of clostridial species.' In *The Anaerobic Bacteria*. p. 85. Ed. by V. Fredette. Montreal; Institute of Microbiology and Hygiene

Batty, I. and Walker, P. D. (1963a). 'The differentiation of *Clostridium septicum* and *Clostridium chauvoei* by the use of fluorescent-labelled antibodies.' *Journal of Pathology and Bacteriology*, **85**, 517

Batty, I. and Walker, P. D. (1963b). 'Fluorescent-labelled clostridial antisera as specific stains.' *Bulletin de l'Office International des Epizooties*, **59**, 1499

Batty, I. and Walker, P. D. (1965). 'Colonial morphology and fluorescent-labelled antibody staining in the identification of species of the genus *Clostridium*.' *Journal of applied Bacteriology*, **28**, 112

Batty, I. and Walker, P. D; (1966). 'Colonial morphology and fluorescent-labelled antibody staining in the identification of species of the genus *Clostridium*.'

In *Identification Methods for Microbiologists.* p. 81. Ed. by B. M. Gibbs and F. A. Skinner. London; Academic Press

Batty, I. and Walker, P. D. (1967). 'The use of the fluorescent-labelled antibody technique for the detection and differentiation of bacterial species.' *Symposia Series in Immunobiological Standardization,* **4,** 73

Beerens, H. and Castell, M. M. (1960). 'Action of bile on the growth of certain Gram-negative anaerobic bacteria.' *Annales de l'Institut Pasteur,* **99,** 454 (French)

Beerens, H., Castel, M. M. and Fievez, L. (1962). 'Classification of the *Bacteroidaceae.*' *International Congress for Microbiology.* p. 120. Montreal

Beerens, H. and Tahon-Castel, M. (1965). *Infections humaines à bactéries anaérobies non-toxigénes.* Brussels; Presses Academiques Europeenes

Berry, J. A. (1933). 'Detection of microbial lipase by copper soap formation.' *Journal of Bacteriology,* **25,** 433

Bessey, O. A. and King, C. G. (1934). 'Proteolytic and deaminizing enzymes of *Clostridium sporogenes* and *Clostridium histolyticum.*' *Journal of infectious Diseases,* **54,** 123

Blazevic, D. J. (1975). 'Evaluation of the modified broth-disk method for determining antibiotic susceptibilities of anaerobic bacteria.' *Antimicrobial Agents and Chemotherapy,* **7,** 721

Breed, R. S., Murray, E. G. D. and Smith, N. R. (1957). Bergey's *Manual of Determinitive Bacteriology,* 7th edn. London; Bailliere, Tindall and Cox

Bricknell, K. S., Sutter, V. L. and Finegold, S. M. (1975). 'Detection and identification of anaerobic bacteria.' In *Gas Chromatographic Applications in Microbiology and Medicine.* p. 251. Ed. by B. M. Mitruka. New York; Wiley

Brock, D. W. and Georg, L. K. (1969). 'Determination and analysis of *Actinomyces israelii* serotypes by fluorescent-antibody procedures.' *Journal of Bacteriology,* **97,** 581

Butler, H. M. (1945). 'Bacteriological studies of *Clostridium welchii* infections in man.' *Surgery, Gynaecology* and *Obstetrics,* **81,** 475

Charles, A. B. and Barrett, F. C. (1963). 'Detection of volatile fatty acids produced by obligate Gram-negative anaerobes.' *Journal of medical laboratory Technology,* **20,** 266

Coons, A. H., Creech, H. J. and Jones, R. N. (1941). 'Immunological properties of an antibody containing a fluorescent group.' *Proceedings of the Society of Experimental Biology, New York,* **47,** 200

Cowan, S. T. (1974). *Manual for the Identification of Medical Bacteria.* 2nd edn. London; Cambridge University Press

Dack, G. M. (1940). 'Non-spore-forming anaerobic bacteria of medical importance.' *Bacteriological Reviews,* **4,** 227

Dornbusch, K., Nord, C-E. and Olsson, B. (1975). 'Antibiotic susceptibility testing of anaerobic bacteria by the standardized disc diffusion method with special reference to *Bacteroides fragilis.*' *Scandinavian Journal of Infectious Diseases,* **7,** 59

Dowell, V. R. (1974). 'Methods and techniques for identification: Invited discussion.' In *Anaerobic Bacteria: Role in Disease.* p. 59. Ed. by A. Balows, R. M. DeHaan, V. R. Dowell and L. B. Guze. Springfield; Thomas

Duerden, B. I., Holbrook, W. P., Collee, J. G. and Watt, B. (1976). 'The characterization of clinically important Gram-negative anaerobic bacilli by conventional bacteriological tests.' *Journal of applied Bacteriology,* **40,** 163

Eggerth, A. H. and Gagnon, B. D. (1933). 'The bacteroides of human faeces.' *Journal of Bacteriology,* **25,** 389

Eijkman, C. (1901). 'On enzymes in bacteria and fungi.' *Zentralblatt fur Bakteriologie, Parasitenkunde, Infektionskrankheiten und Hygiene* (Abteilung I), **29**, 841

Ellner, P. D. and O'Donnell, E. D. (1971). 'A selective differential medium for histotoxic clostridia.' *American Journal of clinical Pathology*, **56**, 197

Ellner, P. D., Granato, P. A. and May, C. B. (1973). 'Recovery and identification of anaerobes: a system suitable for the routine clinical laboratory.' *Applied Microbiology*, **26**, 904

Fiévez, L. (1963). 'Comparative study of strains of *Sphaerophorus necrophorus* isolated from man and animals.' Brussels; Presses Académiques Européenes (French)

Finegold, S. M., Harada, N. E. and Miller, L. G. (1967). 'Antibiotic susceptibility patterns as aids in classification and characterization of Gram-negative anaerobic bacilli.' *Journal of Bacteriology*, **94**, 1443

Finegold, S. M. and Miller, L. G. (1968). 'Susceptibility to antibiotics as an aid in classification of Gram-negative anaerobic bacilli.' *The Anaerobic Bacteria.* p. 139. Ed. by V. Fredette. Montreal; Montreal University

Frazier, W. C. (1926). 'A method for the detection of changes in gelatin due to bacteria.' *Journal of infectious Diseases*, **39**, 302

Frobisher, M. (1933). 'Some pitfalls in bacteriology.' *Journal of Bacteriology*, **25**, 565

Garrod, L. P., Lambert, H. P., O'Grady, F. and Waterworth, P. M. (1973). *Antibiotic and Chemotherapy.* 4th edn. Edinburgh; Churchill Livingstone

Gonzalez, C. and Sierra, G. (1961). 'Lipolytic activity of some anaerobic bacteria.' *Nature, London*, **189**, 601

Gorbach, S. L., Mayhew, J. W., Bartlett, J. G., Thadepalli, H. and Onderdonk, A. B. (1974). 'Rapid diagnosis of *Bacteroides fragilis* infections by direct gas liquid chromatography of clinical specimens.' *Clinical Research*, **2**, 442A

Gorbach, S. L., Mayhew, J. W., Bartlett, J. G., Thadepalli, H. and Onderdonk, A. B. (1976). 'Rapid diagnosis of anaerobic infections by direct gas-liquid chromatography of clinical specimens.' *Journal of Clinical Investigation*, **57**, 478

Gordon, J. (1954). 'The use of meta-amino-acetanilide hydrochloride for the demonstration of hydrogen peroxide production by bacteria.' *Journal of Pathology and Bacteriology*, **68**, 645

Gordon, J. and McLeod, J. W. (1940). 'A simple and rapid method of distinguishing *Cl. novyi (B. oedematiens)* from other bacteria associated with gas gangrene.' *Journal of Pathology and Bacteriology*, **50**, 167

Guillaumie, J., Beerens, H. and Osteux, R. (1956). 'Paper chromatography of C_1 to C_6 aliphatic volatile acids. Its application to the identification of the anaerobic bacteria.' *Annales de l'Institut Pasteur de Lille*, **8**, 13 (French)

Hayward, N. J. (1941). 'Rapid identification of *Cl. welchii* by the Nagler reaction.' *British medical Journal*, **1**, 811, 916

Hayward, N. J. (1942). 'The use of benzidine media for the identification of *Clostridium oedematiens.' Journal of Pathology and Bacteriology*, **54**, 379

Hayward, N. J. (1943). 'The rapid identification of *Cl. welchii* by Nagler tests in plate cultures.' *Journal of Pathology and Bacteriology*, **55**, 285

Holdeman, L. V. and Moore, W. E. C. (1975). *Anaerobe Laboratory Manual.* Blacksburg; Virginia Polytechnic Institute Anaerobe Laboratory

Holmberg, K. and Forsum, U. (1973). 'Identification of *Actinomyces, Arachnia, Bacterionema, Rothia* and *Propionibacterium* species by defined immunofluorescence.' *Applied Microbiology*, **25**, 834

Holt, R. J. (1971). 'The detection of bacterial esterases and lipases.' *Medical Laboratory Technology*, **28**, 208

Howie, J. W. and Kirkpatrick, J. (1934). 'Observations on bacterial capsules as demonstrated by a simple method.' *Journal of Pathology and Bacteriology*, **39**, 165

Ingham, H. R., Selkon, J. B., Codd, A. A. and Hale, J. H. (1970). 'The effect of carbon dioxide on the sensitivity of *Bacteroides fragilis* to certain antibiotics *in vitro.' Journal of clinical Pathology*, **23**, 259

James, A. T. and Martin, A. J. P. (1952). 'Gas-liquid partition chromatography. A technique for the analysis of volatile materials.' *Analyst*, **77**, 915

Jones, A. and Richards, T. (1952). 'Night blue and victoria blue as indicators in lipolysis media.' *Proceedings of the Society of Applied Bacteriology*, **15**, 82

Kwok, Y. Y., Tally, F. P., Sutter, V. L. and Finegold, S. M. (1975). 'Disc susceptibility testing of slow-growing anaerobic bacteria.' *Antimicrobial Agents and Chemotherapy*, **7**, 1

McClung, L. S. and Toabe, R. (1947). 'Egg-yolk plate reaction for presumptive diagnosis of *Clostridium sporogenes* and certain species of gangrene and botulinum groups.' *Journal of Bacteriology*, **53**, 139

McClung, L. S., Heidenreich, P. and Toabe, R. (1945). 'The Nagler reaction for recognition of *Clostridium novyi (Cl. oedematiens).' Journal of Bacteriology*, **50**, 715

Macfarlane, R. G., Oakley, C. L. and Anderson, C. G. (1941). 'Haemolysis and the production of opalescence in serum and lecitho-vitellin by the a-toxin of *Clostridium welchii.' Journal of Pathology and Bacteriology*, **52**, 99

McLeod, J. W. (1928). In *The Newer Knowledge of Bacteriology and Immunology*. p. 213. Ed. by Jordan, E. O., and Falk, I. S. Chicago; University of Chicago Press

Martin, W. J., Gardner, M. and Washington, J. A. (1972). '*In vitro* antimicrobial susceptibility of anaerobic bacteria isolated from clinical specimens.' *Antimicrobial Agents and Chemotherapy*, **1**, 148

Marymont, J. H., Holdeman, L. V. and Moore, W. E. C. (1972). 'Anaerobic bacteriology in the clinical laboratory.' *American Journal of medical Technology*, **38**, 441

Miles, E. M. and Miles, A. A. (1947). 'The lecithinase of *Clostridium bifermentans* and its relation to the a-toxin of *Clostridium welchii.' Journal of general Microbiology*, **1**, 385

Mitruka, B. M., Jonas, A. M., Alexander, M. and Kundargi, R. S. (1973). 'Rapid differentiation of certain bacteria in mixed populations by gas-liquid chromatography.' *Yale Journal of Biology and Medicine*, **46**, 104

Moore, W. E. C. (1970). 'Relationships of metabolic products to taxonomy of anaerobic bacteria.' *International Journal of systematic Bacteriology*, **20**, 535

Moore, W. E. C., Cato, E. P. and Holdeman, L. V. (1966). 'Fermentation patterns of some *Clostridium* species.' *International Journal of systematic Bacteriology*, **16**, 383

Moore, W. E. C., Cato, E. P. and Holdeman, L. V. (1969). 'Anaerobic bacteria of the gastrointestinal flora and their occurrence in clinical infections.' *Journal of infectious Diseases*, **119**, 641

Moore, W. E. C. and Holdeman, L. V. (1972). 'Identification of anaerobic bacteria.' *American Journal of clinical Nutrition*, **25**, 1306

Moussa, R. S. (1958). 'Antigenic formulae for the genus *Clostridium.' Nature, London*, **181**, 123

Moussa, R. S. (1959). 'Antigenic formulae for *Clostridium septicum* and *Clostridium chauvoei.' Journal of Pathology and Bacteriology*, **77**, 341

Nagler, F. P. O. (1939). 'Observations on a reaction between the lethal toxin of *Cl. welchii* (Type A) and human serum.' *British Journal of experimental Pathology*, **20**, 473

Nagler, F. P. O. (1944). 'Bacteriological diagnosis of gas gangrene due to *Clostridium oedematiens.*' *Nature, London*, **153**, 496

Nagler, F. P. O. (1945). 'A cultural reaction for the early diagnosis of *Clostridium oedematiens* infection.' *Australian Journal of experimental Biology and medical Science*, **23**, 59

Oakley, C. L. (1955). 'Bacterial toxins and classification.' *Journal of general Microbiology*, **12**, 344

Oakley, C. L. (1962). 'Soluble bacterial antigens as discriminants in classification.' *Twelfth Symposium of the Society for General Microbiology*. Cambridge; Cambridge University Press

Oakley, C. L. (1967). 'Lesions produced by intradermal injection of clostridial toxins.' In *The Anaerobic Bacteria*. p. 209. Ed. by V. Fredette. Montreal; Institute of Microbiology and Hygiene

Oakley, C. L., Warrack, G. H. and Clarke, P. H. (1947). 'The toxins of *Clostridium oedematiens (Cl. novyi).*' *Journal of general Microbiology*, **1**, 91

Omata, R. R. and Disraely, M. N. (1956). 'A selective medium for oral fusobacteria.' *Journal of Bacteriology*, **72**, 677

Peach, S. (1975). 'Antibiotic-disc tests for rapid identification of non-sporing anaerobes.' *Journal of clinical Pathology*, **28**, 388

Petrie, G. F. and Steabben, D. (1943). 'Specific identification of the chief pathogenic clostridia of gas gangrene.' *British medical Journal*, **1**, 377

Phillips, I. and Warren, C. (1975). 'Susceptibility of *Bacteroides fragilis* to spectinomycin.' *Journal of antimicrobial Chemotherapy*, **1**, 91

Phillips, K. D. (1976). 'A simple and sensitive technique for determining the fermentation reactions of non-sporing anaerobes.' *Journal of applied Bacteriology*, **41**, 325

Phillips, K. D., Tearle, P. V. and Willis, A. T. (1976). 'Rapid diagnosis of anaerobic infections by gas-liquid chromatography of clinical material.' *Journal of clinical Pathology*, **29**, 428

Prévot, A. R. (1957). *Manual of Classification and Determination of Anaerobic Bacteria.* 3rd edn. Paris; Masson (French)

Prévot, A. R. (1966). *Manual for the Classification and Determination of the Anaerobic Bacteria.* 1st American Edn. Translated by V. Fredette. Philadelphia; Lea and Febiger

Reed, G. B. and Orr, J. H. (1941). 'Rapid identification of gas gangrene anaerobes.' *War Medicine, Chicago*, **1**, 493

Rosenblatt, J. E. and Schoenknecht, F. (1972). 'Effect of several components of anaerobic incubation on antibiotic susceptibility test results.' *Antimicrobial Agents and Chemotherapy*, **1**, 433

Rotilie, C. A., Fass, R. J., Prior, R. B. and Perkins, R. L. (1975). 'Microdilution technique for antimicrobial susceptibility testing of anaerobic bacteria.' *Antimicrobial Agents and Chemotherapy*, **7**, 311

Sapico, F. L., Kwok, Y. Y., Sutter, V. L. and Finegold, S. M. (1972). 'Standardized antimicrobial disc susceptibility testing of anaerobic bacteria: *In vitro* susceptibility of *Clostridium perfringens* to nine antibiotics.' *Antimicrobial Agents and Chemotherapy*, **2**, 320

Schaeffer, A. B. and Fulton, M. (1933). 'A simplified method of staining endospores.' *Science, New York*, **77**, 194

Schwabacher, H. and Lucas, D. R. (1947). '*Bacterium melaninogenicum* – a misnomer.' *Journal of general Microbiology,* **1,** 109

Sebald, M. and Prévot, A. R. (1960). 'Qualitative and quantitative studies on the lipolytic activity of 86 strains of clostridia.' *Annales de l'Institute Pasteur,* **99,** 386 (French)

Seiffert, G. (1939). 'A reaction in human serum with *Cl. perfringens* toxin.' *Zeitschrift fur Immunitatsforschung und experimentelle Therapie,* **96,** 515 (German)

Shimada, K., Sutter, V. L. and Finegold, S. M. (1970). 'Effect of bile and desoxychocolate on Gram negative anaerobic bacteria.' *Applied Microbiology,* **20,** 737

Sierra, G. (1957). 'A simple method for the detection of lipolytic activity and some observations on the influence of the contact between cells and fatty substrates.' *Antonie van Leeuwenhoek,* **23,** 15

Slanetz, L. W. and Rettger, L. F. (1933). 'A systematic study of the fusiform bacteria.' *Journal of Bacteriology,* **26,** 599

Slifkin, M. and Hercher, H. J. (1974). 'Paper chromatography as an adjunct in the identification of anaerobic bacteria.' *Applied Microbiology,* **27,** 500

Smith, L. DS. (1955). *Introduction to the Pathogenic Anaerobes.* Chicago; University of Chicago Press

Snyder, M. L., Bullock, W. W. and Parker, R. B. (1967). 'Morphology of Gram-positive filamentous bacteria identified in dental plaque by fluorescent antibody technique.' *Archives of oral Biology,* **12,** 1269

Stalons, D. R. and Thornsberry, C. (1975). 'Broth-dilution method for determining the antibiotic susceptibility of anaerobic bacteria.' *Antimicrobial Agents and Chemotherapy,* 7, 15

Stanbridge, T. N. and Preston, N. W. (1969). 'The motility of some *Clostridium* species.' *Journal of general Microbiology,* **55,** 29

Stauffer, L. R., Hill, E. O., Holland, J. W. and Altemeier, W. A. (1975). 'Indirect fluorescent antibody procedure for the rapid detection and identification of *Bacteroides* and *Fusobacterium* in clinical specimens.' *Journal of clinical Microbiology,* **2,** 337

Sutter, V. L. and Finegold, S. M. (1971). 'Antibiotic disc susceptibility tests for rapid presumptive identification of Gram-negative anaerobic bacilli.' *Applied Microbiology,* **21,** 13

Sutter, V. L. and Finegold, S. M. (1973). 'Anaerobic bacteria: Their recognition and significance in the clinical laboratory.' *Progress in clinical Pathology,* **5,** 219

Sutter, V. L., Kwok, Y. Y. and Finegold, S. M. (1972). 'Standardized antimicrobial disc susceptibility testing of anaerobic bacteria. I. Susceptibility of *Bacteroides fragilis* to tetracycline.' *Applied Microbiology,* **23,** 268

Suzuki, S., Ushijima, T. and Ichinose, H. (1966). 'Differentiation of *Bacteroides* from *Sphaerophorus* and *Fusobacterium.*' *Japanese Journal of Microbiology,* **10,** 193

Taylor, A. L. (1929). 'The anaerobic streptococci.' In *A System of Bacteriology.* Vol. 2, p. 136. London; HMSO

Tittsler, R. P. and Sandholzer, L. A. (1936). 'The use of semi-solid agar for the detection of bacterial motility.' *Journal of Bacteriology,* **31,** 575

Watt, B. and Brown, F. V. (1975). 'Sensitivity testing of anaerobes on solid media.' *Journal of antimicrobial Chemotherapy,* **1,** 440

Wilkins, T. D. and Appleman, M. D. (1976). 'Review of methods for antibiotic susceptibility testing of anaerobic bacteria.' *Laboratory Medicine,* 7, 12

Wilkins, T. D., Holdeman, L. V., Abramson, I. J. and Moore, W. E. C. (1972). 'Standardized single-disc method for antibiotic susceptibility testing of anaerobic bacteria.' *Antimicrobial Agents and Chemotherapy,* **1,** 451

Wilkins, T. D. and Thiel, T. (1973). 'Modified broth-disk method for testing the antibiotic susceptibility of anaerobic bacteria.' *Antimicrobial Agents and Chemotherapy,* **3,** 350

Williams, K. (1971). 'Some observations on *Clostridium tetani.*' *Medical laboratory Technology,* **28,** 399

Willis, A. T. (1960a). 'Observations on the Nagler reaction of some clostridia.' *Nature, London,* **185,** 943

Willis, A. T. (1960b). 'The lipolytic activity of some clostridia.' *Journal of Pathology and Bacteriology,* **80,** 379

Willis, A. T. (1962). 'Some diagnostic reactions of clostridia.' *Laboratory Practice,* **11,** 526

Willis, A. T. (1969). *The Clostridia of Wound Infection.* London: Butterworths

Willis, A. T. and Gowland, G. (1962). 'Some observations on the mechanism of the Nagler reaction.' *Journal of Pathology and Bacteriology,* **83,** 219

Willis, A. T. and Hobbs, G. (1957). 'A modified Nagler medium.' *Nature, London,* **180,** 92

Willis, A. T. and Hobbs, G. (1958). 'A medium for the identification of clostridia producing opalescence in egg yolk emulsions.' *Journal of Pathology and Bacteriology,* **75,** 299

Willis, A. T. and Hobbs, G. (1959). 'Some new media for the isolation and identification of clostridia.' *Journal of Pathology and Bacteriology,* **77,** 511

Willis, A. T. and Williams, K. (1970). 'Some cultural reactions of *Clostridium tetani.*' *Journal of medical Microbiology,* **3,** 291

4

Characteristics of the Pathogenic and Related Clostridia

DEFINITION

The clostridia are anaerobic or micro-aerophilic bacilli, usually staining Gram-positively and producing spores that commonly distend the organism. Some species decompose protein or ferment carbohydrates, or have both activities. Some produce exotoxins and are pathogenic for man. Useful information on the systematics of the clostridia is to be found in the publications of Smith (1955), Prévot (1966), Smith and Holdeman (1968), Willis (1969), Ajl, Kadis and Montie (1970), Buchanan and Gibbons (1974) and Holdeman and Moore (1975).

CLASSIFICATION

In *Table 4.1* are listed the main species of clostridia, classified according to their proteolytic and saccharolytic properties, as shown by their action on milk agar on the one hand, and on the sugars, glucose, maltose, lactose and sucrose, on the other.

These clostridia are the ones most likely to be encountered in clinical material, some of them much more commonly than others. With the general exception of *Cl. botulinum*, which is the causal organism of botulism in man and animals, all the clostridia listed are associated with wound infections. The most important of these are *Cl. tetani*, the causal organism of tetanus, and *Cl. perfringens, Cl. novyi, Cl. septicum,*

Cl. histolyticum, Cl. sordellii, Cl. bifermentans and *Cl. sporogenes,* those most commonly associated with gas gangrene. The other species are less frequently encountered and many of them are non-pathogenic.

Table 4.1 Division of some clostridia according to their saccharolytic and proteolytic properties

Saccharolytic and proteolytic	*Proteolytic but non-saccharolytic*	*Saccharolytic but non-proteolytic*	*Non-saccharolytic and non-proteolytic*
Cl. sporogenes	*Cl. histolyticum*	*Cl. perfringens*	*Cl. cochlearium*
Cl. bifermentans	*Cl. botulinum* type G	*Cl. barati*	*Cl. tetani*
Cl. sordellii		*Cl. butyricum*	
Cl. botulinum type A type B type F		*Cl. tertium* *Cl. fallax*	
		Cl. chauvoei	
		Cl. septicum	
		Cl. sphenoides	
		Cl. novyi (A–D)	
		Cl. botulinum type B	
		type C	
		type D	
		type E	
		type F	
		Cl. cadaveris	

Some characteristics of the commonly encountered clostridia are shown in *Table 4.2*, from which the following useful items of information emerge:

1. Gelatinase activity is not a particularly useful discriminating feature.
2. Clostridia that are lactose fermenters are never strongly proteolytic.
3. Apart from *Cl. cochlearium*, which is entirely inactive and is rarely encountered, *Cl. tetani* is the only non-saccharolytic, non-proteolytic species.

Table 4.2 Reactions of some commonly encountered clostridia

Organism	*Glucose*	*Maltose*	*Lactose*	*Sucrose*	*Gelatinase activity*	*Indole*	*Milk agar digestion*	*Egg yolk agar*		*Patho-genicity for laboratory animals*
								Lecithinase C activity	*Lipase activity*	
Cl. perfringens (A–E)	+	+	+	+	+	−	−	+	−	±
Cl. barati	+	+	+	+	−	−	−	+	−	−
Cl. tertium	+	+	+	+	−	−	−	−	−	−
Cl. fallax	+	+	+	+	−	−	−	−	−	±
Cl. carnis	+	+	+	+	−	−	−	−	−	+
Cl. chauvoei	+	+	+	+	+	−	−	−	−	+
Cl. septicum	+	+	+	−	+	−	−	−	−	+
Cl. paraputrificum	+	+	+	+	−	−	−	−	−	−
Cl. sphenoides	+	+	+	−	−	+	−	−	−	−
Cl. botulinum (A, B, F)	+	+	−	+	+	−	+	−	+	+
(C, D, E)	+	+	−	±	+	−	−	−	+	+
Cl. novyi (A)	+	+	−	−	+	−	−	+	+	+
(B)	+	+	−	−	+	−	−	+	−	+
(C)	+	+	−	−	+	−	−	−	−	−
(D)	+	−	−	−	±	+	−	+	−	+
Cl. bifermentans	+	+	−	−	+	+	+	+	−	−
Cl. sordellii	+	+	−	−	+	+	+	+	−	±
Cl. sporogenes	+	+	−	−	+	−	+	−	+	±
Cl. cadaveris	+	−	−	−	+	±	−	−	−	−
Cl. difficile	+	−	−	−	−	−	−	−	−	+
Cl. cochlearium	−	−	−	−	−	−	−	−	−	−
Cl. tetani	−	−	−	−	+	+	−	−	−	±
Cl. histolyticum	−	−	−	−	+	−	+	−	−	+
Cl. lentoputrescens	−	−	−	−	+	−	+	−	−	−

4. *Cl. histolyticum* and *Cl. botulinum* type G are the only species that are non-saccharolytic but strongly proteolytic; *Cl. botulinum* type G is rarely encountered.
5. Lecithinase C-producing species are *Cl. perfringens* (all types), *Cl. novyi* (types A, B and D), *Cl. bifermentans, Cl. sordellii* and *Cl. barati*.
6. Lipolytic species are *Cl. botulinum* (all types), *Cl. sporogenes* and *Cl. novyi* type A.
7. *Cl. novyi* type A is the only species that produces both a lecithinase C *and* a lipase.
8. It is useful to note here that all the clostridia are motile except *Cl. perfringens*, and that all are non-capsulated except *Cl. perfringens*.

CLOSTRIDIUM TETANI

Clostridium tetani is widely distributed in soil, especially cultivated soil, and in the intestinal tracts of man and animals.

Morphology

In young cultures the tetanus bacillus stains Gram-positively and appears most commonly in a delicate filamentous form. Individual bacilli are usually about 5 × 0.4 μm in size. Incubation at 37 °C for at least 48 h is required for *Cl. tetani* to produce its fully developed spherical terminal spores which give it a characteristic 'drum stick' appearance. Immature spores are oval and terminal, and resemble those of sporulating cultures of *Cl. tertium* and *Cl. cochlearium*. Spore-free forms of *Cl. tetani* have no distinctive features. In older cultures containing fully mature sporing rods, the organism not infrequently stains Gram-negatively. *Cl. tetani* may be stained by the fluorescent labelled antibody technique (Batty and Walker, 1964, 1967).

Cultural characteristics

Cl. tetani is a strict anaerobe, growing only in the absence of oxygen, and being rapidly killed on exposure to the air. Cultures of this organism have a 'burnt-organic' smell.

Isolated surface colonies are difficult to obtain, since *Cl. tetani* tends to spread as a fine rhizoidal film over the surface of solid media, especially in the presence of blood, and on moist plates. The advancing edge of the swarming growth is finely filamentous with curled projections.

The swarming of *Cl. tetani* is a characteristic and distinctive feature which must be sought carefully. It is easily overlooked on account of its extreme fineness and delicacy. Not only is it of diagnostic importance, but it is also of value in obtaining pure cultures of *Cl. tetani* from mixtures with other organisms (*see* p. 119). A laterally inoculated agar plate is completely covered by swarming growth in 24–36 h. Non-motile variants of *Cl. tetani*, however, give rise to discrete colonies which show no tendency to swarm.

Swarming growth of *Cl. tetani* may be prevented by growing the organism on firm agar media (2–3% agar), or on ordinary agar media containing commercial horse tetanus antitoxin (Willis and Williams, 1970; Williams and Willis, 1970; Williams, 1971). On concentrated agar media, colonies are 2–4 mm in diameter after 48 h incubation, irregularly circular to coarsely rhizoidal in shape, translucent, with a granular surface and an ill-defined edge. Discrete colonies of a much more compact structure are obtained when the organism is cultured on 1.5% agar media containing 40–60 units per ml of commercial horse tetanus antitoxin. The antitoxin may be added to the cooled molten medium before plates are poured, or more conveniently and economically, it may be spread over the surface of plates before inoculation. The formation of discrete colonies in the presence of horse tetanus antitoxin is doubtless due to the presence of agglutinating antibody, since strains of *Cl. tetani* are known to possess a common O antigen, but may have type-specific H antigens. Tetanus antitoxin preparations that contain no agglutinating antibody, e.g. human antitetanus immunoglobulin, do not prevent swarming growth. Inhibition of swarming by tetanus antitoxin is specific, and is therefore of diagnostic value. It is conveniently demonstrated in a half-antitoxin fresh blood agar plate.

On horse blood agar growth is often accompanied by α-haemolysis, later passing into β-haemolysis, due to the production of a haemolysin, tetanolysin, a toxin distinct from the neurotoxin, tetanospasmin. Haemolysis has been suggested as a means of identifying *Cl. tetani* (Lowbury and Lilly, 1958). These authors advocated the use of half-antitoxin fresh blood agar plates, using undiluted therapeutic tetanus antitoxin; toxigenic strains of *Cl. tetani* produced haemolysis which was inhibited by the antiserum. But the claim that this is of diagnostic value is unsound for a number of reasons, the most important being that tetanolysin is a typical oxygen-labile haemolysin. Its haemolytic activity is inhibited, not only by homologous antitoxin but also by antisera prepared against other oxygen-labile haemolysins (for example, the

θ-toxin of *Cl. perfringens* and streptolysin O), and by normal serum and cholesterol. Further, haemolysis due to other oxygen-labile haemolysins is inhibited by tetanus antitoxin.

On heated blood agar, good growth is obtained without any change developing in the medium. On egg yolk agar there is no opalescence or pearly layer. In cooked meat broth good growth is obtained after 24 h incubation in the ordinary incubator. The meat is not digested but sometimes shows slight blackening after some weeks on the bench.

Biochemical characteristics

Cl. tetani is non-saccharolytic and non-proteolytic, but produces a gelatinase. Some rare aberrant strains ferment glucose. Most strains produce indole, but none produces hydrogen sulphide. A rennin-like enzyme is formed which produces zones of diffuse opacity about areas of growth on milk agar, due to precipitation of casein. Among the commonly encountered clostridia, this property appears to be shared only with *Cl. novyi* types A and B. Some strains of *Cl. tetani* produce a deoxyribonuclease.

Metabolic products detected by gas liquid chromatography are acetic, propionic and butyric acids, ethanol and butanol.

Antigenic structure

Differentiation by flagellar antigen into at least 10 types has been demonstrated (MacLennan, 1939). Neurotoxin is formed by all types and is antigenically identical in all. Non-toxigenic variants occur, and are not infrequently isolated from wounds in cases of clinical tetanus.

Pathogenicity

Cl. tetani is pathogenic for man by virtue of its exotoxin (neurotoxin) production. Susceptible laboratory animals include the mouse, guinea pig and rabbit. Washed spores or bacilli are harmless, unless their inoculation is associated with the production of a local nidus of necrosis. After the atraumatic inoculation of washed organisms, clinical tetanus may be induced at a later date by injuring the inoculated area.

Experimental tetanus

The mouse is a suitable laboratory animal for the demonstration of the toxigenicity of *Cl. tetani*. Two animals are used for each test. A protected animal is prepared by subcutaneous injection with 0.5 ml of tetanus antitoxin containing 1500 units per ml at least 1 h before it is inoculated with the virulent organism. Both the protected and unprotected animals are then inoculated intramuscularly in the right hind limb with 0.25 ml of the supernatant from a 48-h cooked meat broth culture of the organism. It may be necessary to include 2% calcium chloride in the inoculum to initiate necrosis. After an incubation period of a few hours, signs of tetanus develop in the unprotected mouse. In ascending tetanus the first evidence of disability is usually that the inoculated leg tends to slip backwards as the animal progresses. Later the limb becomes slightly abducted and the ankle is extended. The leg gradually becomes more extended until it is quite stiff from the spasm of opposing muscles. The tail of the mouse becomes stiff, and gradually the leg on the opposite side is affected. Involvement of the trunk muscles leads to hyperextension or lateral flexion of the spine and finally the forelimbs become spastic. During this period the slightest stimulus induces a generalized spasm.

In mixed ascending and descending tetanus, which is produced by a larger inoculum, local tetanus of the injected limb is followed rapidly by generalized spasms, and death occurs in 18–24 h. Descending tetanus in small laboratory animals is very rare unless special routes of inoculation are used. If very large doses of toxin are injected, the animal may die without the development of any of the classic signs of tetanus intoxication.

At post mortem there is a slight hyperaemia at the site of inoculation. The internal organs show little change. The animal protected with tetanus antitoxin shows no evidence of disease.

Toxicology of Cl. tetani

Cultures and culture filtrates of *Cl. tetani* may contain the neurotoxin (tetanospasmin) and a haemolysin (tetanolysin). The production of the neurotoxin is not related to that of the haemolysin, for the latter may be produced by non-toxigenic strains (Kerrin, 1930).

Tetanolysin

Tetanolysin lyses the erythrocytes of many animals, including those of the horse and rabbit. It is a heat-labile, oxygen-labile haemolysin, and is antigenically related to other oxygen-labile haemolysins, such as those of *Cl. perfringens* (θ-toxin), *Cl. novyi* (δ-toxin) and *Streptococcus pyogenes* (streptolysin O). An antiserum against any one of these oxygen-labile haemolysins neutralizes to some extent the haemolytic activity of all the others. Tetanolysin, and indeed all oxygen-labile haemolysins, must therefore be regarded as 'non-specific'. In culture, tetanolysin is produced during the period of active growth; but it is not present in old cultures, owing to its rapid inactivation.

Tetanospasmin

Tetanospasmin is alone responsible for the characteristic features of clinical tetanus in man and animals. It is purely neurotoxic and is without demonstrable specific biochemical activity (Pillemer and Wartman, 1947). Pillemer, Wittler and Grossberg (1946) produced a purified form of the toxin which had a potency of 67×10^6 mouse MLD/mg N (mouse minimum lethal dose per milligram of nitrogen). Tetanospasmin and *Cl. botulinum* neurotoxin thus share the attribute of being the most potent poisons known (*see* p. 315). In culture, the neurotoxin does not appear until the phase of active growth is over, and its production is maximal at a temperature of 35 °C. It is relatively heat-labile, being destroyed at 65 °C in 5 min, but is oxygen-stable. The inoculation of culture filtrates into the tissues of laboratory animals produces the characteristic chain of events described earlier. Administration of the neurotoxin by other routes, however, such as oral, rectal and conjunctival, produces no effect. The pharmacological effects of tetanus toxin have been reviewed by Wright (1955). Though different strains of *Cl. tetani* vary greatly in their ability to produce neurotoxin, that produced by different strains is antigenically and pharmacologically the same. Hence, an antitoxin prepared by inoculating animals with tetanus formol–toxoid, combines with and neutralizes the neurotoxin from all toxigenic strains of *Cl. tetani*. Tetanospasmin has been the subject of a review by van Heyningen and Mellanby (1971).

Main features for recognition of Cl. tetani

1. Spreading growth in surface cultures. Inhibition of swarming by equine antitoxic serum is specific and diagnostic.
2. Filamentous appearance of the organism in young culture; drum-stick appearance in older cultures.
3. Absence of saccharolytic and proteolytic activity.
4. Production of indole (most strains).

5. Production of diffuse opacity in milk agar (3% agar).
6. Volatile metabolic products.
7. Animal inoculation and protection tests.

Isolation of Cl. tetani

Despite the fact that it is a strict anaerobe, *Cl. tetani* is a comparatively easy organism to isolate. A modified Fildes' technique (Fildes, 1925) is used. The material suspected of containing the organism is inoculated into cooked meat broth and incubated for 2–4 days. In the case of pathological material such as excised wounds, the tissue should be finely minced before inoculation, some of which is also inoculated directly on to plates for anaerobic incubation. The cooked meat broth culture is then subcultured to a suitable plate medium such as fresh or heated blood agar, the inoculum being applied near the edge of the medium only, and the plate incubated anaerobically for 18–24 h. If *Cl. tetani* is present it will swarm over the plate leaving unwanted contaminants behind. Subculture is made from the outermost limit of the spreading edge of growth, which, after 18–24 h incubation is usually half to two thirds of the way across the plate. It may sometimes be necessary to repeat this procedure two or three times before a pure culture is obtained. Plate cultures once inoculated should be rendered anaerobic as rapidly as possible, otherwise *Cl. tetani*, if present, may be killed by exposure to the air.

Clearly, this method of isolation, dependent as it is on the motility of the organism, is of no value for the isolation of non-motile variants (type VI strains). These are separated from mixtures with other organisms by subculture of the enrichment culture to plates, followed by colony selection. The use of neomycin sulphate in the plate medium serves to inhibit or suppress the growth of many aerobic organisms.

Heating to 80 °C for 10 min may be useful for destroying non-sporing contaminants, but this treatment should be used with caution, for not infrequently it results in the destruction of *Cl. tetani* as well, despite microscopic evidence of the presence of spores. The safest practice is always to employ duplicate cultures, only one of which is heated.

CLOSTRIDIUM BOTULINUM

Cl. botulinum is unevenly but widely distributed throughout the world, although all the seven toxicological types do not find an equal geographical representation. The natural habitat of the organism is soil.

Our knowledge of the occurrence of the organism in nature is derived indirectly from reported outbreaks of botulism in man and animals, and directly from studies specifically aimed at determining the tellural distribution of the different toxicological types. Although human and animal botulism are comparatively rare diseases, it is not surprising to find that their incidence and type in different parts of the world is related to the local prevalence and type of *Cl. botulinum* in the soil.

In the United Kingdom there is a low incidence of *Cl. botulinum* in soil; the common indigenous varieties are types A and B. Type C is responsible for outbreaks of botulism in mink, wild fowl and broiler chickens, but has not been encountered in British soil except in association with such epizootic episodes. Types B, C, D, and E have been demonstrated in the mud of lakes and waterways of London (Smith and Moryson, 1975a, b), and types B, C, E and F have been shown to occur in farmed trout in Great Britain (Cann, Taylor and Hobbs, 1975). Type G has not been encountered in the British Isles.

Strains of types A and B are probably the commonest and most widely distributed in the rest of the world, occurring in America, China, Hawaii, and in continental Europe and the USSR. In the North American continent these strains are most prevalent in soils of the Rocky Mountain system and of adjoining territory both in Canada and along the American Pacific seaboard. Types C, D, E and F are also encountered in North America and Europe. Type C is found in Asia, Australia, Africa and South America, type D in Australia and Africa, and type E in Asia and South America. The geographical distribution of *Cl. botulinum* type E is peculiarly regional, and apart from these few exceptional areas, its spores are probably sparsely distributed. It occurs most commonly in the soils of the catchment area of the Baltic Sea, in Japan, especially in relation to the island of Hokkaido, and along coastal areas of Alaska, Canada and Russia. Type G was first isolated by Gimenez and Ciccarelli (1967; 1970a, b) from Argentinian soil; nothing is known of its wider distribution, and the organism has not been implicated in human or animal disease.

Morphology

In young cultures *Cl. botulinum* stains Gram-positively. The bacilli are large, stout rods, about 4×1 μm in size, with straight sides and rounded ends. Spores are oval, centrally or subterminally placed, and distend the bacillary body. Although free spores are often numerous, sporulation may be sparse. There is nothing characteristic about the morphological appearance of this organism. The fluorescent labelled antibody staining technique may be of value in differentiating between the different toxicological types of *Cl. botulinum* (Walker and Batty, 1964; Batty and Walker, 1967).

Cultural characteristics

Cl. botulinum is a strict anaerobe. The species is an ill-defined group of organisms, of which seven types (A–G) are recognized. Not only do these differ from one another in their cultural characteristics, but different strains of the same type also show some variation. The different types are identified by their toxin production, and this is the only reliable characteristic on which a diagnosis of species and type can be made. Variability in cultural characteristics and toxin production is often exhibited by a single strain.

Colonies of *Cl. botulinum* on solid media are irregularly circular and measure about 3 mm in diameter after 48 h incubation. The colonies are translucent, with a granular surface and a lobular or an indefinite spreading edge. A tendency for growth to spread over the surface of solid media is sometimes marked. On horse blood agar growth is usually associated with haemolysis which may be coextensive with the colony or may be larger. On heated blood agar proteolytic strains (types A, B, F and G) produce partial clearing of the medium. On egg yolk agar all types of *Cl. botulinum*, except type G, produce a restricted opalescence and pearly layer due to lipolysis. On milk agar types A and G, and ovolytic strains of types B and F produce zones of clearing in the medium, due to proteolytic activity, while non-ovolytic strains of these types, and types C, D and E are non-proteolytic.

In cooked meat broth good growth develops after 24–48 h incubation. Sometimes a butyrous scum is produced on the surface of the broth. The meat particles are not attacked by types C, D, or E, but are digested after some days by type A and G strains, and by ovolytic strains of types B and F.

Biochemical characteristics

Types A, B, E and F ferment glucose, maltose and sucrose; types C and D ferment glucose and maltose, but not sucrose; type G is non-saccharolytic. *Cl. botulinum* is thus exceptional among the pathogenic clostridia in that it commonly ferments sucrose, but not lactose (*Table 4.3*). All types produce gelatinase and hydrogen sulphide, but none produces indole.

Table 4.3 Some properties of different toxicological types of *Cl. botulinum*

Cl. botulinum type	*Fermentation of*				*Gelatinase (gelatine agar)*	*Proteinase on milk agar*	*Lipase on egg yolk agar*	*Haemolysis on horse blood agar*	*Pathogenic for men*
	Glucose	*Maltose*	*Lactose*	*Sucrose*					
Type A	+	+	—	+	+	+	+	+	+
Type B									
Proteolytic	+	±	—	+	+	+	+	+	+
Non-proteolytic	+	±	—	+	+	—	+	+	+
Type C	+	+	—	—	+	—	+	+	—
Type D	+	+	—	—	+	—	+	+	—
Type E	+	±	—	+	±	—	+	+	+
Type F									
Proteolytic	+	+	—	+	+	+	+	+	+
Non-proteolytic	+	±	—	+	+	—	+	+	+
Type G	—	—	—	—	+	+	—	—	Not known

± = Strains vary

Metabolic products detected by gas liquid chromatography are as follows:

1. All type A strains and proteolytic strains of types B and F produce predominantly acetic and butyric acids, with smaller amounts of propionic, isobutyric and isovaleric acids, and propyl, isobutyl, butyl and isoamyl alcohols (cf. *Cl. sporogenes*).
2. All type E strains and non-proteolytic strains of types B and F produce acetic and butyric acids only.
3. All type C and D strains produce predominantly acetic, propionic and butyric acids.
4. Type G strains produce predominantly acetic acid with lesser amounts of isobutyric, butyric, isovaleric and lactic acids.

Mayhew and Gorbach (1975) suggested that gas liquid chromatographic analysis of foods for short-chain fatty acids may provide presumptive evidence of *Cl. botulinum* contamination. Unfortunately, *Cl. sporogenes* produces almost identical metabolic products, so that final identification must still depend on specific toxin assay (*see* p. 126).

Antigenic structure

The seven types of *Cl. botulinum* produce antigenically distinct neurotoxins, so that an antitoxin to one type does not react with the toxins of the other six. There are, however, some minor antigenic relationships between the neurotoxins of types C and D (Mason and Robinson, 1935), and between type E and F (Dolman and Murakami, 1961). This antigenic specificity is used in animal protection tests for determining the type of the organism. In a study of the heat-stable somatic antigens of *Cl. botulinum* types A–F, Walker and Batty (1964) found that the different toxicological types fell into three distinct serological groups. Types A, B and F are serologically related, and are distinct from types C, D and E. Similarly, types C and D are related serologically, and are distinguishable from type E. They noted that some limited cross reactions occurred between strains of *Cl. botulinum* and *Cl. sporogenes*. The somatic serology of type G has not been studied.

Pathogenicity

Cl. botulinum types A, B, E and F cause botulism in man, types C and D are incriminated in outbreaks of botulism in birds and other animals, while no outbreaks in man or animals have been attributed to type G.

The organisms themselves are essentially saprophytes. They do not multiply easily in the body, and are pathogenic largely by virtue of their exotoxin production. The disease is almost always produced naturally by ingestion of the preformed toxin, usually in association with but few organisms, and is thus a unique type of bacterial toxaemia; it is a pure intoxication. Rarely, *Cl. botulinum* may cause a true wound infection from which the typical syndrome of botulism develops (*see* p. 316).

Experimental botulism

Botulism may be demonstrated in the mouse following the intraperitoneal inoculation of a fluid culture or culture supernatant. The supernatant is prepared by centrifugation from a 3–5 day old cooked meat broth culture which has been incubated at 30 °C. A pair of mice is injected, one of the pair having been protected with a polyvalent botulinum antitoxin. After an incubation period of a few hours, the earliest symptom developed is usually dyspnoea, the respiration being mainly costal. This is evidenced by an indrawing of the abdomen which gives a wasp-waist appearance, and by laboured 'bellows' breathing. Thereafter various flaccid paralyses develop, or a generalized paralysis may result so that the animal lies motionless with limbs outstretched. Death usually ensues in 18–24 h, but may be delayed up to 4 days. The specifically protected animal remains unaffected.

At post mortem the internal organs are congested, and thromboses and haemorrhages are frequently present.

Toxicology

Cl. botulinum species are characterized by the production of highly potent, heat-stable exotoxins (neurotoxins) which, unlike all other clostridial exotoxins, excepting the ϵ- and ι-toxins of *Cl. perfringens* are absorbed from the alimentary tract. There are seven main types of *Cl. botulinum* (types A–G) which are distinguished from one another by the antigenic difference of their neurotoxins. The neurotoxins of the different types are pharmacologically similar, but antigenically distinct (compare types C and D, and types E and F, on p. 123). In culture, the toxin is produced slowly and in maximal amount at a temperature of 30 °C. It is destroyed by boiling, and is readily toxoided by formaldehyde treatment. The pharmacological effects of botulinum neurotoxin have been reviewed by Wright (1955), and a general review of the toxins has been published by Boroff and DasGupta (1971).

In egg yolk emulsion, all types of *Cl. botulinum*, except type G, produce opalescence and a pearly layer, which is due to a lipase (Willis, 1960a, b). Most strains of types A, B, F and G produce proteinases, while types C, D and E do not. Most strains produce a gelatinase, and, with the exception of type G, most strains also produce an oxygen-labile haemolysin similar to *Cl. perfringens* θ-toxin and tetanolysin.

The production by *Cl. botulinum* of its neurotoxin is the only way by which this organism, especially strains of A, B and F, can be distinguished with certainty from *Cl. sporogenes*. Consequently, both the recognition and typing of *Cl. botulinum* depend finally on the demonstration of the neurotoxin, and of the toxin type. A proteolytic, non-toxigenic strain of *Cl. botulinum* would be indistinguishable from *Cl. sporogenes*.

Main features for recognition of Cl. botulinum

1. Cultural and biochemical characters. These do not distinguish The proteolytic strains from *Cl. sporogenes,* although colonies of *Cl. sporogenes* are usually rougher, more opaque and more rhizoidal than those of *Cl. botulinum*.
2. Volatile metabolic products.
3. Animal inoculation and protection tests.
4. Fluorescent labelled antibody staining.

Isolation of Cl. botulinum

Two basic methods have been used for determining the presence of *Cl. botulinum* in soil and other samples. Isolation of the organism in pure culture, although ideal, presents special difficulties due partly to the wide variety and abundance of other spore-formers (both aerobic and anaerobic) that are also commonly present, and partly to the exacting requirements of some strains of *Cl. botulinum*. More commonly, enrichment techniques are employed in which the aim is to promote growth and toxin production by *Cl. botulinum* in mixed culture; the presence of the organism in the soil or other samples is then inferred from the demonstration of toxicity of this culture in animal inoculation and protection tests. Failure to demonstrate *Cl. botulinum* in a sample by this method does not imply that the organism is absent; it means merely that the toxin was not present in demonstrable quantity.

For the isolation of the organism from mixtures with other bacteria advantage is taken of the considerable heat and alcohol-resistance of its spores. An aliquot of the sample, homogenized if necessary, is mixed with an equal volume of absolute ethanol in a screw-capped bottle and incubated at 37 °C for 1 h, with occasional shaking. One ml volumes of the alcohol-treated material are then inoculated into each of three tubes of cooked meat broth, one of which is set aside for incubation; the other two are heated at 80 °C for 10 and 20 min respectively. A similar set of cooked meat broth cultures is set up using untreated material as the inoculum. All six enrichment cultures are incubated in the anaerobic jar for 3–5 days at 30 °C. Although 30 °C is not the optimal temperature for growth of *Cl. botulinum*, some strains produce little or no toxin at temperatures above 30 °C. The presence or absence of botulinum toxin in the culture is then determined by animal inoculation experiments (*see below*); if no toxin is demonstrated it can be assumed for practical purposes that *Cl. botulinum* is absent from the culture.

If, on the other hand, toxin is present, subcultures to horse blood agar and egg yolk agar are made. After anaerobic incubation for 36–48 h, colonies are selected for identification. It is to be noted that many strains of *Cl. botulinum* are very sensitive to oxygen, so that inoculated plates should be placed under anaerobic conditions as quickly as possible. The addition of neomycin sulphate (50 μg/ml) to plate culture media helps to suppress the growth of surviving aerobic contaminants. It should be noted, however, that even low concentrations of neomycin may be inhibitory to some strains of *Cl. botulinum* type E (Spencer, 1969). Colony selection is aided by using egg yolk agar, since all types of *Cl. botulinum*, except type G, produce a restricted opalescence and pearly layer on this medium. Culture on lactose egg yolk milk agar differentiates between the proteolytic and non-proteolytic variants of the organism. On this medium, the presence of a non-proteolytic non-lactose fermenting lipolytic colony is strong presumptive evidence of *Cl. botulinum* types C, D or E, or of a non-proteolytic variant of types B or F.

Demonstration of botulinum toxin in food, pathological specimens and culture-supernatants

The bacteriological diagnosis of food poisoning due to *Cl. botulinum* is based on the demonstration of toxin in the food consumed, or in the intestinal contents or blood of the patient. The method which follows is also applicable to culture supernatants.

Preliminary testing

The food, stomach contents or faeces is homogenized with an equal weight of sterile saline, and allowed to stand overnight at 4 °C. The suspension is centrifuged at 3000 rpm for 30 min, and the supernatant retained. Since the toxicity of type E toxin is enhanced by trypsinization, toxicity tests are performed with both trypsinized and untrypsinized supernatant. Trypsinization is effected by mixing together 9 parts of supernatant with 1 part of trypsin solution, and incubating at 37 °C for 1 h. The trypsin solution used is 1% Difco 1:250 trypsin in distilled water. The trypsinized and untrypsinized supernatants are each divided into two portions, and one of each is heated at 100 °C for 10 min.

From each sample of food, stomach contents or faeces, there are thus prepared four extracts for animal inoculation – untreated supernatants, heated and unheated; and trypsinized supernatant, heated and unheated.

Penicillin is added to each extract to give a final concentration of about 100 units per ml (0.4 ml extract + 0.1 ml penicillin containing 1000 units per ml).

Three pairs of mice are then inoculated intraperitoneally with untrypsinized extract as follows (*Table 4.4*).

Pair A is protected with polyvalent botulinum antitoxin (A, B and E antitoxins – Pasteur Institute) and is injected with 0.5 ml of the unheated material.
Pair B, unprotected, is similarly inoculated with 0.5 ml of the unheated material.
Pair C, unprotected, is inoculated with 0.5 ml of the heated material.

Three similar pairs of mice are inoculated in the same way with the trypsinized material.

In the summary of these inoculation procedures shown in *Table 4.4*, five possible results are indicated, which may be interpreted as follows:

Result 1. An uncomplicated result indicating the presence of botulinum toxin. Unprotected mice receiving the unheated sample die, while protected mice, and those receiving the heated sample survive.
Result 2. The increased toxicity of some toxins (notably type E) by trypsinization may cause death of the protected animals inoculated with the trypsinized sample. This is easily checked by repeating the tests with saline dilutions of the trypsinized sample.

Table 4.4 Scheme of mouse inoculation for detection of *Cl. botulinum* toxin in culture-supernatants, and in extracts of food, stomach contents and faeces

Paired mice inoculated with extract	*Inoculum*		*Some possible results*[1]				
	Extract	*Penicillin (1000 units/ml)*	*1*	*2*	*3*	*4*	*5*
Untrypsinized extract							
Pair A (polyvalent botulinum antitoxin A, B, and E)	0.4ml – unheated	0.1ml	–	–	+	+	–
Pair B (unprotected)	0.4ml – unheated	0.1ml	+	+	+	+	–
Pair C (unprotected)	0.4ml – heated	0.1ml	–	–	–	+	–
Trypsinized extract							
Pair A (polyvalent botulinum antitoxin A, B, and E)	0.4ml – unheated	0.1ml	–	+	+	+	–
Pair B (unprotected)	0.4ml – unheated	0.1ml	+	+	+	+	–
Pair C (unprotected)	0.4ml – heated	0.1ml	–	–	–	+	–

[1] Interpretation of results:
1. Botulism
2. Botulism – inadequate protection for trypsinized extract
3. Botulism – potent toxin, or toxin of type not represented in polyvalent antitoxin (i.e. C, D, or F); sometimes due to tetanus intoxication
4. Not botulism – non-specific reaction
5. Not botulism

Result 3. If none of the protected mice survive, the toxin may be of type C, D or F, or a potent toxin of one of the types tested. This is clarified by repeating the untrypsinized test series (a) with the passively immunized mice protected with types C, D and E antitoxins, and (b) with saline dilutions of the sample. Sometimes *Result 3* may be due to tetanus intoxication, especially if the material under test is a culture supernatant. This problem of 'non-specific' tetanus intoxication can be avoided by preliminary protection of all mice with commercial tetanus antitoxin. It is pertinent to note here that some botulinum antitoxins may contain appreciable amounts of tetanus antitoxin, so that the protection they afford is not restricted to or specific for botulism. This can cause immense confusion if the sample under test contains tetanus neurotoxin.
Result 4. Death of all animals, including those receiving the heated samples indicates the presence of some non-specific toxic substance.
Result 5. Survival of all animals indicates the absence of botulinum toxin from the material.

Blood serum samples do not require trypsinization, and are not heated before testing. Penicillin is added to the serum as described above for food extracts, and the serum is then inoculated intraperitoneally into two pairs of mice, one pair having been protected with the polyvalent botulinum antitoxin. Death of the unprotected pair with paralytic symptoms, with survival of the protected pair is diagnostic of botulism. Since the intraperitoneal injection of normal human serum into mice sometimes causes early non-specific death, the volume of serum should not exceed 0.5 ml. The slight dilution of the serum with penicillin solution helps to avoid these non-specific deaths.

Type specificity testing

Once it has been established that *Cl. botulinum* toxin is present, the antigenic type of toxin is determined. Here, it is convenient to use sample–antitoxin mixtures for inoculation into the test animal; the two pairs of unprotected control animals are included as before, and receive respectively 0.5 ml of heated and unheated sample.

The sample–antitoxin mixtures are prepared in small test tubes as follows: 1.2 ml of fluid sample is mixed with 0.3 ml of monovalent type specific antitoxin (Pasteur Institute) diluted to contain 500 units of activity per ml with penicillin solution (1000 units per ml). The mixture is incubated at 37 °C for 30 min. Volumes of 0.5 ml of this mixture are injected intraperitoneally into each of the two test mice.

THE HISTOTOXIC CLOSTRIDIA OF INFECTED WOUNDS

The organisms associated with gas gangrene may be divided into: (1) the actively pathogenic species, *Cl. perfringens, Cl. novyi, Cl. septicum* and *Cl. histolyticum* (also *Cl. chauvoei* in animals); and (2) the species that are often non-pathogenic but which nevertheless exert a harmful effect in the presence of each other or of the pathogenic species – these include *Cl. sporogenes, Cl. fallax, Cl. sordellii, Cl. bifermentans* and *Cl. tertium*.

CLOSTRIDIUM PERFRINGENS

Synonym – Cl. welchii. This organism is widely distributed in nature: in soil, sewage, water, and so on, and in the intestinal tracts of man and animals. Two types (A and C) occur in man; type D has been recorded on two occasions.

Morphology

Cl. perfringens is a strongly Gram-positive rod, about 4 × 1.5 μm, straight with parallel sides and rounded ends. Human type C strains are rather larger than this, and may exhibit filaments and swollen forms. Most strains of *Cl. perfringens* under conditions of rapid growth produce many very short forms, some of which are almost coccal in appearance. Spores are rarely seen in artificial culture and their absence is one of the characteristic features of *Cl. perfringens*. Sporulation is said to be favoured by an alkaline environment and by the absence of fermentable carbohydrates. Contrary to popular belief, many strains do not sporulate readily, if at all, on inspissated serum or in alkaline egg albumen media. A good yield of spores is produced by some strains in the magnesium sulphate peptone medium described by Ellner (1956), in the peptone starch medium of Duncan and Strong (1968), and in the glucose ion-exchange resin medium described by Clifford and Anellis (1971) (*see* p. 48). In the sporulating strains seen by the author the spores were large, oval and central and distended the organism. Occasionally a wild strain of *Cl. perfringens* is encountered which sporulates readily on ordinary media, and sporulating mutant strains have been induced by treatment with acridine orange or nitrosoguanidine (Sebald and

Cassier, 1969; Duncan, Strong and Sebald, 1972). Capsules are formed in the animal body but are not usually observed in artificial culture.

Cultural characteristics

Cl. perfringens is not a strict anaerobe, growing readily in the presence of small amounts of oxygen. It is one of the most rapidly growing anaerobes, surface growth often being detectable after only 4–6 h incubation; growth in deep broth may be evident after only 2 h. Surface colonies are circular, about 2–4 mm in diameter after 24 h incubation, convex, semitranslucent, smooth and with an entire edge. Less commonly, colonies are umbonate, with radial striations and a crenated or scalloped edge. There is no tendency for growth to spread over the surface of the medium. Some human type C strains produce rough colonies with characteristic thorn-like outgrowths.

On horse blood agar many strains of *Cl. perfringens* produce zones of complete haemolysis due to the production of θ-toxin (oxygen-labile haemolysin). Partial haemolysis, however, is frequently seen among type A strains, and absence of haemolysis is characteristic of some heat-resistant food poisoning type A strains, and of some type C strains, since they do not produce any θ-toxin. Partial haemolysis on horse blood agar is due to the α-toxin (lecithinase C), which is relatively inactive against the red cells of the horse. The addition of 0.5% $CaCl_2$ to horse blood agar renders the erythrocytes more susceptible to the action of α-toxin, so that colonies are surrounded by a wide zone of partial haemolysis; strains producing θ-toxin, in addition to α-toxin, produce two zones of haemolysis ('target' haemolysis) – a narrow zone of complete haemolysis due to θ-toxin, and a much wider peripheral zone of partial haemolysis (Evans, 1945). It is common to see reported isolates of *Cl. perfringens* (especially those from food poisoning outbreaks, *see* p. 323) as α-haemolytic, or β-haemolytic, or α- and β-haemolytic. As with all bacterial cultures on blood agar, these descriptions refer to the type of haemolysis (partial or complete), and not to the specific toxin that is causing the lysis. (It is unfortunate and confusing that Greek alphabetical nomenclature has been used to denote both toxin type and variety of haemolysis, since the two are unrelated.)

On heated blood agar a few strains produce evidence of slight proteolytic activity, but this is never very pronounced. On egg yolk agar *Cl. perfringens* produces diffuse opalescence, which is inhibited by *Cl. perfringens* antitoxin; no pearly layer is produced. In addition to these egg yolk reactions, cultures on lactose egg yolk milk agar show lactose fermentation, but no marked proteolytic activity.

In cooked meat broth growth is evident in a few hours; a fair amount of gas is produced and the meat particles are turned pink; there is no digestion. In ordinary milk medium rapid fermentation of the lactose occurs, with the subsequent development of a characteristic 'stormy clot' reaction.

Biochemical characteristics

All types ferment glucose, maltose, lactose and sucrose, and are gelatinase producers. They are indole-negative and hydrogen sulphide positive. Major products of metabolism are acetic and butyric acids; butanol is sometimes also produced.

Antigenic structure

Studies of the somatic serology of *Cl. perfringens* show that there is an extreme degree of heterogeneity within the species, so that agglutination and other tests are of little value in subdividing the species. Despite these difficulties, slide agglutination tests are of importance in the epidemiological study of outbreaks of food poisoning due to *Cl. perfringens* type A (Hobbs *et al.*, 1953).

Cl. perfringens is differentiated into five serological types (A–E) according to the kinds of exotoxins which it produces. The toxins are antigenic, and antitoxic sera are used in the routine typing of strains. The α-toxin, a lecithinase C, is produced by all types of *Cl. perfringens* and is antigenically related to the lecithinase of *Cl. bifermentans* and *Cl. sordelli* (*see* Chapter 3).

Pathogenicity

There is considerable variation in the pathogenicity of different strains. Type A strains are the ones responsible for gas gangrene and food poisoning in man. Human type C strains are the causal organisms of enteritis necroticans, and type A has been incriminated in necrotizing colitis.

Experimental Cl. perfringens gas gangrene

A guinea pig is inoculated in the right hind leg intramuscularly with 1 ml of a fresh 24-h cooked meat broth culture; death occurs in 24–48 h. In the course of the first few hours marked swelling develops in the injected limb and there may be crepitation from gas formation. The oedema spreads up over the abdomen and sometimes reaches as far as the axillary region. The disease may be prevented by protecting the animal with *Cl. perfringens* antitoxin before inoculation of the culture.

At post mortem there is extensive blood-stained fluid oedema, often with some gas in the tissues. Characteristically the oedema fluid contains a considerable amount of free fat (*see* Macfarlane and MacLennan, 1945). The muscles of the inoculated limb are friable and pale pink, but there is no putrefaction. The internal organs show little change, but the adrenal glands are often deep red or mottled.

Cl. perfringens invades the blood stream in the course of the disease and is easily recoverable from the heart blood and from the spleen.

Toxicology of Cl. perfringens

Cl. perfringens produces multiple exotoxins, at least 12 different soluble antigens being recognized. According to the kinds and amounts of toxins produced, different strains of *Cl. perfringens* are divided into five types, A–E (*Table 4.5*).

Though individual toxins are commonly identified by their particular effects, it will be noted from *Table 4.5* that some toxins produced by *Cl. perfringens* have identical effects; both ϵ- and ι- toxins, for example, are lethal and necrotizing, and both are activated by trypsin. Distinction in such cases is accomplished by neutralization tests with mono-specific antitoxic sera. The type to which any particular strain of *Cl. perfringens* is assigned depends on the sorts and amounts of lethal toxins which it produces. Thus, type A strains produce predominantly α-toxin, type B strains β- and ϵ-toxins, type C strains β- and δ-toxins, type D strains ϵ-toxin and type E ι-toxin. However, the distinction between these different types of *Cl. perfringens* is not always clear cut. Thus, some strains may lose their ability to produce a characteristic toxin and are then referred to as degraded types. As with other toxin-producing clostridia, non-toxigenic variants of *Cl. perfringens* occur. The early literature on the toxins of *Cl. perfringens* was comprehensively reviewed by Oakley (1943). More recent reviews have been published by Ispolatovskaya (1971) and Hauschild (1971).

Table 4.5 The distribution of antigens and heat-resistance amongst the different types of *Cl. perfringens*

		Major antigens[1]				*Minor antigens*[1]								
		α	β	ϵ	ι	γ	σ	η	θ	κ	λ	μ	ν	
Type	*Occurrence (country where first described)*	*Lethal; necrotizing; Ca-dependent; lecithinase*	*Lethal; necrotizing*	*Lethal; necrotizing*	*Lethal; necrotizing*	*Lethal*	*Lethal; haemolytic*	*Lethal*	*Haemolytic; oxygen-labile*	*Collagenase*	*Proteinase*	*Hyaluronidase*	*Deoxy-ribonuclease*	*Heat-resistance*
A	1. Gas gangrene of man and animals Intestinal commensal, man and animals Putrefactive processes, soil, etc. (United States)	+++	0	0	0	0	0	+	++	++	0	++	++	0
	2. Food Poisoning (Britain)	+++	0	0	0	—	0	—	+	++	0	+	+++	+++
B	1. Lamb dysentry Enterotoxaemia of foals (Britain)	+++	+++	+++	0	++	0[2]	—	++	0	+++	+++	++	0
	2. Enterotoxaemia of sheep and goats (Iran)	+++	+++	+++	0	—	—	—	+++	+++	0	0	+	0

C	1. Enterotoxaemia ('struck') of sheep (Britain)	+++	+++[2]	0	0	++	+++	–	+++	+++	0	0	++	0
	2. Enterotoxaemia of calves, lambs (United States)	+++	+++	0	0	–	0	–	+++	+++	0	0	++	0
	3. Enterotoxaemia of piglets (Britain)	+++	+++	0	0	–	0	–	++	++	0	+	+++	0
	4. Necrotic enteritis of man (formerly type F) (Germany)	+++	+++	0	0	+++	0	–	0	0	0	0	+++	+++
	5. Necrotic enteritis of man (Papua–New Guinea)	+++	+++	0	0	–	0	–	++	++	0	++	–	0
D	Enterotoxaemia of sheep, lambs, goats, cattle and possibly man (Australia)	+++	0	+++	0	–	0	–	+++	++	++	++	++	0
E	Sheep and cattle, pathogenicity doubtful (Britain)	+++	0	0	+++	–	0	–	+++	+++	+++	+	++	0

+++ = Produced by most strains ++ = Produced by some strains + = Produced by few strains 0 = Not produced by any – = Not tested

[1] The major antigens are those defining the type and predominantly responsible for pathogenicity; the minor antigens are of a lower order of toxicity and of little or no importance in pathogenicity.
[2] Occasionally the presence of this antigen must be assumed from the production of the appropriate antitoxin by hyperimmunized horses.
[3] Type F has been abandoned and the strains included in it transferred here.

(Reproduced from Sterne and Warrack (1964), *J. Path. Bact.*, 88, 279, by courtesy of the authors and Editor)

Cl. perfringens α-toxin

Though this toxin is produced by all types of *Cl. perfringens*, it is usually produced in greatest amount by type A strains, of which it is the principal lethal toxin. In addition to its lethal effect, it also shows necrotizing, lecithinase C and haemolytic activity. It is heat-stable, and is reversibly inactivated by substances which remove calcium ions.

The lecithinase C activity of α-toxin was first observed by Nagler (1939) and Seiffert (1939), who noticed that toxic filtrates of *Cl. perfringens* produced an opalescence in human serum. A similar but stronger reaction is produced in egg yolk emulsions (Macfarlane, Oakley and Anderson, 1941). Macfarlane and Knight (1941) established the identity of α-toxin as a lecithinase, by demonstrating a quantitative breakdown of lecithin by the toxin. The attack on lecithin is inhibited by α-antitoxin (Hayward, 1941, 1943), and this is made use of in half-antitoxin egg yolk agar or human serum agar plates for the identification of *Cl. perfringens* (*see* p. 131). The lecithinases C of *Cl. bifermentans* and *Cl. sordellii* are partially neutralized by *Cl. perfringens* α-antitoxin and are therefore antigenically related to α-toxin (Miles and Miles, 1947). *Cl. perfringens* α-toxin shows no activity against lysolecithin, and is inactivated by lecithin, ox brain cerebroside and sphingomyelin (Zamecnik, Brewster and Lipman, 1947; Gordon, Turner and Dmochowski, 1954; Gowland and Willis, 1961).

The haemolytic activity of α-toxin is merely a reflection of its lecithinase effect. Macfarlane (1950) showed that lysis of erythrocytes was brought about by the α-toxin acting on the phospholipins of the red cell surface. The erythrocytes of most laboratory animals are attacked, excepting those of the goat and horse. Haemolysis produced by strains of *Cl. perfringens* growing on horse blood agar is not, therefore, due to the α-toxin; it is, in fact, the result of θ-toxin activity (*see below*).

The α-toxin activity of a toxin preparation is most easily determined by incubating the material with an egg yolk suspension (van Heyningen, 1941). The mixture becomes turbid as lecithinase activity proceeds, and the degree of turbidity produced depends on the amount of α-toxin present. The reaction is inhibited by α-antitoxin. A similar *in vitro* determination is made by using lysis of sheep or mouse erythrocytes as the indicating effect. If the α-toxin alone is being sought, then the red cells of the mouse are more conveniently used, since they are moderately resistant to lysis by θ- and δ-toxins. *In vivo* tests for the lethal and necrotizing activities of α-toxin may also be used but are less convenient than the *in vitro* methods.

Cl. perfringens β-toxin

Cl. perfringens β-toxin is produced only by *Cl. perfringens* types B and C. It is a necrotizing and lethal toxin, but is not haemolytic. It is heat-labile, and is easily toxoided with formaldehyde. Its maximum concentration occurs in young cultures, and it disappears from culture filtrates more than a few days old. It is identified by the intradermal inoculation of depilated albino guinea pigs, in which it produces a purplish necrotic area. The β-reactions produced by type B filtrates are very irregular in shape due to the presence of hyaluronidase (μ-antigen) in the filtrates. Hyaluronidase is not present in type C filtrates, however, and the necrotic lesions due to β-toxin in these are smaller and more nearly circular. β-toxin is distinguished from the other necrotoxins of *Cl. perfringens* (α-, ε- and ι-toxins) by the methods described by Oakley and Warrack (1953).

Cl. perfringens δ-toxin

δ-toxin is lethal and haemolytic. It is active against the erythrocytes of the sheep, but not those of the horse or rabbit. It is found in young cultures of *Cl. perfringens* types B and C, but disappears rapidly after active growth has ceased. The presence of δ-toxin in culture filtrates is demonstrated by haemolytic activity, after α- and θ-toxins have been inhibited by appropriate antisera.

Cl. perfringens θ-toxin

θ-toxin is an oxygen labile haemolysin, active against the red cells of the sheep and horse, but not every active against those of the mouse. It is produced in greatest amount by type C strains, but is also found in culture filtrates of types A, B, D and E. θ-toxin is antigenically related to other oxygen-labile haemolysins, such as streptolysin O and tetanolysin, and is inhibited to some extent by antisera to these haemolysins. Normal sera from a variety of animals inhibit the action of θ-toxin (and other oxygen-labile haemolysins), as does cholesterol, due to the presence of lipoid material. Lipoid-free antisera must therefore be used when θ-toxin is being sought. Since θ-toxin is reversibly inactivated by oxidation, it is most easily detected in culture filtrates in the presence of a reducing agent such as sodium thioglycollate (Oakley and Warrack, 1953).

Cl. perfringens ε- and ι-toxins

Cl. perfringens ε- and ι-toxins differ from the other toxins of *Cl. perfringens* in that they are both absorbed from the intestinal tract, a property which is shared with them only by *Cl. botulinum* neurotoxin. Further, they are both formed as *prototoxins* which require proteolytic digestion for activation. ε- and ι-prototoxins are produced during the period of active growth of the organism and may subsequently become activated by the proteolytic enzymes of the growing organism. Activation may also be effected by treating toxic material with 5% trypsin. ε-toxin is produced only by *Cl. perfringens* types B and D, while ι-toxin is produced only by type E strains. Both toxins are lethal and necrotizing.

ε- and ι-toxins are detected in culture filtrates by intradermal testing in guinea pigs. Trypsinized material is used to ensure activation of the prototoxins. Necrosis due to residual traces of α- and β-toxins is prevented by neuralization with the appropriate antisera (Oakley and Warrack, 1953).

Cl. perfringens κ- and λ-toxins

κ-toxin is a collagenase and gelatinase, which also exhibits lethal and necrotizing effects. It is produced by *Cl. perfringens* types A, C and E, and by some strains of type D. It attacks native collagen, collagen paper and azocoll (Oakley, Warrack and Warren, 1948), but has no effect on haemoglobin or casein.

λ-toxin is a proteolytic enzyme which attacks gelatin and azocoll, but not collagen paper (Delaunay, Guillaumie and Delaunay, 1949). It also differs from κ-toxin in that it attacks casein and haemoglobin. λ-toxin is produced by *Cl. perfringens* types B and E, and by some strains of type D. The literature on these toxins was reviewed by Bidwell (1950).

Other soluble substances produced by Cl. perfringens

γ- and η-toxins are lethal toxins without other demonstrable activity, and both require very extensive immunological investigations for their detection. They are of no importance in typing.

μ- and υ-antigens are a hyaluronidase and deoxyribonuclease respectively. μ-antigen is estimated by the ACRA test (Oakley and Warrack, 1951), in which horse synovial fluid is used as the hyaluronic acid substrate. υ-antigen is also estimated by the ACRA test, using sodium deoxyribonucleate as the substrate.

A variety of other soluble substances has been identified in culture filtrates of *Cl. perfringens*, and includes enzymes that destroy blood group substances, neuraminidase, fibrinolysin, and a 'non-$\alpha\theta\delta$-haemolysin'.

Details of the routine typing of *Cl. perfringens* are given by Oakley and Warrack (1953), Brooks, Sterne and Warrack (1957) and Willis (1969). The soluble substances produced by *Cl. perfringens* have been reviewed by Willis (1969).

Cl. perfringens enterotoxin

The enterotoxin of *Cl. perfringens* is an endotoxic component of the sporulating cells of some strains of the organism, and is the factor responsible for *Cl. perfringens* food poisoning in man. Our present knowledge of this endotoxin, which is the result of the extensive studies of Hauschild and his colleagues in Canada, has been reviewed by Hauschild (1973, 1974). The enterotoxin is formed exclusively in sporulating cells of *Cl. perfringens*, and is located in the vegetative part of the cell, from which it is released when the vegetative remnants are lysed. It has lethal and emetic properties, causes fluid accumulation in ligated intestinal loops, increases the permeability of blood capillaries, and causes cutaneous erythema in the guinea pig and rabbit. It is inactivated by heat at 60 °C, and by treatment with proteolytic enzymes. The purified toxin has all the characteristics of a protein, and has a toxicity for mice of the order of 2000 MLD/mg N. It is antigenic, and although antitoxin to it protects rabbits against its lethal action, circulating antibody has no neutralizing effect on enterotoxin in the intestine. *Cl. perfringens* enterotoxin is produced by food poisoning strains of type A, and type C_5 strains that are associated with 'pig-bel' in New Guinea (Skjelkvale and Duncan, 1975).

Main features for recognition of Cl. perfringens

1. Morphological appearances and absence of spores.
2. Culture on half-antitoxin lactose egg yolk agar (diagnostic).

3. Colonial appearances and sugar fermentation reactions.
4. Animal inoculation and protection tests.
5. Volatile metabolic products.

Isolation of Cl. perfringens

Cl. perfringens is an èasy organism to isolate, owing to the rapidity of its growth and its tolerance to low concentrations of oxygen. Material containing the organism is inoculated directly on to horse blood agar, an egg yolk agar medium and into cooked meat broth. After 8–10 h incubation the broth culture is subcultured to horse blood agar and egg yolk agar plates. The incubation time of the enrichment culture may be reduced to 4–6 h if it is incubated in a water bath. It is sometimes advantageous to incorporate neomycin sulphate (100 μg/ml) in the plate culture media to inhibit the growth of aerobic contaminants.

All plate cultures are incubated for 18 h anaerobically, and colonies are then selected for examination. In the event of difficulty from the presence of strict anaerobes that tend to swarm (notably *Cl. tetani* and *Cl. septicum*), incubation of the plates under partially anaerobic conditions will allow growth of *Cl. perfringens* but not of the strict anaerobes. Suitable partial anaerobic conditions are obtained by evacuating the anaerobic jar to –600 mmHg and replacing the air with hydrogen. The catalyst is removed from the jar before use. Note that the haemolytic properties of colonies of *Cl. perfringens* grown in this atmosphere will differ from those obtained under fully anaerobic conditions.

Heating material to destroy non-sporing organisms is not necessary, and is likely to kill any *Cl. perfringens* present, especially if the specimen is of pathological origin. This is due to the fact that *Cl. perfringens* does not sporulate in the tissues; indeed, it is difficult, and often impossible, to induce sporulation of this organism under any circumstances.

In the case of heat-resistant food poisoning strains of *Cl. perfringens* type A, as opposed to heat-sensitive food poisoning strains, isolation from faeces is effected by steaming a tube of cooked meat broth inoculated with the suspected material for 15–30 min, and then incubating at 37 °C overnight. In the intestinal tract food poisoning strains of the organism sporulate abundantly, and the spores of heat-resistant variants survive this preliminary heat treatment. Subcultures are made as already described. It is important to note that resistance to heat can be determined only prior to incubation; once the spores present in the faeces have vegetated it is often impossible to induce sporulation and so to demonstrate heat resistance. Samples of food should not be heated before cultivation, as all spores have usually germinated. The bacteriological diagnosis of food poisoning due to *Cl. perfringens* is discussed on p. 324.

CLOSTRIDIUM NOVYI

Synonym – *Cl. oedematiens. Cl. novyi* is widely distributed in the soil. Type D strains are sometimes referred to as *Cl. haemolyticum*.

Morphology

It is a fairly large bacillus, about 5–10 × 1 μm with parallel sides and rounded ends, its long axis straight or slightly curved. It is Gram-positive in young culture, and it sporulates freely. The spores are large, oval, subterminal and distend the organism. There is nothing characteristic about the morphology of this organism, which in young culture often resembles *Cl. perfringens*.

Cultural characteristics

Cl. novyi is a strict anaerobe which dies rapidly on exposure to air. Four types (A–D) are recognized. Strains of type A grow readily on ordinary horse blood agar. Types B, C and D, however, are extremely demanding both in their nutritional requirements and in their tolerance of oxygen. These organisms are conveniently cultured on the neopeptone (NP) glucose blood agar medium of Moore (1968), which contains cysteine and dithiothreitol as essential ingredients (*see* p. 43). Surface colonies are irregularly circular, 2–3 mm in diameter after 48 h incubation, semitranslucent, with a finely lobulated or crenated edge and a finely granular surface. There is a tendency for the colonies of types A and B to spread over the surface of the medium. The colonies of type D (*Cl. haemolyticum*) are often exceedingly small, appearing as mere pin-points after 24 h incubation. Type B strains are more difficult to grow than type A strains, while strains of type D are among the most fastidious clostridia known.

On horse blood agar a zone of β-haemolysis coextensive with the colony is produced, except in the case of type D strains which produce extensive haemolysis in 24 h, and type C strains which are non-haemolytic.

On egg yolk agar media *Cl. novyi* type A strains produce a diffuse lecithinase C opalescence, and a restricted pearly layer due to lipolysis; this pair of reactions on egg yolk agar is diagnostic of *Cl. novyi* type A. Strains of types B and D produce diffuse opalescence only, while type C strains are egg yolk negative. Lactose egg yolk milk agar, used as a half-antitoxin plate with a mixture of *Cl. perfringens* type A and *Cl. novyi* type A antitoxic sera, is diagnostic for type A strains, and differentiates types B and D strains from other egg yolk positive clostridia. The fact that type B organisms grow much more readily on this medium than type D strains affords presumptive evidence of their identity. Type D strains commonly produce large amounts of toxin when grown in fluid media such as cooked meat broth, and this alone may lead to haemolytic and lecithinase effects on the appropriate media when subculture is made. These enzyme effects occur in the absence of growth and may mislead one into thinking that plate subcultures have been successful.

On benzidine blood agar many strains of *Cl. novyi* types A and B show blackening of their colonies on exposure to air following incubation (Gordon and McLeod, 1940). This reaction is paralleled by the bleaching of heated blood agar which also occurs under aerobic conditions. The benzidine reaction, however, is not specific for *Cl. novyi*. Since benzidine is a carcinogenic substance, this medium should not be used for routine purposes.

In cooked meat broth good growth with slight gas production is obtained in 24 h, but type D strains often require 48 h incubation for the production of satisfactory growth.

Biochemical characteristics

Glucose and maltose are fermented by types A and B, but glucose only by type C and D strains (Rutter, 1970). All types produce gelatinase, but do not attack more complex proteins. Hydrogen sulphide is produced, especially by type D strains, which also produce large amounts of indole. Types A, B and C are indole-negative.

Major products of metabolism include propionic and butyric acids, with lesser amounts of acetic and valeric acids.

Toxicology

Oakley, Warrack and Clarke (1947), Macfarlane (1955) and Oakley and Warrack (1959) demonstrated the presence of eight different soluble antigens in culture filtrates of strains of *Cl. novyi*. Oakley and his colleagues showed that the species was easily divisible into three types according to their soluble antigen production, and they suggested that *Cl. haemolyticum* might well be included in *Cl. novyi*. This has proved to be the case, and *Cl. haemolyticum* is now commonly referred to as *Cl. novyi* type D (Oakley and Warrack, 1959). The activities of the soluble antigens and their distribution among the four types of *Cl. novyi* are shown in *Table 4.6*.

Table 4.6 Activities and distribution of toxins among types of *Cl. novyi* (after Oakley and Warrack, 1959)

Designation	*Activities of toxin*	*Presence in* Cl. novyi			
		Type A	*Type B*	*Type C*	*Type D*[1]
α	Necrotizing, lethal	+	+	–	–
β	Haemolytic, necrotizing, lethal, lecithinase	–	+	–	+
γ	Haemolytic, necrotizing, lecithinase	+	–	–	–
δ	Oxygen-labile haemolysin	+	–	–	–
ε	Opalescence in egg-yolk (lipase), pearly layer	+	–	–	–
ζ	Haemolysin	–	+	–	–
η	Tropomyosinase	–	+	–	+
θ	Opalescence in egg yolk	–	?tr.	–	+

[1] *Cl. haemolyticum*

A satisfactory yield of all soluble elements is obtained by culture in the papain digest of horse meat suggested by Macfarlane and Knight (1941), and in Brewer's medium (Hayward, 1943).

Cl. novyi α-*toxin*

The α-toxin is lethal and necrotizing, and is produced by strains of types A and B, but in greatest amount by type B. It is heat-labile, and though it may be toxoided with formaldehyde, it is much more resistant to modification than most toxins. Following the subcutaneous injection of α-toxin into mice, the animals die in 24–48 h. At post mortem there is extensive subcutaneous gelatinous oedema surrounding the site of inoculation. Toxin and antitoxin may be titrated by *in vivo* testing, using toxin–antitoxin mixtures in the usual way.

Cl. novyi β-and γ-toxins

The β- and γ-toxins are both haemolytic lecithinases C, and are the only toxins of importance in typing strains of *Cl. novyi*. They both produce opalescence in egg yolk emulsions at 37 °C, but are hot–cold haemolysins; that is, haemolysis develops only on cooling, following incubation at 37 °C. Both their lecithinase and haemolytic activities are inhibited by the appropriate antitoxic sera.

β-toxin is produced by strains of types B and D, type D strains being very much more active in this respect. Its haemolytic activity is most marked against the erythrocytes of the mouse; those of the sheep and horse are about equally susceptible, but less so than the red cells of the mouse.

γ-toxin is produced only by type A strains. Unlike β-toxin, its haemolytic activity is much more pronounced for the red cells of the horse than for those of the sheep. It is probable that the haemolytic activities of both β- and γ-toxins are, like that of *Cl. perfringens*, a reflection of their lecithinase activities (Taguchi and Ikezawa, 1975).

The production of β-toxin and γ-toxin is most conveniently determined by their lecithinase activities. This is carried out with culture filtrates in test tubes, in which the toxic material is incubated with lecithovitellin (LV) preparation. The production of opalescence at 37 °C, which is inhibited by appropriate antitoxic serum, is indicative of β- or γ-activity. The toxins can be similarly identified in plate cultures of *Cl. novyi*, using half-antitoxin egg yolk agar plates (Oakley, Warrack and Clarke, 1947), or half-antitoxin fresh blood agar plates (Hayward and Gray, 1946; Smith, 1955). In the latter method red cells of the horse are used.

Examination of strains of *Cl. novyi* for the production of β- and γ-toxins is adequate for typing. The techniques of typing by LV and haemolytic testing of filtrates, and by cultural reactions on half-antitoxin egg yolk emulsion agar, are described by Oakley, Warrack and Clarke (1947) (*see also* Oakley and Warrack, 1959; Willis, 1969).

Cl. novyi δ- and ζ- antigens

δ-antigen is an oxygen-labile haemolysin, produced by type A strains only. Its antigenic relationship to other oxygen-labile haemolysins has not been determined. ζ-antigen is a haemolysin formed only by type B strains.

Cl. novyi ε-*antigen*

ε-antigen is also produced only by type A strains. It is a lipase which produces an opalescence in egg yolk emulsions, and is responsible for the pearly layer which develops on the surface of egg yolk agar plate cultures of *Cl. novyi* type A strains (Willis, 1960a, b; Rutter and Collee, 1969). In tube tests, opalescence develops after the incubated tubes have been cooled to room temperature.

Cl. novyi θ-*antigen*

θ-antigen causes opalescence in egg yolk emulsions. It is produced by strains of type D, and probably also by type B strains (Oakley and Warrack, 1959).

Cl. novyi η-*antigen*

η-antigen (Macfarlane, 1955) is a tropomyosinase which is produced by type B and D strains.

Detailed confirmatory studies concerned with the lecithinase C, lipase and haemolytic activities of *Cl. novyi* were published by Rutter and Collee (1969).

Antigenic structure

All strains of *Cl. novyi* types A, B and C share two somatic antigens in varying proportions. Although there has been some dispute about the somatic relationship of type D to the other types, Batty and Walker (1964, 1967) have shown that fluorescent labelled antiserum prepared against type B organisms stains strains of all types. The somatic fluorescent straining technique is of particular value for the rapid and early recognition of *Cl. novyi* in pathological material, in tissue sections and in artificial culture (*see* Roberts, Guven and Worrall, 1970; Williams and Moe, 1973).

Pathogenicity

Cl. novyi types A and B cause gas gangrene in man and animals, while types B and D are responsible for other diseases in animals. Type A strains are the only ones likely to be encountered by the clinical bacteriologist.

Experimental Cl. novyi gas gangrene

A guinea pig is inoculated intramuscularly in the right hind leg with 1 ml of a fresh 24-h cooked meat broth culture. It may be found necessary to include 1% calcium chloride with the inoculum to initiate the infection. Death occurs 24–48 h after inoculation. The infection produces a profound toxaemia, and the development of massive oedema spreading from the site of inoculation.

At post mortem there is an extensive thick, gelatinous oedema, usually colourless, but sometimes blood-stained, extending from the inoculated limb up over the abdomen. The inoculated muscles are red and softened. There is no gas formation. The organism is recoverable from heart-blood. The disease may be prevented by preliminary protection of the guinea pig with *Cl. novyi* antitoxin. If the mouse is used as the experimental animal, the oedema is so extensive that the animal is flattened and is twice its normal width across the pelvic region, so that it assumes the shape of a flattened pear.

Although *Cl. novyi* type D, which causes bacillary haemoglobinuria in cattle, is never likely to be encountered by the clinical bacteriologist, it is nevertheless worth noting its effect on inoculation into laboratory animals. Death of the animal occurs in 24–48 h. At post mortem there is little haemorrhagic subcutaneous oedema and the muscles are reddened; and the bladder may be filled with haemoglobin-containing urine. The organism is recoverable from the heart-blood and from the spleen. Haemoglobinuria is most constantly produced in rabbits and mice.

Main features for recognition of Cl. novyi

1. Reactions on half-antitoxin lactose egg yolk milk agar (species specific, and type specific for type A strains).
2. Sugar fermentation reactions.
3. Volatile products of metabolism.
4. Animal inoculation and protection tests.

Isolation of Cl. novyi

Cl. novyi is a relatively difficult organism to grow, and its isolation from pathological material or from soil is often difficult and slow. Strains of type A are the easiest to isolate, while those of type D are the most difficult. All plate media should be freshly poured from pre-reduced stock. Alternatively, but less satisfactorily, routine plate media may be pre-reduced in the anaerobic jar. When types B, C and D are being sought, the cysteine dithiothreitol medium of Moore (1968) must be employed. The plates are inoculated immediately and placed under anaerobic conditions without delay. So sensitive is *Cl. novyi* to free oxygen, that unless these precautions are taken the organism dies. Before culture the material may be heated at 80–100 °C for 10–15 min to destroy vegetative contaminants. The spores of *Cl. novyi* are unaffected by this treatment. Preliminary enrichment of the material in cooked meat broth is often of value, especially following heat treatment. Incubation of all cultures should be for 48 h in the anaerobic jar.

The intramuscular inoculation of a guinea pig with the ground-up pathological material may eliminate unwanted contaminants. One ml of ground suspension in 1% calcium chloride is inoculated; the organism may be recovered from the infected muscle at the site of inoculation or from the heart-blood.

Incorporation of neomycin sulphate into culture media serves to suppress the growth of aerobic organisms. Neomycin egg yolk agar may be used for this purpose, since it acts both as a selective and an indicator medium for *Cl. novyi.*

CLOSTRIDIUM SEPTICUM AND CLOSTRIDIUM CHAUVOEI

Synonym – *Cl. chauvoei* is sometimes referred to as *Cl. feseri.*

These two organisms are dealt with together because their characteristics are similar in many ways. Indeed, there are good reasons for regarding them as two types of a single species, *Cl. septicum* type A (*Cl. septicum*) being pathogenic for man and animals, and *Cl. septicum* type B (*Cl. chauvoei*) being pathogenic for animals only (Moussa, 1959; Princewill, 1970). Both organisms are found chiefly in soil.

Morphology

Cl. chauvoei and *Cl. septicum* are rod-shaped organisms about 3–8 × 0.6 μm, with parallel sides and rounded ends, the axis being straight or slightly curved. Filamentous forms are common, and navicular and citron forms occur, especially in the presence of serum or fresh tissue. They are Gram-positive in young culture. Spores are oval, subterminal and distend the organism. On the peritoneal surface of the liver in infected animals *Cl. chauvoei* occurs singly or in pairs, while *Cl. septicum* tends to form long jointed filaments.

Cultural characteristics

These organisms are strict anaerobes, *Cl. chauvoei* being rather more fastidious than *Cl. septicum*. The growth of *Cl. chauvoei* is greatly favoured by the presence of liver extract in the medium (*see* p. 43).

The colonial characters of the two organisms in surface culture are quite distinctive. Colonies of *Cl. chauvoei* are small, about 1–2 mm in diameter after 48 h incubation, commonly umbonate, irregularly circular, shiny and semitranslucent. The organism shows little or no tendency to swarm over the surface of the medium. Colonies of *Cl. septicum* are sometimes small and discrete with a coarsely rhizoidal edge. Much more commonly, however, the organism swarms over the surface of the medium in a manner similar to that of *Cl. tetani*, although the film of growth is much thicker and shows a coarsely irregular surface. The rhizoidal spreading edge of growth is evident after 24 h incubation; after 48 h, the growth usually completely covers the surface of the agar plate.

On horse blood agar, growth of *Cl. septicum* is associated with β-haemolysis, while *Cl. chauvoei* is commonly non-haemolytic. Haemolysis due to *Cl. chauvoei* is best observed on sheep blood agar. Neither organism produces any conspicuous change on heated blood agar or on egg yolk agar media. In cooked meat broth both organisms grow well in 48 h, with slight gas production, and the meat particles are often turned pink.

Biochemical characteristics

Cl. chauvoei ferments glucose, maltose, lactose and sucrose, while *Cl. septicum* ferments all but sucrose. In addition, *Cl. septicum* attacks salicin, which is not fermented by *Cl. chauvoei.* Both organisms produce hydrogen sulphide, but neither produce indole. Both are gelatinase producers, and both produce deoxyribonuclease.

Major products of metabolism produced by *Cl. septicum* and *Cl. chauvoei* are acetic and butyric acids with lesser amounts of formic acid.

Toxicology

Cl. chauvoei produces an exotoxin complex which is antigenically related to that produced by *Cl. septicum*. Consequently, *Cl. septicum* antitoxin provides not only homologous protection, but also protection against intoxication by *Cl. chauvoei. Cl. chauvoei* antitoxin, however, does not protect against *Cl. septicum* toxaemia, owing to the fact that the latter organism produces some additional toxins (notably *Cl. septicum* α-toxin) which are not elaborated by *Cl. chauvoei.*

Cl. septicum

At least three exotoxins are produced by strains of *Cl. septicum*. The α-toxin is lethal, necrotizing and haemolytic (Bernheimer, 1944; Moussa, 1958). Its lethal effect is demonstrated by intravenous injection into mice; convulsions are followed by paralysis, and death occurs rapidly. At post mortem there is intense capillary engorgement, and interstitial haemorrhages are present in the heart. Haemolytic activity is shown against the red cells of the sheep, and it can be determined by haemolysin tests using *Cl. septicum* antitoxin. The haemolysis produced by cultures growing on horse blood agar is due to the α-toxin.

The α-toxin is an oxygen-stable haemolysin which exhibits an induction period of 5–60 min before lysis of the red cells occurs (compare with the oxygen-labile haemolysin). Haemolysis, which was inhibited by specific anti-toxin, was used by Hayward and Gray (1946) for the identification of strains of *Cl. septicum* and *Cl. oedematiens* type A. *Cl. septicum* α-toxin is antigenically related to *Cl. histolyticum* α-toxin (Guillaumie, Kreguer and Fabre, 1946a, b; Sterne and Warrack, 1962).

The β-toxin is a deoxyribonuclease (Warrack, Bidwell and Oakley, 1951). Its action is demonstrated either by the ACRA test as used for the determination of *Cl. perfringens* υ-antigen, or by the toluidine blue DNA technique described by Princewill and Oakley (1972a). It is specifically inhibited by *Cl. septicum* antitoxin.

Moussa (1958) showed that *Cl. septicum* produced an oxygen-labile haemolysin similar in properties, and antigenically related to, other oxygen-labile haemolysins. Unlike the α-toxin, no induction period precedes lysis of erythrocytes. Moussa referred to this haemolysin as δ-toxin.

Cl. septicum γ-antigen is a hyaluronidase. The organism also produces a haemagglutinin and a neuraminidase (Gadalla and Collee, 1967, 1968).

Cl. chauvoei

Much of our modern knowledge of the soluble substances produced by *Cl. chauvoei* is due to the work of Moussa (1958).

Cl. chauvoei α-toxin is an oxygen-stable haemolysin that is also necrotizing; it appears to be unrelated to *Cl. septicum* α-toxin. The other antigens examined by Moussa are a deoxyribonuclease (β-toxin), a hyaluronidase (γ-antigen), and an oxygen-labile haemolysin (δ-antigen), all of which appear to be antigenically related to the corresponding antigens produced by *Cl. septicum* (*see also* Princewill and Oakley, 1972a, b; 1976).

Antigenic structure

Moussa (1959), in a study of the flagellar, somatic and spore antigens of *Cl. septicum* and *Cl. chauvoei*, divided 37 strains of *Cl. septicum* into two groups on the basis of the 'O' antigen; neither of these groups cross-reacted with *Cl. chauvoei.* He also found that his 37 strains of *Cl. chauvoei* possessed a common 'O' antigen. This work was confirmed by Batty and Walker (1963), who further showed that *Cl. septicum* and *Cl. chauvoei* can be identified with certainty, and differentiated from one another, by the use of fluorescent labelled antibodies. This is a matter of particular value to the veterinary bacteriologist, since *Cl. septicum* may be present as a contaminant in lesions of cattle that are primarily due to *Cl. chauvoei*.

Willis and Williams (1972) showed that the swarming growth of *Cl. septicum* in surface culture was prevented by treating plates with a polyvalent *Cl. septicum* 'O' antiserum before inoculation. Such an agglutinating antiserum is easily prepared in the rabbit, and requires the use, as antigen, of only two strains of the organism, selected to represent each of the two serological groups of Moussa (1959).

Pathogenicity

Cl. septicum is pathogenic for man and animals and is associated with gas gangrene in man, while *Cl. chauvoei* is not pathogenic for man but causes disease in some animals, especially the ruminants.

Experimental Cl. septicum and Cl. chauvoei gas gangrene

After intramuscular inoculation of a guinea pig with 1 ml of a fresh 24–48 h cooked meat broth culture, extensive oedema develops and death follows in 24–48 h. There is considerable gas formation in the tissues. The disease may be prevented by protection of the animal with antitoxin.

At post mortem there is extensive blood-stained oedema of a deep red colour, spreading from the site of inoculation up over the abdomen. The muscles about the site of inoculation are very deep red, and there is a considerable amount of gas in the tissues. The muscles are neither friable nor softened. The organisms are readily isolated from the heart-blood, and the liver in particular provides a focus for their multiplication. As has been noted, *Cl. septicum* tends to form long filaments on the peritoneal surface of the liver, a feature not exhibited by *Cl. chauvoei.*

Main features for recognition of Cl. septicum and Cl. chauvoei

1. Direct staining with fluorescent labelled antibodies.
2. Colonial morphology.
3. Sugar fermentation reactions, notably sucrose and salicin.
4. Volatile products of metabolism.
5. Animal pathogenicity and protection tests.
6. Morphological appearance on the peritoneal surface of the liver.

Isolation of Cl. septicum and Cl. chauvoei

Cl. septicum and *Cl. chauvoei* are relatively easy to grow, bearing in mind that the growth of *Cl. chauvoei* is greatly favoured by the addition of liver extract to the medium. Their isolation is accomplished by colony selection from plate cultures. Preliminary enrichment of the material in cooked meat broth may be necessary, and aerobic contaminants may be reduced by the use of neomycin sulphate. Initial heating of the enrichment culture may be used to destroy vegetative organisms. The use of egg yolk agar media for primary plating is not indicated, since these organisms are egg yolk negative. In practice the clinical bacteriologist will rarely, if ever, encounter *Cl. chauvoei,* since this organism is not associated with disease in man. Its isolation from blackleg lesions in cattle is usually not difficult, but may be complicated by contamination with *Cl. septicum* which is a normal inhabitant of the intestinal tracts of herbivorous animals. Under these circumstances it is necessary to prevent the swarming growth of *Cl. septicum* by plating on firm agar media, or by using *Cl. septicum* somatic antiserum plates as described by Willis and Williams (1972). To the medical bacteriologist *Cl. chauvoei* is important only because it has to be distinguished from *Cl. septicum*.

CLOSTRIDIUM HISTOLYTICUM

Cl. histolyticum is widely but sparsely distributed in soil and probably also in the intestinal tracts of man and animals.

Morphology

Cl. histolyticum is a Gram-positive bacillus, about 3–5 × 0.5 μm in size with parallel sides and rounded ends, and usually a straight axis. Filamentous forms are developed in old cultures and under aerobic conditions (*see below*). Spores are large, oval, subterminal and distend the organism. They are readily formed in large numbers, except under aerobic conditions when sporulation does not occur at all. There is nothing characteristic about the appearance of *Cl. histolyticum*.

Cultural characteristics

Cl. histolyticum is not an exacting anaerobe, and grows readily as surface colonies under aerobic conditions, a property which it shares with *Cl. tertium* and *Cl. carnis*. Aerobic colonies, however, are small and are composed predominantly of filamentous and pleomorphic forms, spores being absent. Growth is greatly improved in an anaerobic environment. After 24–48 h anaerobic incubation, surface colonies are roughly circular in shape, about 1 mm in diameter, opaque and greyish-white in colour with a shiny surface and an entire edge. There is no tendency for growth to spread over the surface of the medium. Aerobic colonies are slow to appear; after 48 h incubation they are about 0.5 mm in diameter, hemispherical and transparent, and they do not increase in size with subsequent incubation. On horse blood agar, colonies are surrounded by a narrow zone of haemolysis. On heated blood agar quite extensive proteolysis occurs, with the development of wide zones of partial clearing about the colonies. On lactose egg yolk milk agar the milk is attacked, with resultant partial clearing of the medium; lactose is not fermented, and an egg yolk reaction is not produced. The reactions on lactose egg yolk milk agar are virtually diagnostic. In cooked meat broth good growth is obtained in 24 h, and with subsequent incubation at 37 °C or standing at room temperature vigorous proteolysis of the meat particles occurs.

Biochemical characteristics

Cl. histolyticum ferments no sugars but is strongly proteolytic, attacking both gelatin and more complex proteins. It is the only commonly occurring clostridium that is gelatinase-positive but glucose-negative. It is, indeed, the only common example of a strongly proteolytic, non-saccharolytic clostridium (cf. *Cl. lentoputrescens* and *Cl. subterminale*). Indole is not formed, but hydrogen sulphide is produced in abundance.

The only volatile product of metabolism produced in any amount is acetic acid.

Toxicology

Oakley and Warrack (1950) demonstrated the presence of three separate soluble antigens in culture filtrates of *Cl. histolyticum*. The α-toxin is lethal and necrotizing, and is antigenically related to *Cl. septicum* α-toxin (Guillaumie, Krueger and Fabre, 1946a, b; Sterne and Warrack, 1962); the β-antigen is a collagenase which attacks azocoll and gelatin; the γ-antigen, which is a proteinase activated by reducing agents, does not attack native collagen but attacks azocoll, gelatin and casein. MacLennan, Mandl and Howes (1958) and Oakley and Warrack (1958) demonstrated the presence of a third proteolytic enzyme, δ-antigen. This has since been shown by Oakley and Banerjee (1963) to be an elastase. Methods for the production and assay of elastase in cultures and culture filtrates are described by the latter workers. δ-antigen also attacks azocoll and gelatin, and is partially inhibited by reducing agents. ε-antigen (Hobbs, 1958) is a typical oxygen-labile haemolysin which is neutralized by antisera to other oxygen-labile haemolysins. In addition to these soluble antigens, others are almost certainly produced, as for example, the peptidases described by Mandl, Ferguson and Zaffuto (1957), and the gelatinase described by Mandl and Zaffuto (1958). Webster and her colleagues (1962) claimed that one strain of *Cl. histolyticum*, which they studied, produced at least nine different proteinases.

Pathogenicity

Cl. histolyticum is pathogenic for man and animals. In man it is associated with gas gangrene infections, but it rarely occurs as the sole anaerobe.

Experimental Cl. histolyticum gas gangrene

Following the intramuscular inoculation of 1 ml of a fresh 24-h cooked meat broth culture into the leg of the guinea pig, death occurs in 2–3 days. Oedema develops at the site of inoculation and spreads up over the abdomen; the inoculated muscles begin to break down, the skin ulcerates, and within 24 h the bone of the limb is denuded of soft tissues. Auto-amputation of the limb at the hip joint may occur as a result of digestion of the joint ligaments and capsule.

Main features for recognition of Cl. histolyticum

1. Powerful proteolytic activity but absence of saccharolytic activity. Appearances on lactose egg yolk agar are virtually diagnostic.
2. Action on glucose gelatin.
3. Volatile products of metabolism.
4. Animal inoculation experiments.

Isolation of Cl. histolyticum

Cl. histolyticum is not a difficult organism to grow, and its colonies are easily recognized on lactose egg yolk milk agar. Since it sporulates quite readily in artificial culture, differential heating may be used to eliminate unwanted vegetative organisms from mixed cultures. The use of neomycin sulphate in plate cultures is of value in eliminating aerobic contaminants. *Cl. histolyticum* may be separated from mixtures with other anaerobic organisms by incubating plate subcultures under aerobic or partially anaerobic conditions.

CLOSTRIDIUM TERTIUM

Cl. tertium was so named because it was the third commonest anaerobe encountered by Henry (1916–17) in war wounds. It is a non-pathogen.

Morphology

Cl. tertium is a Gram-positive bacillus, about 3–5 × 0.3 μm in size, usually slightly curved. Sporulation readily occurs, the spores being large, oval and terminal, and distending the bacillary body. Like *Cl. histolyticum, Cl. tertium* grows under aerobic conditions but spores are produced only in the absence of oxygen.

Cultural characteristics

Although *Cl. tertium* is an aerotolerant organism, and grows quite well in the presence of free oxygen, best growth is obtained under anaerobic conditions. Anaerobic colonies are about 1–2 mm in diameter, transparent, domed and round, with an entire or slightly crenated edge. Aerobic colonies are less than 1 mm in diameter, circular, domed and featureless. On horse blood agar no haemolysis is produced, and the organism produces no change in heated blood agar or in egg yolk agar media.

Biochemical characteristics

The organism ferments glucose, maltose, lactose and sucrose, but does not attack gelatin or produce indole.

Major volatile products of metabolism include acetic and lactic acids with lesser amounts of butyric and formic acids.

CLOSTRIDIUM SPOROGENES

Cl. sporogenes is widely distributed in nature – in soil and in the intestinal tract of man and animals. It is one of the commonest clostridia and a very common contaminant of anaerobic cultures.

Morphology

The organism is a strongly Gram-positive bacillus, about 3–6 × 0.5 μm in size. It has parallel sides and rounded ends, and its axis is either straight or slightly curved. Filaments sometimes occur under favourable conditions. Sporulation occurs readily, the spore being oval, subterminal and distending the bacillary body. Free spores are common. Occasional strains of this organism do not form spores freely. There is nothing characteristic about the appearance of *Cl. sporogenes*.

Cultural characteristics

After 48 h incubation surface colonies are 3–5 mm in diameter, umbonate, with an opaque, greyish-white centre, a flattened irregularly circular periphery and highly rhizoidal edge. There is often some tendency for growth to spread over the surface of the medium.

On horse blood agar zones of pseudo-haemolysis are sometimes seen around the colonies; but contrary to general belief, haemolysis does not occur. On heated blood agar partial clearing of the medium is produced, due to proteolytic activity. On lactose egg yolk milk agar the lactose is not fermented. Restricted opalescence and a pearly layer are produced due to lipase activity, both of which are more or less coextensive with growth. Proteolysis of the milk leads to the development of wide zones of clearing about area of growth. Incubation for 48 h is usually required for the full development of these characters. In cooked meat broth good growth is obtained in 24 h. With further incubation, or on standing at room temperature, the meat particles are attacked and partially digested. A butyrous scum of fatty acids sometimes develops on the surface of the broth if the meat is not completely fat-free. Cultures are associated with an odour of skatol.

Biochemical characteristics

Glucose and maltose are fermented, but lactose and sucrose are not attacked. Gelatin is attacked and more complex proteins are also broken down. Indole is not produced, but hydrogen sulphide is formed in large quantities.

Volatile products of metabolism are multiple and complex, a wide range of both acids and alcohols being produced. Major acids may include acetic, propionic, isobutyric, butyric, isovaleric and isocaproic.

Toxicology

Cl. sporogenes does not produce any soluble antigens of clinical significance.

Pathogenicity

Most strains of this organism in pure culture are able to produce local putrefactive changes in inoculated muscles, especially if the muscle tissue is damaged at the time of inoculation; but there is no associated toxaemia. In mixed clostridial infections, *Cl. sporogenes* increases the virulence of pathogenic species, such as *Cl. perfringens* and *Cl. septicum*. Occasionally, apparently truly pathogenic strains of *Cl. sporogenes* have been reported, cultures of which cause death of the guinea pig in 1–2 days following intramuscular inoculation.

Main features for recognition of Cl. sporogenes

1. Sugar fermentation reactions and proteolytic activity.
2. Reactions on lactose egg yolk milk agar virtually diagnostic.
3. Volatile products of metabolism (but cf. *Cl. botulinum, Cl. bifermentans-sordellii* and *Cl. difficile*).
4. Distinction from proteolytic strains of *Cl. botulinum* by pathogenicity tests in animals, and by somatic serological reactions (Walker and Batty, 1964).

Isolation of Cl. sporogenes

As with *Cl. histolyticum*, use may be made of the heat resistance of the spores of *Cl. sporogenes* and of the inhibitory effect of neomycin sulphate against aerobic organisms. The characteristic effects produced by the anaerobe on lactose egg yolk milk agar make this medium useful for both primary and subsequent plate culture. The main concern of the clinical microbiologist is usually to exclude this organism from cultures of other clostridia!

CLOSTRIDIUM BIFERMENTANS AND CLOSTRIDIUM SORDELLII

Cl. bifermentans and *Cl. sordellii* are widely distributed in the soil, and *Cl. bifermentans* is a common inhabitant of the large bowel of man and animals.

Table 4.7 Some biochemical differences between *Cl. bifermentans* and *Cl. sordellii*

Organism	*Glucose*	*Maltose*	*Lactose*	*Mannose*	*Sorbitol*	*Salicin*	*Urease activity*
Cl. bifermentans	+	+	−	+	+	+	−
Cl. sordellii	+	+	−	−	−	−	+

In the past it has been a common practice to regard these organisms as non-pathogenic and pathogenic strains of the single species *Cl. bifermentans*. However, in a study of the *bifermentans–sordellii* complex, Brooks and Epps (1959) showed that the two species, although similar in many important respects, are distinct and easily distinguishable. Thus, *Cl. bifermentans* is urease-negative but ferments mannose, sorbitol and salicin, whereas *Cl. sordellii* is urease-positive and does not attack these fermentable substances (*Table 4.7*). Further, the colonial appearances of the two organisms on horse blood agar containing 3% of agar are distinct: *Cl. sordellii* colonies have crenated or coarsely rhizoidal margins, while the margins of *Cl. bifermentans* colonies are entire or slightly undulate. Not all strains of *Cl. sordellii* are pathogenic, but *Cl. bifermentans* is never pathogenic (*see also* Nakamura *et al.* 1975).

Against the work of Brooks and Epps must be set the findings of Huang (1959) who considered that the biochemical differences between *Cl. sordellii* and *Cl. bifermentans* were too minor to warrant species separation. Huang observed mutation of a urease-positive pathogenic strain of *Cl. sordellii* to a non-pathogenic variant without urease activity and *vice versa*, and also found that *Cl. bifermentans* and *Cl. sordellii* shared some somatic and spore antigens. Walker (1963), on the other hand, found that the two species could be differentiated on the basis of their spore agglutinogens, although differentiation on the basis of their spore precipitinogens was not possible. Walker further noted the absence of cross agglutination and precipitation between the spores of the *Cl. bifermentans–sordellii* group on the one hand, and those of *Cl. sporogenes, Cl. histolyticum Cl. sphenoides* and *Cl. perfringens* types A–D on the other. It seems clear, therefore, that *Cl. bifermentans* and *Cl. sordellii* are rightly regarded as separate species.

Morphology

These species are strongly Gram-positive bacillii about 2–4 × 1μm in size. The organisms sporulate freely in culture, the spores being large and cylindrical, usually centrally placed but only slightly distending the bacillary body. Free spores are common and chains of sporulating organisms are frequently seen. Chain and filament formation in fluid media is characteristic of these organisms.

Cultural characteristics

Cl. bifermentans and *Cl. sordellii* are not exacting anaerobes and are comparatively easy to grow. After 24 h incubation surface colonies are 2–3 mm in diameter, greyish-white, with a low convex surface and an entire or irregular edge; not uncommonly swarming growth tends to develop. The colonial differences between the two species growing on 3% agar plates are noted above. On horse blood agar, colonies are usually surrounded by a narrow zone of haemolysis. On heated blood agar partial clearing of the medium is produced, due to proteolysis. On lactose egg yolk milk agar extensive opalescence is produced, which is inhibited by *Cl. perfringens* type A antitoxin, though the inhibition is seldom complete. There is no pearly layer, lactose is not fermented but the milk is attacked, so that zones of partial clearing are developed around areas of growth. The reactions in this medium, used as a half-antitoxin plate with *Cl. perfringens* type A antitoxin, are diagnostic of the *bifermentans–sordellii* complex. In cooked meat broth good growth is obtained in 24 h, and the meat particles are subsequently partially digested. In this, as in other broth media, a viscous mucoid deposit is produced, a feature peculiar to these organisms.

Biochemical characteristics

Glucose and maltose are attacked, but not lactose or sucrose. Gelatin and more complex proteins are broken down, and hydrogen sulphide is produced. Indole is also formed. The biochemical differences between the two organisms have already been noted.

Major products of metabolism include large amounts of acetic, isobutyric and isovaleric acids, with smaller amounts of propionic and isocaproic acids.

Toxicology

Pathogenic strains of *Cl. sordellii* produce a lethal toxin, against which antitoxic sera can be prepared. All strains of *Cl. sordellii* and *Cl. bifermentans* produce a lecithinase C which is antigenically related to, but not identical with, the lecithinase C (α-toxin) of *Cl. perfringens*. Our present knowledge of the *bifermentans-sordellii* lecithinase C is due to the work of Miles and Miles (1947, 1950). As with other haemolytic lecithinases, lysis of red cells is probably due to the lecithinase activity of the toxin. The erythrocytes of the mouse are very susceptible, those of the rabbit less so, while red cells of the sheep and horse are comparatively resistant.

The proteolytic activity of strains of *Cl. bifermentans* and *Cl. sordellii* indicates that at least one other soluble antigen is produced by them, but this awaits further study (*see* Arsecularatne, Panabokke and Wijesundera, 1969).

Pathogenicity

Intramuscular inoculation of a culture of a pathogenic strain of *Cl. sordellii* into the guinea pig is followed by death of the animal in 1–2 days.

At post mortem there is subcutaneous gelatinous oedema spreading away from the site of inoculation, the inoculated muscles are haemorrhagic and some gas is present in the tissues. Some strains produce severely proteolytic local effects, especially if the muscle tissue is damaged at the time of inoculation. This proteolysis may progress like that of *Cl. histolyticum* infections, so that the bones of the inoculated limb are denuded of soft tissues, and auto-amputation of the limb at the hip joint may occur.

Main features for recognition of the bifermentans–sordellii complex

1. Sugar fermentation reactions and proteolytic activity.
2. Reactions on half-antitoxin lactose egg yolk milk agar (diagnostic).
3. Mucoid deposit and filament or chain formation in broth cultures.
4. Indole production.
5. *Cl. bifermentans* and *Cl. sordellii* are distinguished from one another biochemically; pathogenic strains of *Cl. sordellii* are recognized by animal inoculation tests.
6. Volatile products of metabolism.

Isolation of Cl. bifermentans and Cl. sordellii

Advantage may be taken of the rapid growth of the organisms in cooked meat broth, the heat resistance of their spores, and their characteristic appearance on lactose egg yolk milk agar. Neomycin sulphate may be incorporated in solid media to inhibit the growth of aerobic organisms.

CLOSTRIDIUM FALLAX

Cl. fallax is probably widely distributed in nature, although its natural habitat does not appear to have been determined.

Morphology

It is a Gram-positive bacillus, about 2–5 $\times$ 0.5 μm in size with a straight axis and rounded ends. Spores, which are rarely produced, are centrally or subterminally situated, oval, and distend the organism. Sporulation is favoured by a neutral or alkaline environment and is inhibited by the presence of fermentable carbohydrates. Capsules are said to be formed by some strains on first isolation.

Cultural characteristics

Cl. fallax is a strict anaerobe. After 24–48 h incubation surface colonies are 1–2 mm in diameter, at first flat and transparent, with an irregular edge, later becoming umbonate and more opaque. On horse blood agar small zones of haemolysis are produced by some strains. On heated blood agar no change is produced in the medium, and the organism is egg yolk negative. In cooked meat broth the organism produces abundant growth of the saccharolytic type in 24 h.

Biochemical characteristics

Glucose, maltose, lactose and sucrose are all fermented. Gelatin is not attacked, and hydrogen sulphide and indole are not produced.

Major products of metabolism are lactic, acetic and butyric acids.

Toxicology

A lethal toxin is produced.

Pathogenicity

Cl. fallax is often pathogenic for the mouse and guinea pig when first isolated, but it loses its virulence rapidly in culture. The intramuscular inoculation of a guinea pig with a virulent strain leads to the development of a localized haemorrhagic lesion with gelatinous oedema fluid and some gas formation. The oedema may extend up over the abdomen. Death occurs after about 48 h and is preceded by invasion of the blood stream by the organism.

Isolation of Cl. fallax

Since this organism infrequently produces spores, and since the spores when formed are not very resistant to heat, it is not advisable to use heat as an aid to purifying it from mixtures with other organisms.

CLOSTRIDIUM COCHLEARIUM

Cl. cochlearium is a slender, weakly Gram-positive bacillus, about 3–5 × 0.5 μm in size. Spores, which are not readily produced, are large, oval and terminal, and give the organism the appearance of a spoon. *Cl. cochlearium* is of interest because it is the only commonly occurring clostridium which is chemically almost inert. It is non-saccharolytic, non-proteolytic, non-haemolytic and egg yolk negative. It does not produce indole, but produces small amounts of hydrogen sulphide. It is non-pathogenic for man and laboratory animals. Volatile products of metabolism include a major amount of butyric acid with smaller amounts of acetic and propionic acids.

CLOSTRIDIUM PARAPUTRIFICUM

This organism is a normal inhabitant of the intestinal tract of man and animals. It is most commonly found in the faeces of infants.

Cl. paraputrificum is a long, slender Gram-variable bacillus, about 2–6 × 0.3–0.5 μm in size with a straight or slightly curved axis. It sporulates freely in 2–3 day old cultures, producing large oval terminal spores that distend the bacillary body. It is a strict anaerobe. Colonies are 1–2 mm in diameter after 48 h incubation, translucent, and irregularly circular. The organism produces no change on horse blood agar, heated blood agar or on egg yolk agar. It is actively saccharolytic, fermenting glucose, maltose, lactose and sucrose, but is entirely non-proteolytic. Neither indole nor hydrogen sulphide is produced.

Volatile products of metabolism are major amounts of lactic, acetic and butyric acids, with lesser amounts of formic acid.

Cl. paraputrificum is non-pathogenic for man and laboratory animals.

CLOSTRIDIUM LENTOPUTRESCENS

Cl. lentoputrescens is widely distributed in soil, and in the intestinal tract of man.

It is a long, slender Gram-positive bacillus, 7–9 × 0.3–0.5 μm in size, with a straight or slightly curved axis. Both on solid and in fluid media it shows a marked tendency to develop filamentous forms, which are seen characteristically in older cultures as masses of tangled threads. Spores, which are developed slowly but abundantly in cooked meat broth, are large, spherical and terminally situated.

Cl. lentoputrescens is a strict anaerobe. Colonies are 2–3 mm in diameter after 3–4 days' incubation, semitranslucent, flat and irregularly circular, with a filamentous edge. The surface of the colony is finely furrowed which gives it a ground-glass appearance. The organism produces narrow zones of pseudohaemolysis on horse blood agar. It is strongly, but slowly, proteolytic, attacking both gelatin and more complex proteins such as heated blood and milk. It is entirely non-saccharolytic (cf. *Cl. histolyticum* and *Cl. subterminale*) and is egg yolk negative. It produces hydrogen sulphide but not indole. Major volatile products of metabolism are lactic, acetic and butyric acids with lesser amounts of formic and propionic.

Cl. lentoputrescens is non-pathogenic for man and laboratory animals.

CLOSTRIDIUM CADAVERIS

Cl. cadaveris occurs in soil and in the intestinal tracts of man and animals. It was formerly known as *Cl. capitovale*.

It is a slender Gram-positive bacillus, about 2.0 × 0.5–0.8 μm in size, with a slightly curved axis. Spores, which are readily produced are oval and terminal, and distend the bacillary body. The organism is a strict anaerobe. Colonies appear as tiny 'dewdrops' less than 1 mm in diameter after 48 h incubation. It is non-haemolytic on horse blood agar and is without effect on heated blood agar, milk agar, and egg yolk agar. Gelatin is attacked, hydrogen sulphide is formed, and some strains produce indole. Glucose is fermented, but maltose, lactose and sucrose are not attacked.

Major products of metabolism are acetic and butyric acids with smaller amounts of propionic, isobutyric and isovaleric acids.

Cl. cadaveris is non-pathogenic for man and laboratory animals.

CLOSTRIDIUM SPHENOIDES

Cl. sphenoides is probably widely distributed in nature.

It is a weakly Gram-positive bacillus, about 2–5 × 0.4–0.6 μm in size, usually with a slightly curved axis. The organism is characteristically fusiform in shape, often with pointed ends, a morphological appearance that is unique among the spore-forming anaerobes. Spores, which are readily formed, are large, spherical and terminally situated; developing spores distort the bacillary body in such a way that the sporulating cell is at first wedge-shaped, tapering from the spore to the distal pointed end of the cell.

Cl. sphenoides is a strict anaerobe. Colonies are about 1 mm in diameter after 24 h incubation, irregularly circular and semitranslucent. It is haemolytic on horse blood agar, but is without effect on heated blood agar. It is egg yolk negative. The species is entirely non-proteolytic, but is moderately saccharolytic, fermenting glucose, maltose and lactose. Indole is produced, but hydrogen sulphide is not formed.

Major products of metabolism are acetic and formic acids.

Cl. sphenoides is non-pathogenic for man and laboratory animals.

CLOSTRIDIUM CARNIS

Cl. carnis is a rarely encountered organism, probably sparsely distributed in soil, although evidence of its natural distribution is lacking.

It is a strongly Gram-positive bacillus, about 1.5–4.5 × 0.5 μm in size, which tends to pleomorphism in old cultures. Abundant spores are formed in the absence of fermentable carbohydrate; they are large, oval and subterminal and slightly distort the bacillary body.

Cl. carnis is aerotolerant, but less so than *Cl. histolyticum* and *Cl. terium*, producing barely visible pin-point colonies in 24 h under aerobic conditions. After 48 h aerobic incubation colonies are about 0.5 mm in diameter, flat, transparent discs with a lobulated margin; spores are not produced under aerobic conditions. After 48 h anaerobic incubation, colonies are about 1.5 mm in diameter, flat, greyish and irregularly circular. It is non-haemolytic on horse blood agar, entirely non-proteolytic, but is actively saccharolytic, fermenting glucose, maltose, lactose and sucrose. It is egg yolk negative, and produces neither indole nor hydrogen sulphide.

Major products of metabolism are acetic, butyric and formic acids.

Cl. carnis is probably non-pathogenic for man, but causes death of the guinea pig following intramuscular injection.

CLOSTRIDIUM DIFFICILE

Cl. difficile is most commonly encountered in the faeces of young infants, and has recently been reported to occur in the genitourinary tracts of a large proportion of patients suspected of having venereal infections (Hafiz *et al.*, 1975).

It is a long slender Gram-positive bacillus, about 6–8 × 0.5 μm in size, which produces large, oval, subterminal spores that distend the bacillary body. It is a strict anaerobe. Colonies are 2–3 mm in diameter after 48 h incubation, slightly raised, white, opaque and circular, with an entire margin. It is non-haemolytic on horse blood agar, is entirely non-proteolytic, and is egg yolk negative. It ferments glucose, but not maltose, lactose or sucrose. Indole and hydrogen sulphide are not formed.

Products of fermentation are multiple and complex, and include small amounts of acetic, isobutyric, isovaleric, valeric, butyric and isocaproic acids.

Cl. difficile is unusual in that it is tolerant to cresol, which it produces during growth (Hafiz and Oakley, 1976). Advantage may be taken of this for its isolation from mixtures with other organisms. Paracresol (0.2%) added to an enrichment broth allows selective growth of *Cl. difficile* (Hafiz *et al.* 1975).

CLOSTRIDIUM SUBTERMINALE

Cl. subterminale is probably widely distributed in soil.

It is a Gram-positive bacillus about 2–5 × 0.5–1 μm in size, which readily forms oval, subterminal spores that distend the bacterial cell. It is a strict anaerobe. Colonies are 2–3 mm in diameter after 24 h incubation, greyish white, with a low convex surface and an irregular edge. Small zones of haemolysis are produced on horse blood agar. It is entirely non-saccharolytic, but is actively proteolytic, attacking gelatin, heated blood and milk (cf. *Cl. histolyticum* and *Cl. lentoputrescens*). On egg yolk agar media it produces a faint diffuse opacity, similar in appearance to that of a weak lecithinase C reaction, but it is not lipolytic. The organism is indole-negative.

The major product of metabolism is acetic acid with smaller amounts of butyric, isobutyric and isovaleric acids.

Cl. subterminale is non-pathogenic for man and laboratory animals.

CLOSTRIDIUM BARATI

Cl. barati is probably widely but sparsely distributed in nature. It is a Gram-positive bacillus, about 1–3 × 0.5 μm in size, which shows a tendency to filament formation. Spores are sparse and subterminally situated.

It is entirely non-proteolytic, but is actively saccharolytic, fermenting glucose, maltose, lactose, sucrose, salicin and aesculin. On egg yolk agar media it produces a faint lecithinase C reaction, but is not lipolytic. It is indole negative.

Major products of metabolism are butyric, lactic and acetic acids, with a smaller amount of formic acid.

The organism is not pathogenic.

REFERENCES

Ajl, S. J., Kadis, S. and Montie, T. C. (1970). *Microbial Toxins*, Vol 1. New York; Academic Press

Arseculаratne, S. N., Panabokke, R. G. and Wijesundera, S. (1969). 'The toxins responsible for the lesions of *Clostridium sordellii* gas gangrene.' *Journal of medical Microbiology*, **2**, 37

Batty, I. and Walker, P. D. (1963). 'Differentiation of *Clostridium septicum* and *Clostridium chauvoei* by the use of fluorescent labelled antibodies.' *Journal of Pathology and Bacteriology*, **85**, 517

Batty, I. and Walker, P. D. (1964). 'The identification of *Clostridium novyi* (*Clostridium oedematiens*) and *Clostridium tetani* by the use of fluorescent labelled antibodies.' *Journal of Pathology and Bacteriology*, **88**, 327

Batty, I. and Walker, P. D. (1967). 'The use of the fluorescent labelled antibody technique for the detection and differentiation of bacterial species.' In *International Symposium of Immunological Methods of Biological Standardization, Royaumont 1965*. Vol. 4, p. 73 Basel; Karger

Bengston, I. A. (1922). 'Preliminary note on a toxin-producing anaerobe isolated from the larvae of *Lucilia caesar.*' *Public Health Reports*, **37**, 164

Bernheimer, A. W. (1944). 'Parallelism in the lethal and haemolytic activity of the toxin of *Clostridium septicum.*' *Journal of experimental Medicine*, **80**, 309

Bidwell, E. (1950). 'Proteolytic enzymes of *Clostridium welchii.*' *Biochemical Journal*, **46**, 589

Boroff, D. A. and DasGupta, B. R. (1971). 'Botulinum toxin.' In *Microbial Toxins Vol IIA*. p. 1. Ed. by S. Kadis, T. C. Montie and S. J. Ajl. New York; Academic Press

Brooks, M. E. and Epps, H. B. G. (1959). 'Taxonomic studies of the genus *Clostridium: Clostridium bifermentans* and *Cl. sordellii.*' *Journal of general Microbiology*, **21**, 144

Brooks, M. E., Sterne, M. and Warrack, G. H. (1957). 'A re-assessment of the criteria used for type differentiation of *Clostridium perfringens.*' *Journal of Pathology and Bacteriology*, **74**, 185

Buchanan, R. E. and Gibbons, N. E. (1974). *Bergey's Manual of Determinative Bacteriology*, 8th edn Baltimore; Williams and Wilkins

Cann, D. C., Taylor, L. Y. and Hobbs, G. (1975). 'The incidence of *Clostridium botulinum* in farmed trout raised in Great Britain.' *Journal of applied Bacteriology,* **39,** 331

Clifford, W. J. and Anellis, A. (1971). '*Clostridium perfringens.* I. Sporulation in a biphasic glucose-ion-exchange resin medium.' *Applied Microbiology,* **22,** 856

Delaunay, M., Guillaumie, M. and Delaunay, A. (1949). 'Studies on collagen, I. Bacterial collagenases.' *Annales de l'Institute Pasteur.* **76,** 16 (French)

Dolman, C. E. (1964). 'Botulism as a world health problem.' In *Botulism: Proceedings of a Symposium.* p. 5. Ed. by. K. H. Lewis and K. Cassel. Cincinnati; US department of Health

Dolman, C. E. and Murakami, L. (1961). '*Clostridium botulinum* type F with recent observations on other types.' *Journal of infectious Diseases,* **109,** 107

Duncan, C. L. and Strong, D. H. (1968). 'Improved medium for sporulation of *Clostridium perfringens.*' *Applied Microbiology,* **16,** 82

Duncan, C. L., Strong, D. H. and Sebald, M. (1972). 'Sporulation and enterotoxin production by mutants of *Clostridium perfringens.*' *Journal of Bacteriology,* **110,** 378

Ellner, P. D. (1956). 'A medium promoting rapid quantitative sporulation in *Clostridium perfringens.*' *Journal of Bacteriology,* **71,** 495

Evans, D. G. (1945). 'The *in-vitro* production of α-toxin, θ-haemolysin and hyaluronidase by strains of *Cl. welchii* type A, and the relationship of *in-vitro* properties to virulence for guinea pigs.' *Journal of Pathology and Bacteriology,* **57,** 75

Fildes, P. (1925). 'Tetanus. I. Isolation, morphology and cultural reactions of *B. tetani.*' *British Journal of experimental Pathology,* **6,** 62

Gadalla, M. S. A. and Collee, J. G. (1967). 'The nature and properties of the haemagglutinin of *Clostridium septicum.*' *Journal of Pathology and Bacteriology,* **93,** 255

Gadalla, M. S. A. and Collee, J. G. (1968). 'The relationship of the neuraminidase of *Clostridium septicum* to the haemagglutinin and other soluble products of the organism.' *Journal of Pathology and Bacteriology,* **96,** 169

Gimenez, D. F., and Ciccarelli, A. S. (1967). 'A new type of *Cl. botulinum.*' In *Botulism 1966. Proceedings of the Fifth international Symposium on Food Microbiology; Moscow, July, 1966.* p. 455. Ed. by M. Ingram and T. A. Roberts. London; Chapman and Hall

Gimenez, D. F. and Ciccarelli, A. S. (1970a). 'Studies on strain 84 of *Clostridium botulinum.*' *Zentralblatt für Bakteriologie, Paraseitenkunde, Infektionskrankheiten und Hygiene (Abteilung I) Originale,* **215,** 212

Gimenez, D. F. and Ciccarelli, A. S. (1970b). 'Another type of *Clostridium Botulinum.* 'Zentralblatt für Bakteriologie, Paraseotenkunde, Infektionskrankheiten und Hygiene (Abteilung I) Originale,* **215,** 221

Gordon, J. and Mcleod, J. W. (1940). 'A simple and rapid method of distinguishing *Cl. novyi* (*B. oedematiens*) from other bacteria associated with gas gangrene.' *Journal of Pathology and Bacteriology,* **50,** 167

Gordon, J., Turner, G. C. and Dmochowski, L. (1954). 'The inhibition of the alpha and theta toxins of *Clostridium welchii* by lecithin.' *Journal of Pathology and Bacteriology,* **67,** 605

Gowland, G., and Willis, A. T. (1961). 'Resistance of lysolecithin to the action of *Clostridium welchii* alpha-toxin.' *Journal of Pathology and Bacteriology,* **81,** 542

Guillaumie, M., Kreguer, A. and Fabre, M. (1946a). 'New investigation on titration of *Cl. septicum* antitoxin.' *Annales de l'Institue Pasteur,* **72,** 814 (French)

Guillaumie, M., Kreguer, A. and Fabre, M. (1946b). 'Properties of *Cl. histolyticum* antitoxin. Antigens elaborated by different strains of *Cl. histolyticum* and *Cl. septicum.' Annales de l'Institute Pasteur,* **72,** 818 (French)

Gunnison, J. B., Cummings, J. R. and Meyer, K. F. (1935). '*Clostridium botulinum* type E.' *Proceedings of the Society of Experimental Biology and Medicine,* **35,** 278

Hafiz, S., McEntegart, M. G., Morton, R. S. and Waitkins, S. A. (1975). '*Clostridium difficile* in urogenital tract of males and females.' *Lancet,* **1,** 420

Hafiz, S. and Oakley, C. L. (1976). '*Clostridium difficile*; Isolation and characteristics.' *Journal of medical Microbiology,* **9,** 129

Hauschild, A. H. W. (1971). '*Clostridium perfringens* toxins types B, C, D and E.' In *Microbial Toxins*, Vol IIA, p. 159. Ed. by S. Kadis, T. C. Montie and S. J. Ajl. New York; Academic Press

Hauschild, A. H. W. (1973). 'Food poisoning by *Clostridium perfringens.' Canadian Institute of Food Science and Technology,* **6,** 106

Hauschild, A. H. W. (1974). 'Enterotoxin of *Clostridium perfringens. 'In Anaerobic Bacteria: Role in Disease*. p. 149. Ed. by A. Balows, R. M. DeHaan, V. R. Dowell and L. B. Guze. Springfield; Charles C. Thomas

Hayward, N. J. (1941). 'Rapid identification of *Cl. welchii* by the Nagler reaction.' *British medical Journal,* **1,** 811, 916

Hayward, N. J. (1943). 'The rapid identification of *Cl. welchii* by Nagler tests in plate cultures.' *Journal of Pathology and Bacteriology,* **55,** 285

Hayward, N. J. and Gray, J. A. B. (1946). 'Haemolysin tests for the rapid identification of *Cl. oedematiens* and *Cl. septicum.' Journal of Pathology and Bacteriology,* **58,** 11

Henry, H. (1916–17). 'An investigation of the cultural reactions of certain anaerobes found in wounds.' *Journal of Pathology and Bacteriology,* **21,** 344

Heyningen, W. E. van (1941). 'The biochemistry of the gas gangrene toxins. I. Estimation of the α-toxin of *Cl. welchii* type A.' *Biochemical Journal,* **35,** 1246

Heyningen, W. E. van. and Mellanby, J. (1971). 'Tetanus toxin.' In *Microbial Toxins Vol IIA*, p. 69. Ed. by S. Kadis, T. C. Montie and S. J. Ajl. New York; Academic Press

Hobbs, B. C., Smith, M. E., Oakley, C. L., Warrack, G. H. and Cruickshank, J. C. (1953). '*Clostridium welchii* food poisoning.' *Journal of Hygiene, Cambridge,* **51,** 75

Hobbs, G. (1958). 'The soluble antigens produced by *Clostridium histolyticum*.' *PhD Thesis,* University of Leeds

Holdeman, L. V. and Moore, W. E. C. (1975). *Anaerobe Laboratory Manual;* V.P.I. Anaerobe Laboratory. Blacksburg; Virginia Polytechnic Institute and State University

Huang, C. T. (1959). 'Comparison of *Clostridium bifermentans* and *Clostridium sordellii*.' *PhD Thesis,* University of Leeds

Ispolatovskaya, M. V. (1971). 'Type A *Clostridium perfringens* toxin. 'In *Microbial Toxins Vol IIA*. p. 109. Ed. by S. Kadis, T. C. Montie and S. J. Ajl. New York; Academic Press

Kerrin, J. C. (1930). 'Studies in the haemolysin produced by atoxic strains of *B. tetani*.' *British Journal of experimental Pathology,* **11,** 153

Leuchs, J. (1910). 'Contributions to the knowledge of the toxins and antitoxins of *Bacillus botulinum*.' *Zeitschrift für Hygiene und Infectionskrankheiten,* **65,** 55 (German)

Lowbury, E. J. L. and Lilly, H. A. (1958). 'Contamination of operating-theatre air with *Cl. tetani*.' *British medical Journal,* **2,** 1334

Macfarlane, M. G. (1950). 'Biochemistry of bacterial toxins; Variation in haemolytic activity of immunologically distinct lecithinase towards erythrocytes from different species.' *Biochemical Journal,* **47,** 270

Macfarlane, M. G. (1955). '*Clostridium oedematiens* η-antigen, an enzyme decomposing tropomyosin.' *Biochemical Journal,* **61,** 308

Macfarlane, M. G. and Knight, B. C. J. G. (1941). 'The biochemistry of bacterial toxins. I. The lecithinase activity of *Cl. welchii* toxins.' *Biochemical Journal,* **35,** 884

Macfarlane, R. G. and MacLennan, J. D. (1945). 'The toxaemia of gas gangrene.' *Lancet,* **2,** 328

Macfarlane, R. G., Oakley, C. L. and Anderson, C. G. (1941). 'Haemolysis and the production of opalescence in serum and lecitho-vitellin by the α-toxin of *Clostridium welchii.' Journal of Pathology and Bacteriology,* **52,** 99

MacLennan, J. D. (1939). 'The serological identification of *Cl. tetani.*' *British Journal of experimental Pathology,* **20,** 372

MacLennan, J. D., Mandl, I. and Howes, E. L. (1958). 'New proteolytic enzymes from *Clostridium histolyticum* filtrates.' *Journal of general microbiology,* **18,** 1

Mandl, I. and Zaffuto, S. F. (1958). 'Serological evidence for a specific *Clostridium histolyticum* gelatinase.' *Journal of general Microbiology,* **18,** 13

Mandl, I., Ferguson, L. T. and Zaffuto, S. F. (1957). 'Exopeptidases of *Clostridium histolyticum.' Archives of Biochemistry,* **69,** 565

Mason, J. H. and Robinson, E. M. (1935). 'The antigenic components of the toxins of *Cl. botulinum* types C and D.' *Onderstepoort Journal of Veterinary Science,* **5,** 65

Mayhew, J. W. and Gorbach, S. L. (1975). 'Rapid gas chromatographic technique for presumptive detection of *Clostridium botulinum* in contaminated food.' *Applied Microbiology,* **29,** 297

Meyer, K. F. and Gunnison, J. B. (1929). 'South African cultures of *Clostridium botulinum* and *parabotulinum*. XXXVII. With a description of *Cl. botulinum* type D, n. sp.' *Journal of infectious Diseases,* **45,** 106

Miles, E. M. and Miles, A. A. (1947). 'The lecithinase of *Clostridium bifermentans* and its relation to the α-toxin of *Clostridium welchii.*' *Journal of general Microbiology,* **1,** 385

Miles, E. M. and Miles, A. A. (1950). 'The relation of toxicity and enzyme activity in the lecithinases of *Clostridium bifermentans* and *Cl. welchii.*' *Journal of general Microbiology,* **4,** 22

Moeller, V. and Scheibel, I. (1960). 'Preliminary report on the isolation of an apparently new type of *Cl. botulinum.*' *Acta Pathologica et Microbiologica Scandinavica,* **40,** 80

Moore, W. B. (1968). 'Solidified media suitable for the cultivation of *Clostridium novyi* type B.' *Journal of general Microbiology,* **53,** 415

Moussa, R. S. (1958). 'Complexity of toxins from *Clostridium septicum* and *Clostridium chauvoei.*' *Journal of Bacteriology,* **76,** 538

Moussa, R. S. (1959). 'Antigenic formulae for *Clostridium septicum* and *Clostridium chauvoei.' Journal of Pathology and Bacteriology,* **77,** 341

Nagler, F. P. O. (1939). 'Observations on a reaction between the lethal toxin of *Cl. welchii* (Type A) and human serum.' *British Journal of experimental Pathology,* **20,** 473

Nakamura, S., Shimamura, T., Hayashi, H. and Nishida, S. (1975). 'Reinvestigation of the taxonomy of *Clostridium bifermentans* and *Clostridium sordellii.*' *Journal of medical Microbiology,* **8,** 299

Oakley, C. L. (1943). 'Toxins of *Clostridium welchii;* Critical review.' *Bulletin of Hygiene, London,* **18,** 781

Oakley, C. L. and Banerjee, N. G. (1963). 'Bacterial elastases.' *Journal of Pathology and Bacteriology,* **85,** 489

Oakley, C. L. and Warrack, G. H. (1950). 'The alpha, beta and gamma antigens of *Clostridium histolyticum.' Journal of general Microbiology,* **4,** 365

Oakley, C. L. and Warrack, G. H. (1951). 'The ACRA test as a means of estimating hyaluronidase, deoxyribonuclease and their antibodies.' *Journal of Pathology and Bacteriology,* **63,** 45

Oakley, C. L. and Warrack, G. H. (1953). 'Routine typing of *Clostridium welchii.' Journal of Hygiene, Cambridge,* **51,** 102

Oakley, C. L. and Warrack, G. H. (1958). 'The cysteine-activated proteinase (δ-antigen) of *Clostridium histolyticum.' Journal of general Microbiology,* **18,** 9

Oakley, C. L. and Warrack, G. H. (1959). 'The soluble antigens of *Clostridium oedematiens* type D (*Cl. haemolyticum*).' *Journal of Pathology and Bacteriology,* **78,** 543

Oakley, C. L., Warrack, G. H. and Clarke, P. H. (1947). 'The toxins of *Clostridium oedematiens* (*Cl. novyi*).' *Journal of general Microbiology,* **1,** 91

Oakley, C. L., Warrack, G. H. and Warren, M. E. (1948). 'The kappa and lambda antigens of *Clostridium welchii.*' *Journal of Pathology and Bacteriology,* **60,** 495

Pillemer, L. and Wartman, W. B. (1947). 'The clinical behaviour, incubation period, and pathology of tetanus induced in white Swiss mice by injection of crystalline tetanal toxin.' *Journal of Immunology,* **55,** 277

Pillemer, L., Wittler, R. and Grossberg, D. B. (1946). 'Isolation and crystallization of tetanus toxin.' *Science,* **103,** 615

Prévot, A. R. (1966). *Manual for the Classification and Determination of the Anaerobic Bacteria*; 1st American edn. Philadelphia; Lea and Febiger

Princewill, T. J. T. (1970). 'Deoxyribonucleases and hyaluronidases of *Clostridium septicum* and *Clostridium chauvoei.*' *Journal of medical laboratory Technology,* **27,** 276

Princewill, T. J. T. and Oakley, C. L. (1972a). 'The deoxyribonuleases and hyaluronidases of *Clostridium septicum* and *Cl. chauvoei*. I. An agar plate method for testing for deoxyribonuclease.' *Medical laboratory Technology,* **29,** 243

Princewill, T. J. T. and Oakley, C. L. (1972b). 'The deoxyribonucleases and hyaluronidases of *Clostridium septicum* and *Cl. chauvoei*. II. An agar plate method for testing for hyaluronidase.' *Medical laboratory Technology,* **29,** 255

Princewill, T. J. T. and Oakley, C. L. (1976). 'Deoxyribonuleases and hyaluronidases of *Clostridium septicum* and *Clostridium chauvoei.*' *Medical laboratory Science,* **33,** 105

Roberts, R. S., Guven, S. and Worrall, E. E. (1970). '*Clostridium oedematiens* in the livers of healthy sheep.' *Veterinary Record,* **86,** 628

Rutter, J. M. (1970). 'A study of the carbohydrate fermentation reactions of *Clostridium oedematiens* (*Cl. novyi*).' *Journal of medical Microbiology,* **3,** 283

Rutter, J. M. and Collee, J. G. (1969). 'Studies on the soluble antigens of *Clostridium oedematiens* (*Cl. novyi*).' *Journal of medical Microbiology,* **2,** 395

Seabald, M. and Cassier, M. (1969). 'Sporulation and toxigenicity in mutant strains of *Clostridium perfringens.*' In *Spores IV.* p. 306. Ed. by L. L. Campbell. Bethesda; American Society for Microbiology

Seddon, H. R. (1922). 'Bulbar paralysis in cattle due to the action of a toxicogenic bacillus, with a discussion on the relationship of the condition to forage poisoning (botulism).' *Journal of comparative Pathology and Therapeutics*, **35**, 147

Seiffert, G. (1939). 'A reaction in human serum with *Cl. perfringens* toxin.' *Zeitschrift für Immunitatsforschung und experimentelle Therapie*, **96**, 515 (German)

Skjelkvale, R. and Duncan, C. L. (1975). 'Enterotoxin formation by different toxigenic types of *Clostridium perfringens*.' *Infection and Immunity*, **11**, 563

Smith, G. R. and Moryson, C. J. (1975a). 'Fish farms and botulism.' *British medical Journal*, **3**, 301

Smith, G. R. and Moryson, C. J. (1975b). '*Clostridium botulinum* in the lakes and waterways of London.' *Journal of Hygiene, Cambridge*, **75**, 371

Smith, L. DS. (1955). *Introduction to the Pathogenic Anaerobes*. Chicago; University of Chicago Press

Smith, L. DS. and Holdeman, L. V. (1968). *The Pathogenic Anaerobic Bacteria* Springfield; Thomas

Spencer, R. (1969). 'Neomycin-containing media in the isolation of *Clostridium botulinum* and food poisoning strains of *Clostridium welchii*.' *Journal of applied Bacteriology*, **32**, 170

Sterne, M. and Warrack, G. H. (1962). 'The interactions between *Clostridium septicum* and *Clostridium histolyticum* toxins and antitoxins.' *Journal of Pathology and Bacteriology*, **84**, 277

Taguchi, R. and Ikezawa, H. (1975). 'Phospolipase C from *Clostridium novyi* type A.' *Biochemica et Biophysica Acta*, **409**, 75

Theiler, A. and Robinson, E. M. (1927). *11th and 12th Reports, Director of Veterinary Education and Research*. Part II, Section 5; p. 1201. Union of South Africa; Department of Agriculture

Walker, P. D. (1963). 'The spore antigens of *Clostridium sporogenes, Cl. bifermentans* and *Cl. sordellii*.' *Journal of Pathology and Bacteriology*, **85**, 41

Walker, P. D. and Batty, I. (1964). 'Fluorescent studies in the genus *Clostridium* II. A rapid method for differentiating *Clostridium botulinum* types A, B and F and types C, D and E.' *Journal of applied Bacteriology*, **27**, 140

Warrack, G. H., Bidwell, E. and Oakley, C. L. (1951). 'The beta-toxin (deoxyribonuclease) of *Cl. septicum*.' *Journal of Pathology and Bacteriology*, **63**, 293

Webster, M. E., Altieri, P. L., Conklin, D. A., Berman, S., Lowenthal, J. P. and Gochenour, R. B. (1962). 'Enzymatic debridement of third-degree burns on guinea pigs by *Clostridium histolyticum* proteinases.' *Journal of Bacteriology*, **83**, 602

Williams, H. E. and Moe, V. G. (1973). '*Clostridium oedematiens* infection in cattle in Trinidad, West Indies.' *Veterinary Record*, **92**, 451

Williams, K. (1971). 'Some observations on *Clostridium tetani*.' *Medical laboratory Technology*, **28**, 399

Williams, K. and Willis, A. T. (1970). 'A method of performing surface viable counts with *Clostridium tetani*.' *Journal of medical Microbiology*, **3**, 639

Willis, A. T. (1960a). 'Observations on the Nagler reaction of some clostridia.' *Nature, London*, **185**, 943

Willis, A. T. (1960b). 'The lipolytic activity of some clostridia.' *Journal of Pathology and Bacteriology*, **80**, 379

Willis, A. T. (1969). *Clostridia of Wound Infection*. London; Butterworths

Willis, A. T. and Williams, K. (1970). 'Some cultural reactions of *Clostridium tetani*.' *Journal of medical Microbiology*, **3**, 291

Willis, A. T. and Williams, K. (1972). 'Prevention of swarming of *Clostridium septicum.' Journal of medical Microbiology,* **5**, 493

Wright, G. P. (1955). 'The neurotoxins of *Clostridium botulinum* and *Clostridium tetani.*' *Pharmacological Reviews,* **7**, 413

Zamecnik, P. C., Brewster, L. E. and Lipmann, F. (1947). 'A manometric method for measuring the activity of the *Cl. welchii* lecithinase and a description of certain properties of this enzyme.' *Journal of experimental Medicine,* **85**, 381

5

The Non-Clostridial Anaerobes

Much of our modern understanding of the non-clostridial anaerobes is due to the meticulous systematic and taxonomic studies carried out by Professor W. E. C. Moore and his colleagues at the Virginia Polytechnic Institute Anaerobe Laboratory in the USA. The results of their extensive observations on all varieties of anaerobes have been put together and published in the invaluable *Anaerobe Laboratory Manual* (Holdeman and Moore, 1975). Fortunately too, Professors Moore and Holdeman are major contributors to the eighth edition of Bergey's Manual (Buchanan and Gibbons, 1974), for which they have written the sections dealing with *Bacteroides, Fusobacterium, Propionibacterium* and *Eubacterium*.

BACTEROIDES

The bacteroides are Gram-negative non-sporing bacilli. They are strict anaerobes, and many species are obligate parasites of man and other animals, not normally occurring outside the body except perhaps in sewage. They differ from species of *Fusobacterium* in a number of general ways. They are commonly actively saccharolytic but do not produce indole or threonine deaminase. More specifically they are distinguished from species of *Fusobacterium* by their inability to produce significant amounts of butyric acid from peptone or glucose; some exceptional strains of *Bacteroides* that do produce butyric acid

Table 5.1 Differential characteristics of some *Bacteroides* species

Species	*Aesculin hydrolysis*	*Glucose*	*Maltose*	*Lactose*	*Sucrose*	*Growth in bile*	*Indole*	*Resistant to*			*Other*
								Penicillin	*Kanamycin*	*Rifampicin*	
B. fragilis	+	+	+	+	+	+	−/+	+	+	+	
B. melaninogenicus	−/+	+/−	+/−	+/−	+/−	−	+/−	−	+	−	Proteolytic, black fluorescent colonies
B. oralis	+	+	+	+	+	−	−	−	+	−	
B. capillosus	+	+	−	−	−	+	−	+	+	−	
B. praeacutus	−	−	−	−	−	+	−	+			
B. corrodens	−	−	−	−	−	+	−	−	−	−	Oxidase +ve; pitting of agar

also produce isobutyric and isovaleric acids which fusobacteria do not. Twenty-two species of *Bacteroides* are described, of which five are of animal origin, two have been isolated from termites, and the remaining 15 are associated with man.

The species most commonly implicated as significant pathogens in human infections are *Bacteroides fragilis* and *B. melaninogenicus*, both of which are divided into a number of sub-species. Of the other 13 species of bacteroides that may be isolated from clinical material, the most commonly encountered are *B. corrodens, B. oralis, B. capillosus* and *B. praeacutus.* Some differential characters of these species are summarized in *Table 5.1*. Not infrequently (and not surprisingly) strains of bacteroides are encountered for which no specific identification can be made.

It is useful to note here the following general rules about particular species of bacteroides:

1. *B. fragilis* is by far the most commonly encountered pathogen in human infections; unlike most other species it is resistant to penicillin G.
2. *B. melaninogenicus* is also commonly encountered in human infections. It is the only bacteroides which is proteolytic, which produces black pigmented colonies on horse blood agar, and whose colonies show red fluorescence in ultraviolet light.
3. *B. corrodens* is the only species that causes pitting of agar media, and is the only oxidase-positive bacteroides.

Bacteroides fragilis

B. fragilis occurs normally in the large bowel of man and animals where it forms part of the predominantly anaerobic faecal flora; it is also found in the normal vagina and less frequently in the mouth. It is the commonest cause of all anaerobic infections in man (*see for example* Zabransky, 1970).

Five sub-species of *B. fragilis* have been described – *fragilis, vulgatus, distasonis, ovatus* and *thetaiotaomicron*. Sub-species *fragilis* is the one most often implicated in clinical infections, although it is an uncommon isolate from faecal material (Werner and Pulverer, 1971; Werner, 1974; Sonnenwirth, 1974). Sub-species *vulgatus* is the one most commonly encountered in normal faeces. There is some uncertainty about the taxonomic validity of the various sub-species of *B. fragilis*, and it seems likely that sub-species *fragilis* will in future be placed in a separate species. *B. fragilis* is the type species of the genus *Bacteroides*.

Morphology

B. fragilis is a Gram-negative bacillus. Cells from surface colonies and from fluid media that contain no fermentable substances are about 0.4 × 3–5 μm in size, regularly shaped, with a straight or slightly curved axis and rounded ends. In glucose broth cultures pleomorphism is usually marked with filamentous and curved forms, and staining is markedly irregular. Individual cells may contain one or more unstained vacuoles that either slightly distort the bacillary body when they may be mistaken for clostridial spores, or cause gross distension and distortion of the bacillus.

Cultural characteristics

B. fragilis is an obligate anaerobe that grows well on horse blood agar. Colonies are low convex circular domes, 1–3 mm in diameter after 24–48 h incubation, and semitranslucent or greyish-white in colour. Most strains are non-haemolytic but occasional strains are frankly haemolytic. In fluid media growth of *B. fragilis* is not inhibited by 20% bile; indeed, its growth may be actually stimulated, but it is often difficult to distinguish between the turbidity due to enhanced growth and that caused by acid precipitation of the bile. Failure of bile to inhibit its growth is an important feature that distinguishes *B. fragilis* from most other members of the genus. Growth of *B. fragilis* is favoured by haemin, in the absence of which atypical or negative biochemical reactions may be given.

Biochemical characteristics

Differences in some biochemical characteristics serve to distinguish between the five sub-species of *B. fragilis* (*Table 5.2*), although, as

Table 5.2 Some differential characteristics of sub-species of *B. fragilis*

Sub-species	*Rhamnose*	*Trehalose*	*Mannitol*	*Indole*
ovatus	+	+	+	+
thetaiotaomicron	+	+/–	–	+
distasonis	+/–	+	–	–
vulgatus	+	–	–	–
fragilis	–	–	–	–

noted above, there is some debate as to the validity or clinical usefulness of this subdivision (Johnson, 1973). Among strains of *B. fragilis* sub-species *fragilis* a variety of carbohydrates is fermented with the production of acid and some gas. Gelatin is not attacked, indole, hydrogen sulphide and lipase are not produced, and nitrate is not reduced. The organism hydrolyses aesculin.

Major products of glucose metabolism are acetic and succinic acids; lesser amounts of lactic, isovaleric, formic and propionic acids are also produced.

Antibiotic resistance

Most strains of *B. fragilis* are resistant to benzylpenicillin (2 units/ml), neomycin (1mg/ml) and kanamycin (1mg/ml). Its resistance to penicillin distinguishes *B. fragilis* from most other bacteroides (Finegold, Harada and Miller, 1967; Finegold and Miller, 1968). The organism is sensitive to erythromycin (60 μg) and rifampicin (15 μg).

Bacteroides melaninogenicus

B. melaninogenicus occurs normally in the human mouth, large bowel and vagina, and is implicated in a variety of infections. Three sub-species are recognized on the basis of saccharolytic and proteolytic activity—*melaninogenicus, intermedius* and *asaccharolyticus* (Sawyer, Macdonald and Gibbons, 1962). It is one of the few pigment-producing anaerobes of medical importance. Strains of sub-species *melaninogenicus* are uncommonly encountered; sub-species *asaccharolyticus* appears to reside mainly in the gut, and *intermedius* in the mouth (Holdeman and Moore, 1975; Williams *et al.*, 1975). Oral strains of the organism appear to be increasingly prevalent with age; by the early 'teens' it can be isolated from most mouths, especially from the gingival crevice region (Bailit, Baldwin and Hunt, 1964; Kelstrup, 1966).

A useful little review of *B. melaninogenicus* was published by Hart (1969).

Morphology

There is nothing especially characteristic about the cellular morphology of *B. melaninogenicus*. The cells from surface colonies are short Gram-negative rods, commonly more coccal than coccobacillary in appearance. Bacillary forms are straight or slightly curved, and measure about 0.4 × 0.6–1 μm. Mild pleomorphism sometimes occurs in broth cultures, the cells of some strains being longer, vacuolated and showing irregular staining; short filaments are sometimes seen.

Cultural characteristics

B. melaninogenicus is an obligate anaerobe that grows well but slowly on horse blood agar. After 2–3 days' incubation colonies are 0.5–3 mm in diameter, circular, convex or umbonate, opaque and grey, brown or black in colour; many strains are β-haemolytic. Blackening of the colony, which commences at its centre on about the third day of incubation, extends towards the periphery to produce a shiny jet black colony after 5–6 days. The development of pigmentation is often associated with death of the culture. The black pigmentation is due to a haematin derivative (Schwabacher, Lucas and Rimington, 1947), and is consequently developed only on blood-containing media. Pigmentation develops more rapidly on media containing lysed blood than on media containing whole erythrocytes.

B. melaninogenicus is often difficult to grow and maintain in pure culture although it usually grows well in the presence of other organisms such as *E. coli* and *B. fragilis*. This is probably a reflection of the requirement of *B. melaninogenicus* for vitamin K, since a variety of other organisms, both aerobic and anaerobic, synthesize napthoquinones (Gibbons and Engle, 1964).

The colonies of many strains of *B. melaninogenicus* show a brick red fluorescence in ultraviolet light. Fluorescence develops early, before black pigmentation appears, although young pigmented colonies also fluoresce. In old, fully pigmented cultures fluorescence is often lost. Fluorescence is probably due to a diffusable substance since both the medium about fluorescent colonies, and closely adjacent colonies of other bacterial species (e.g. *B. fragilis*) also show excitation under ultraviolet light. The fluorescent substance can be removed from the organism by methanol extraction.

Many strains of *B. melaninogenicus* require vitamin K (menadione, 0.1 μg/ml) in addition to haemin (1.0 μg/ml) for growth (Gibbons and Macdonald, 1960).

Biochemical characteristics

Differences in biochemical characteristics serve to distinguish between the three sub-species of *B. melaninogenicus* – sub-species *melaninogenicus, intermedius* and *asaccharolyticus* (*Table 5.3*). No strains of

B. melaninogenicus reduce nitrates, and none grows in the presence of 20% bile.

Strains of *B. melaninogenicus* sub-species *melaninogenicus* hydrolyse a range of fermentable substances including glucose, maltose, lactose, sucrose, cellobiose, starch and aesculin; they produce gelatinase, but are

Table 5.3 Some differential characteristics of sub-species of *B. melaninogenicus*

Sub-species	*Indole*	*Proteolysis (milk agar)*	*Starch hydrolysis*	*Glucose*	*Aesculin hydrolysis*	*Lactose*
asaccharolyticus	+	+	–	–	–	–
intermedius	+	+	+	+	–	–
melaninogenicus	–	–	+	+	+	+

otherwise non-proteolytic, and they do not form hydrogen sulphide or indole. *B. melaninogenicus* sub-species *asaccharolyticus* is entirely without saccharolytic activity but is strongly proteolytic, hydrolysing gelatin, casein, coagulated serum and egg, collagen and azocol; it also produces both hydrogen sulphide and indole. *B. melaninogenicus* sub-species *intermedius* has proteolytic and limited saccharolytic activity. It attacks gelatin and more complex proteins, and hydrolyses glucose and starch. Indole is produced, but not hydrogen sulphide.

The three sub-species of *B. melaninogenicus* also show differences in the end products of their metabolism. *B. melaninogenicus* sub-species *melaninogenicus* produces predominantly succinic acid and some lactic acid, sub-species *intermedius* produces predominantly succinic and acetic acids, with lesser amounts of isobutyric and isovaleric acids, while sub-species *asaccharolyticus* produces mainly acetic and butyric acids, with smaller amounts of isobutyric and isovaleric acids. Other substantial differences between sub-species of *B. melaninogenicus* (cell wall composition, malate dehydrogenase mobility and deoxyribonucleic acid base composition) led Williams *et al.* (1975) to suggest that it may be inappropriate to retain sub-species *asaccharolyticus* and *intermedius* within the same species.

Antibiotic resistance

All strains of *B. melaninogenicus* are sensitive to benzylpenicillin (2 units), erythromycin (60 μg) and rifampicin (15 μg). The organism is resistant to neomycin (1mg), and sometimes to kanamycin (1mg) (Sutter and Finegold, 1971).

Pathogenicity

B. melaninogenicus is not uncommonly implicated in anaerobic infections in man. Although monomicrobial infections with it do occur, the organism is found much more commonly in association with other species of bacteroides, with anaerobic and carbon dioxide-dependent cocci, and with fusobacteria. From extensive *in vivo* studies carried out at the Forsyth Dental Infirmary in Boston, it has been concluded that *B. melaninogenicus* plays a dominant and probably an essential role in a number of mixed anaerobic infections that have been derived from the indigenous bacterial flora (Macdonald, 1962; Macdonald, Socransky and Gibbons, 1963; Gibbons, 1974). *B. melaninogenicus* is most frequently encountered in infections of the female genital tract and of the orofacial region (Dormer and Babett, 1972; *Journal of Antimicrobial Chemotherapy*, 1975). It is of some interest to note that *B. melaninogenicus* is the only strongly proteolytic member of the indigenous oral flora of man, and it has been suggested that the organism may be related for this reason to chronic periodontal disease (Macdonald and Gibbons, 1962). There appears to be no clear evidence, however, that *B. melaninogenicus* is an essential component of the dental plaque which is the direct precursor of chronic periodontitis (Hardie, 1974).

Bacteroides corrodens

B. corrodens as originally described by Eiken (1958) included both anaerobic and facultative organisms. Today, *Bacteroides corrodens* refers to the obligate anaerobes, while the facultative strains have been transferred to the genus *Eikenella* (Jackson *et al.*, 1971).

The principal habitat of *B. corrodens* is probably the oropharynx, but *B. corrodens* and *E. corrodens* have frequently been confused with one another. A useful comparative study of the two organisms has been published by James and Robinson (1974).

Morphology

B. corrodens is a Gram-negative bacillus, about $0.5 \times 1–2\ \mu m$ in size, with an occasional tendency to chain formation. The organism has no distinctive morphological characteristics.

Cultural characteristics

B. corrodens is an obligate anaerobe. It grows well, but slowly, on horse blood agar to produce small (1 mm diameter), non-haemolytic colonies after 4–5 days' incubation. Young (48-h) colonies are barely discernable with the naked eye, and present the appearance of pin-point depressions in the agar. Mature colonies are circular with entire or slightly undulating margins, low convex or umbonate, semitranslucent and greyish-white in colour; colonies sometimes have thin spreading edges. Classically, colonies of fresh isolates cause pitting of the agar – either the colonies grow in depressions in the medium or they produce circumferential depressions around the colonies; pitting is best developed after 7 days' incubation (Robinson and James, 1974). The ability to pit agar media is commonly lost by stock laboratory strains, and is by no means always associated with fresh isolates. The mechanism of pitting is not known.

Biochemical characteristics

B. corrodens is entirely non-saccharolytic and non-proteolytic. Nitrate is reduced to nitrite, and urease and oxidase are produced, but neither indole nor hydrogen sulphide is formed. Like *B. fragilis*, the organism grows in the presence of 20% bile. Aesculin is not hydrolysed.

Antibiotic resistance

B. corrodens is sensitive to benzylpenicillin (2 units), neomycin (1mg), kanamycin (1mg), erythromycin (60 μg) and rifampicin (15 μg).

Pathogenicity

Few observations have been made. Because strains of *Eikenella corrodens* are more commonly found in clinical specimens than *Bacteroides corrodens*, the literature relating to the clinical significance of '*B. corrodens*' prior to 1972 should be interpreted with caution.

Bacteroides oralis

B. oralis occurs in man as part of the normal oral flora. It is also encountered in infections of the upper respiratory tract and female genital tract, although its pathogenetic significance is not known.

Morphology

B. oralis is a Gram-negative bacillus, about 0.5 × 1–2 μm in size, and with a marked tendency to chain formation. In broth cultures that contain a fermentable carbohydrate short filaments are commonly produced.

Cultural characteristics

B. oralis is a strict anaerobe. It grows well on horse blood agar producing discrete circular colonies that are 0.5–2 mm in diameter after 48 h incubation. Colonies are entire, shiny and translucent, and are non-haemolytic.

Biochemical characteristics

B. oralis attacks a variety of fermentable substances including glucose, maltose, lactose, sucrose, starch and aesculin. Unlike *B. fragilis* it does not attack arabinose or xylose. The organism is non-proteolytic, does not reduce nitrates, and produces neither indole nor hydrogen sulphide. It does not grow in the presence of 20% bile, but its growth is slightly enhanced by haemin and by Tween 80. Major end products of metabolism are succinic and acetic acids.

Antibiotic resistance

B. oralis is resistant to kanamycin (1mg), but is sensitive to benxyl-penicillin (2 units), erythromycin (60 μg), neomycin (1mg) and rifampicin (15 μg).

Bacteroides capillosus

B. capillosus occurs as part of the normal faecal flora of man, and is probably also a normal inhabitant of the vagina. It has been encountered in a variety of soft tissue infections.

Morphology

B. capillosus is a Gram-negative bacillus, the cells of which vary considerably in size, from about 0.5–1.5 μm wide and 1.5–8.0 μm in length. Pleomorphism is often marked, especially in fluid media, with bent filaments, distorted bacilli and irregular staining.

Cultural characteristics

B. capillosus is a strict anaerobe. Surface colonies on horse blood agar are 0.5–2 mm in diameter after 48 h incubation, circular, entire, low convex, greyish and translucent. The organism is non-haemolytic.

Biochemical characteristics

B. capillosus is entirely non-proteolytic and only weakly saccharolytic, attacking only glucose with the formation of acid. Aesculin is also hydrolysed. Nitrate is not reduced, and neither indole nor hydrogen sulphide is formed. The growth of the organism is mildly enhanced by 20% bile, by Tween 80 and by haemin. Major products of metabolism are succinic acid and a smaller amount of acetic acid; small amounts of lactic and formic acids are produced by some strains.

Antibiotic resistance

B. capillosus is resistant to benzylpenicillin (2 units), kanamycin (1mg) and neomycin (1mg), but sensitive to erythromycin (60 μg) and rifampicin (15 μg).

Bacteroides praeacutus

B. praeacutus is a normal inhabitant of the large bowel of man, and has been implicated in a variety of soft tissue infections.

Morphology

B. praeacutus is a Gram-negative bacillus that shows considerable variation in size, ranging from 0.5–2 μm wide and 2–12 μm long. Pleomorphism is common with short chains, filaments and swollen forms.

Cultural characteristics

B. praeacutus is a strict anaerobe. Forty-eight hour old colonies on horse blood agar are small (0.5 mm diameter), circular and flat with

Table 5.4 Differential characteristics of some *Fusobacterium* species

Species	*Aesculin hydrolysis*	*Glucose*	*Maltose*	*Lactose*	*Sucrose*	*Growth in bile*	*Indole*	*Resistant to*			*Other*
								Penicillin	*Kanamycin*	*Rifampicin*	
F. mortiferum	+	+	+	+	—	+	—	—	—	+	
F. varium	—	+	—	—	—	+	+	—	—	+	
F. nucleatum	—	+/—	—	—	—	—	+	—	—	—	
F. necrophorum	—	+/—	—	—	—	—	+	—	—	—	Lipolytic

scalloped or diffuse margins, greyish and semitranslucent. It is non-haemolytic.

Biochemical characteristics

B. praeacutus is entirely non-saccharolytic. Gelatinase is produced, but more complex proteins are not attacked. Nitrate is reduced, and hydrogen sulphide produced, but indole is not formed. The organism grows in the presence of 20% bile. Major products of metabolism are acetic, butyric, isovaleric and isobutyric acids, together with some propionic acid.

FUSOBACTERIUM

The fusobacteria are Gram-negative non-sporing bacilli. They are strict anaerobes, and like the bacteroides many species are obligate parasites of man and other animals. They are distinguished from the bacteroides by their weakly saccharolytic activity, by their production of threonine deaminase and indole, and more specifically by their formation of butyric acid as a major product of metabolism. Sixteen species are described, all of which have been isolated at one time or another from human sources. The four species most commonly implicated as significant human pathogens are *F. necrophorum, F. nucleatum, F. mortiferum* and *F. varium*. Some differential characters of these species are summarized in *Table 5.4*.

Fusobacterium necrophorum

An organism with an extremely chequered taxonomic background (it was the *Sphaerophorus necrophorus,* and *Sphaerophorus funduliformis* of earlier times) *F. necrophorum* is a well recognized pathogen of man and of a wide variety of animals (Simon and Stovell, 1969). It occurs normally in the human upper respiratory and intestinal tracts. The characteristics of the organism have been reported upon in some detail by Aalbaek (1971).

Morphology

F. necrophorum is a Gram-negative bacillus with a tendency to pleomorphism and irregular staining. Cells are about 0.5 × 5–10 μm in size, the larger ones commonly being curved; filament formation is not uncommon. Fusiform swelling of bacilli is sometimes seen but free spheroids are rarely developed. In pathological material pleomorphism may be the predominant feature, coccal, bacillary and filamentous forms all being represented.

Cultural characteristics

F. necrophorum is an obligate anaerobe that grows well on horse blood agar. Colonies are 1–2 mm in diameter after 48 h incubation, circular with scalloped or diffuse margins, high convex or umbonate, with a ridged surface and semitranslucent or opaque, and yellowish in colour. Many strains of the organism are haemolytic, and many are lipolytic on egg yolk agar.

Biochemical characteristics

F. necrophorum is generally non-saccharolytic although some strains ferment glucose weakly. Some strains produce a gelatinase, but more complex proteins are not attacked. Nitrate is not reduced, but indole and hydrogen sulphide are produced. The organism usually does not grow in the presence of 20% bile. Lactate and threonine are deaminated to propionate.

Major products of metabolism are butyric acid, and lesser amounts of acetic and propionic acids.

Antibiotic resistance

The organism is sensitive to benzylpenicillin (2 units), kanamycin (1mg), neomycin (1mg), erythromycin (60 μg) and rifampicin (15 μg); it is resistant to vancomycin (5 μg).

Pathogenicity

F. necrophorum produces an endotoxic lipopolysaccharide that is lethal dermonecrotic and pyrogenic (Hofstad and Kristoffersen, 1971; Warner *et al.*, 1975). A lipase is also produced. The organism is a cause of anaerobic infections in man, especially in relation to, or derived from the upper respiratory tract.

F. necrophorum causes a variety of infections in animals, including bovine hepatic abscess and foot rot in sheep.

Fusobacterium nucleatum

F. nucleatum is the type species of the genus. It is a normal inhabitant of the human oropharynx, and is involved in various anaerobic soft tissue infections, especially in relation to the respiratory tract.

Morphology

F. nucleatum is a Gram-negative bacillus of classic fusiform appearance, the cells being spindle-shaped, sharply pointed and occasionally with central or eccentrically placed swellings. The cells are about 1 × 5–10 μm in size. Free spheroids are commonly seen.

Cultural characteristics

F. nucleatum is an obligate anaerobe that grows well on horse blood agar. Colonies are 1–2 mm in diameter after 48 h incubation, irregularly circular, low convex, greyish, glistening and translucent. Colonies often show an iridescent flecking when viewed by transmitted light. The organism is usually non-haemolytic.

Biochemical characteristics

F. nucleatum is biochemically fairly inactive. Most strains weakly ferment fructose, and some weakly attack glucose. The organism is otherwise non-saccharolytic, and is entirely non-proteolytic. Threonine is converted to propionate, and indole is produced, but nitrate is not reduced and hydrogen sulphide is not formed. The organism does not grow in the presence of 20% bile.

Major products of metabolism include butyric acid, with lesser amounts of acetic, propionic and succinic acids.

Antibiotic resistance

The organism is sensitive to benzylpenicillin (2 units), kanamycin (1mg), rifampicin (15 μg), neomycin (1mg) and erythromycin (60 μg). It is resistant to vancomycin (5 μg).

Fusobacterium varium

F. varium is a normal inhabitant of the human large intestine, and has been implicated in a variety of soft tissue infections in man.

Morphology

F. varium is a small Gram-negative bacillus, about 0.5 × 1–2 μm in size; coccobacillary forms are not uncommon.

Cultural characteristics

F. varium is a strict anaerobe that grows satisfactorily on horse blood agar. After 48 h incubation colonies may appear as mere pin-points or may measure up to 1 mm in diameter. Colonies are circular flat discs, greyish and translucent, and are non-haemolytic.

Biochemical characteristics

Most strains of *F. varium* weakly ferment glucose, fructose and mannose; some strains also weakly hydrolyse galactose. The organism is entirely non-proteolytic; indole is produced, but nitrate is not reduced and hydrogen sulphide is not formed. Threonine is deaminated to propionate, and the organism grows quite well in the presence of 20% bile.

Main products of metabolism are butyric, lactic and acetic acids, all produced in substantial quantity.

Antibiotic resistance

The organism is sensitive to benzylpenicillin (2 units) and kanamycin (1mg), but resistant to erythromycin (60 μg), rifampicin (15 μg) and vancomycin (5 μg).

Fusobacterium mortiferum

F. mortiferum forms part of the normal bacterial flora of the human gastrointestinal and female genital tracts, and is implicated in a variety of soft tissue infections.

Morphology

F. mortiferum is a Gram-negative, highly pleomorphic bacillus. The cells are extremely variable in size – from 0.5–2 μm wide and 1–10 μm in length. Many cells are grossly irregular in shape, and some are distorted with centrally or eccentrically placed swellings; large 'ball' forms are not uncommon.

Cultural characteristics

F. mortiferum is an obligate anaerobe that grows well on horse blood agar. Forty-eight hour old colonies are 1–2 mm in diameter, irregularly circular, low convex or slightly umbonate, greyish and semitranslucent. The organism is non-haemolytic.

Biochemical characteristics

F. mortiferum is moderately saccharolytic, weakly fermenting glucose, maltose, lactose, mannose, salicin and cellobiose. It is entirely non-proteolytic, and does not reduce nitrate or produce indole or hydrogen sulphide. Aesculin is hydrolysed, threonine deaminase is produced and the organism grows well in the presence of 20% bile.

Major products of metabolism are butyric and acetic acids, with lesser amounts of propionic acid.

Antibiotic resistance

F. mortiferum is sensitive to benzylpenicillin (2 units) and to kanamycin (1mg), but it is resistant to erythromycin (60 μg), rifampicin (15 μg) and vancomycin (5 μg).

ACTINOMYCES AND ARACHNIA

Strains of *Arachnia propionica* and of 'anaerobic' *Actinomyces* are virtually indistinguishable from one another by conventional tests; they differ, however, in the nature of their products of metabolism, for *Arachnia propionica* produces predominantly propionic acid, while *Actinomyces* species produce predominantly acetic acid but not propionic acid. Other definitive features that distinguish *Arachnia* from

Table 5.5 Some differential characteristics of *Actinomyces* and *Arachnia*

Organism	*Growth in* *Air*	*Air+CO_2*	*ANO_2+CO_2*	*'Spider' microcolony*	*Red macrocolony on blood agar*	*Starch hydrolysis (wide zone)*	*Aesculin hydrolysis*	*Serological group*	*Propionic acid produced*
A. israelii	—	slight	+	+	—	—	+	D	—
A. odontolyticus	+	+	+	—	+	—	+/—	E	—
A. eriksonii	—	—	+	—	—	—	—	C	—
A. bovis	slight	+	+	—	—	+	+	B	—
A. naeslundii	+	+	+	—	—	—	+/—	A	—
Arachnia propionica	slight	slight	+	+	—	—	—		+

Actinomyces include differences in cell wall composition, and in the serological specificity of the organisms.

Arachnia propionica and species of *Actinomyces* occur as part of the indigenous oral flora of mammals. *Arachnia propionica, A. israelii, A. eriksonii* and *A. odontolyticus* occur normally only in the human mouth; *A. bovis* appears to be normally confined to the bovine mouth.

As a group the *Actinomyces* have been aptly described as 'facultative but preferentially anaerobic organisms' (Cross and Goodfellow, 1973), whose growth is usually enhanced by carbon dioxide. Some differential characteristics of *Actinomyces* and *Arachnia*, including differences in their atmospheric requirements, are summarized in *Table 5.5*.

Rapid and specific serological methods of differentiation are provided by fluorescent antibody, gel diffusion and cell wall agglutination tests, and by analysis of cell wall components (Brock and Georg, 1969a, b; Boone and Pine, 1968; Slack, Landfried and Gerencser, 1969; Bowden and Hardie, 1973; Holmberg and Forsum, 1973). Fluorescent antibody techniques provide the most rapid and specific approach, on the basis of which six serological groups are recognized and are as follows: group A, *A. naeslundii*; group B, *A. bovis;* group C, *A. eriksonii;* group D, *A. israelii;* group, E. *A. odontolyticus*; and group F, *A. viscosus*.

The systematics of *Arachnia propionica* are so closely similar to those of *A. israelii* that a separate description here would be redundant (*see* Buchanan and Pine, 1962; Pine and Georg, 1969; Brock *et al.*, 1973). *A. odontolyticus* and *A. naeslundii* are frankly facultatively anaerobic organisms, while *A. bovis* is carbon dioxide-dependent rather than anaerobic, and is not encountered in clinical material.

The *Actinomyces* and related organisms have been excellently reviewed by Georg (1970, 1974), Bowden and Hardie (1973) and Slack and Gerencser (1975).

Actinomyces israelii

A. israelii is an obligate parasite of man, where its predominant habitat is the mouth. It is the commonest cause of human actinomycosis. Laboratory animals are suceptible to infection, especially when the organism is injected intraperitoneally.

Morphology

A. israelii is a Gram-positive bacillus, about 1 μm in width, that may present as long filaments, long and short curved bacilli, and as coccoid, clubbed and tapered forms, any of which may show true branching. In artificial culture rough colonies are usually composed largely of branching cells. It is generally true to say that, although the cellular morphology of different species of *Actinomyces* is very similar, *A. israelii* conforms more to the accepted picture of branching filaments, while other species tend towards rod formation and a 'diphtheroid' appearance. *A. israelii* falls into the fluorescent antibody serological group D.

Cultural characteristics

A. israelii does not grow under ordinary atmospheric conditions but grows slightly in air plus 10% carbon dioxide. Optimal growth is obtained under anaerobic conditions with added carbon dioxide.

Microcolonies on brain heart infusion agar are 0.03–0.06mm in diameter after 24 h incubation. Viewed through the plate microscope they are seen to be composed of branching filaments of varying length that originate from a single central point – these are referred to as 'spider' colonies. With continued incubation macrocolonies are developed after 7–14 days. These mature colonies are 0.5–3 mm in diameter, irregularly circular, usually heaped up, white or cream in colour and commonly with a granular or convoluted surface; they have been variously described as molar-tooth, breadcrumb or raspberry colonies. Colonies are commonly adherent to the agar, and although they are sometimes friable, many are hard and may be moved over the agar surface intact. Mature colonies have a similar appearance on horse blood agar and are non-haemolytic.

Biochemical characteristics

The biochemical activity of *A. israelii* shows marked variation from strain to strain (*see* Brock and Georg, 1969a; Bowden and Hardie, 1971). Most strains ferment glucose, maltose, lactose, sucrose, mannose and ribose, and hydrolyse aesculin; starch is not attacked and the organism is entirely non-proteolytic. The major products of metabolism are lactic, succinic and acetic acids.

Actinomyces eriksonii

A. eriksonii is considered by Holdeman and Moore (1975) to be more properly regarded as a species of *Bifidobacterium*. The organism was first recovered and described by Georg *et al.*, (1965). Some strains are slightly pathogenic for the mouse on intraperitoneal inoculation.

Morphology

A. eriksonii is a Gram-positive bacillus about 1 × 2–4 μm in size, often with clubbed or bifid ends. Short, highly branched filaments sometimes occur. Cells tend to be larger in cultures grown in the presence of a fermentable carbohydrate. The organism falls into the fluorescent antibody group C.

Cultural characteristics

A. eriksonii is an obligate anaerobe. On brain–heart infusion agar mature colonies are 0.5–1 mm in diameter after 7–14 days' incubation, circular, entire, low to high convex and white. 'Spider' microcolonies are not produced and the organism is non-haemolytic on horse blood agar.

Biochemical characteristics

A. eriksonii is a highly saccharolytic organism. Some strains hydrolyse aesculin but the organism in non-proteolytic. The chief products of metabolism are acetic and lactic acids.

Actinomyces odontolyticus

A. odontolyticus is a normal inhabitant of the human oral cavity. The organism is slightly pathogenic for the mouse, but its relationship to human disease has not been established.

Morphology

The morphology of *A. odontolyticus* is very similar to that of *A. eriksonii*. For a detailed description see Batty (1958). The organism falls into the fluorescent antibody serological group E.

Cultural characteristics

A. odontolytics is a facultative anaerobe. On brain–heart infusion agar, macrocolonies are 1–2 mm in diameter after 7–14 days' incubation, irregularly circular, high convex or umbonate, opaque and white.

On horse blood agar there may be a zone of greening around the colonies when they resemble green-forming streptococci. Colonies on horse blood agar are dark red when viewed by transmitted light, due to a pigment that seems to be best developed under aerobic conditions at room temperature. 'Spider' microcolonies, indistinguishable from those of *A. israelii*, are produced by occasional strains.

Biochemical characteristics

Most strains of *A. odontolyticus* attack a variety of fermentable substances including glucose, maltose, lactose and sucrose. Some strains hydrolyse aesculin, but none is proteolytic. The major products of metabolism are acetic and lactic acids.

Actinomyces bovis

A. bovis is the type species of the genus. It is the cause of bovine actinomycosis but is not a human pathogen, and is unlikely to be encountered in the clinical laboratory. The organism is pathogenic for laboratory animals on intraperitoneal inoculation, but less so than *A. israelii.*

Morphology

The cellular morphology of *A. bovis* is generally similar to that of *A. israelii* except that filamentous forms and branching are much less conspicuous features. Indeed, like other 'non-*israelii*' species of *Actinomyces*, short rod formation is commonly predominant and it may be impossible to distinguish the organism from some other non-sporing Gram-positive anaerobic bacilli. The organism falls into the fluorescent antibody group B.

Cultural characteristics

A. bovis is a carbon dioxide-dependent facultative organism. On brain–heart infusion agar microcolonies are about 0.06 mm in diameter after 24 h incubation, circular, entire and flat, with a smooth or granular appearance. Occasional strains produce filamentous colonies similar to those of *A. israelii.* Macrocolonies are 0.5–1 mm in diameter after 7–14 days' incubation; they are circular, entire, low to high convex, opaque and white and with a smooth or granular surface. Some strains produce heaped-up colonies similar to *A. israelii.* The organism is non-haemolytic on horse blood agar.

Biochemical characteristics

Like *A. israelii, A. bovis* attacks a variety of fermentable substances including glucose, maltose, lactose and sucrose. Aesculin is hydrolysed, and a wide zone of hydrolysis is produced on starch agar. Ribose is not attacked, however, and the organism is non-proteolytic. The major products of metabolism are acetic, lactic and succinic acids.

Actinomyces naeslundii

A. naeslundii is an obligate parasite of man, is found predominantly in the oral cavity, and is occasionally implicated in human infections. The organism is pathogenic for the mouse on intraperitoneal inoculation (Coleman and Georg, 1969).

Morphology

The cellular morphology of *A. naeslundii* is generally similar to that of other 'non-*israelii*' actinomyces, although the presence of short branches is often a predominant feature. The organism belongs to the fluorescent antibody serological group A.

Cultural characteristics

A. naeslundii is a facultative anaerobe. On brain–heart infusion agar microcolonies are about 0.5 mm in diameter after 48 h incubation. Typically the colonies are made up centrally of a dense mass of tangled filaments and diphtheroidal cells, surrounded by a narrow corona of branched filaments. Macrocolonies are 1–3 mm in diameter after 7–14 days' incubation, round with an entire or scalloped margin, convex, white and opaque, and with a smooth or finely granular surface. Occasional colonies present the 'molar-tooth' appearance similar to *A. israelii*. The organism is non-haemolytic on horse blood agar.

Table 5.6 Some differential characters of *Eubacterium*, *Propionibacterium*, *Actinomyces* and *Arachnia*

Organism	*Propionic acid produced*	*Gelatinase produced*	*Catalase produced*	*True branching*	*Growth in air or in air* + CO_2
Eubacterium (*lentum* and *limosum*)	–	–	–	–	–
Propionibacterium (*avidum* and *acnes*)	+	+	+	–	+
Actinomyces	–	–	–	+	+/–
Arachnia	+	–	–	+	+/–

Biochemical characteristics

Like *A. israelii, A. naeslundii* attacks a variety of fermentable substances including glucose, maltose, lactose, sucrose and mannose, but not ribose. Some strains hydrolyse aesculin, but none attacks starch vigorously, and the organism is non-proteolytic. The major products of metabolism are acetic, lactic and succinic acids.

EUBACTERIUM

The genus *Eubacterium* is made up of Gram-positive non-sporing obligate anaerobic bacilli. Most species appear to be associated with the mouth or intestinal tract of man and animals. Of the 28 described species, only two – *E. lentum* and *E. limosum* – are encountered with much frequency in human clinical material. There is little evidence to suggest, however, that these organisms are of pathological significance.

Differentiation among the Gram-positive non-sporing anaerobic bacilli of clinical interest – *Actinomyces, Arachnia, Eubacterium* and *Propionibacterium* – is aided by the criteria summarized in *Table 5.6*.

Eubacterium lentum

The cells of *E. lentum* are 0.5 × 1–2 μm, but coccal forms, occurring in pairs, and short chains are common. Colonies on horse blood agar are 0.5–2 mm in diameter after 48 h incubation, circular, low convex, entire and semitranslucent. The organism is non-haemolytic, is entirely non-saccharolytic and non-proteolytic. It does not produce indole but reduces nitrate. Volatile products of metabolism are minimal or absent.

Eubacterium limosum

The cells of *E. limosum* are about 1 × 4–6 μm in size, and the organism shows some degree of pleomorphism with swollen and club forms. Colonies on horse blood agar are 2 mm in diameter after 48 h incubation, circular, entire, low convex and semitranslucent. The organism is non-haemolytic. A variety of fermentable substances is attacked including glucose, fructose and mannitol. Aesculin is hydrolysed, but indole is not produced and nitrate is not reduced. The organism is entirely non-proteolytic. Principal products of metabolism are acetic and lactic acids, with smaller amounts of butyric acid.

PROPIONIBACTERIUM

The propionibacteria are anaerobic to facultative Gram-positive non-sporing bacilli of typical diphtheroidal cellular morphology. Indeed, these organisms were formerly placed in the genus *Corynebacterium*. They are associated with a variety of habitats including the bodies of man and other animals and with dairy products (*see* Marples and McGinley, 1974). Of the eight described species only two – *P. acnes* and *P. avidum* – are commonly encountered in clinical material.

Propionibacterium acnes

P. acnes is widely distributed on the adult human skin and hair, and in the oropharynx and gastrointestinal tract, a ubiquity that explains its common occurrence as a clinical and laboratory contaminant.

Morphology

P. acnes is a Gram-positive bacillus, about 0.5–1.5 × 1–10 μm in size. The organism shows a typically diphtheroidal pleomorphism with clubs, 'tadpole' forms, short bifid and branched elements and coccal forms.

Cultural characteristics

P. acnes is anaerobic to facultatively anaerobic in its atmospheric requirements. A small proportion of strains grow aerobically on primary isolation. The majority of fresh isolates grow initially only in an anaerobic atmosphere, but may show facultatively anaerobic growth after several transfers. About one-third of strains are found to be carbon dioxide-dependent rather than anaerobic on primary isolation.

Colonies on horse blood agar are about 0.5–2 mm in diameter after 48 h incubation, circular, entire, convex, greyish-white and opaque, with a glistening surface. A few strains are haemolytic.

Biochemical characteristics

P. acnes ferments glucose, fructuse, galactose, mannose and glycerol; maltose, lactose and sucrose are not attacked, and aesculin is not hydrolysed. The organism is proteolytic, producing both a gelatinase and a caseinolytic enzyme; it produces indole and reduces nitrate. About two-thirds of strains show catalase activity after exposure of surface colonies to air for about 30 min. Variability in reactions, which is not uncommon, may be reduced by the addition of 0.025 % Tween 80 to the media; Tween 80 stimulates the growth of the organism. Like other propionibacteria the major products of metabolism are propionic and acetic acids.

P. acnes produces a number of soluble enzymes including a lipase (Hassing, 1971) which is active against triglycerides but inactive against the fats in egg yolk (the organism is egg yolk negative).

Pathogenicity

The ubiquity of *P. acnes* on the normal human body ensures that the organism is commonly recovered from anaerobic cultures of clinical material, most frequently as an incidental contaminant. The organism has aroused considerable interest as a major contributor to the complex pathogenesis of acne (*see for example* Marples *et al.*, 1971, 1973, 1974; Cunliffe and Cotterill, 1973; Cunliffe *et al.*, 1973; Leyden *et al.*, 1974). It has been suggested that the lesions of acne are initiated by pilosebaceous duct obstruction and a local irritant effect of long chain fatty acids derived from bacterial hydrolysis of lipids in the sebum.

Propionibacterium avidum

The distribution of *P. avidum* on the human body is similar to that of *P. acnes*.

Morphology

The cellular morphology of *P. avidum* is generally similar to that of *P. acnes*, except that the cells are somewhat smaller and pleomorphism less marked.

Cultural characteristics

P. avidum is anaerobic to facultatively anaerobic in its atmospheric requirements. After several transfers most strains grow well under ordinary atmospheric conditions. Surface colonies on horse blood agar vary from 'pin-points' to about 0.5 mm in diameter after 48 h incubation. They are circular, entire, high convex, opaque and white, and have a shiny surface. Many strains are haemolytic, and a few produce a lipase opacity on egg yolk agar.

Biochemical characteristics

P. avidum is actively saccharolytic and may be distinguished from *P. acnes* by its ability to ferment maltose and sucrose and to hydrolyse aesculin. The organism is proteolytic, but does not produce indole or reduce nitrate. It is catalase-positive and is able to grow in the presence of 20% bile.

Pathogenicity

Although *P. avidum* is commonly encountered in clinical specimens its role as a pathogen is uncertain.

THE ANAEROBIC COCCI

The anaerobic cocci of clinical interest are the *Veillonella* (Gram-negative cocci), *Peptococcus* (Gram-positive; anaerobic equivalent of *Staphylococcus*), and *Peptostreptococcus* (Gram-positive; anaerobic equivalent of *Streptococcus*). Unfortunately, identification of species is commonly based on variable characteristics, so that delineation of particular isolates at the specific, or even at the generic, level may be difficult. The problem is compounded by the variety of contemporary classifications of the anaerobic cocci that have been published and the small measure of agreement that exists between them (Prévot, 1957; Prévot and Fredette, 1966; Hare, 1967; Buchanan and Gibbons, 1974; Dowell and Hawkins, 1974; Holdeman and Moore, 1975).

Although the pathogenetic significance of the veillonellae is uncertain, there is no doubt that both the peptococci and peptostreptococci are significant pathogens for man. The role of these organisms in clinical infections has been discussed by Weinberg (1974), Lambe, Vroon and Rietz, (1974) and Pien, Thompson and Martin (1972).

VEILLONELLA ALCALESCENS AND VEILLONELLA PARVULA

V. alcalescens and *V. parvula* form part of the normal bacterial flora of the oropharynx, gastrointestinal tract and female genital tract. Although *V. parvula* is encountered much more frequently in clinical material than *V. alcalescens*, there is little convincing evidence to suggest that either organism alone is truly pathogenic.

These organisms are small Gram-negative cocci, about 0.3–0.5 μm in diameter, that occur in pairs, in short chains and in irregular masses. Colonies on horse blood agar are 1–2 mm in diameter after 48 h incubation, circular, entire, convex, greyish white and butyrous. The organisms are non-haemolytic. According to Chow, Patten and Guze (1975) all strains of *V. parvula* and *V. alcalescens* display red fluorescence under uv light immediately after removal from the anaerobic jar; fluorescence is rapidly lost on exposure of colonies to air. Biochemically they are remarkably inactive in conventional tests, but they ferment several organic acids including lactic, pyruvic, fumaric and succinic acids; the products of fermentation of lactate are acetic and propionic acids, with carbon dioxide and hydrogen release. All strains are indole-negative, but reduce nitrate and will produce hydrogen sulphide in the presence of cysteine (Rogosa and Bishop, 1964b). Differentiation between the species is based on differences in their nutritional requirements (Rogosa and Bishop, 1964a).

Media containing sodium lactate (0.85%) and vancomycin (7.5 μg/ml) facilitate isolation of these organisms (Rogosa, 1964).

The veillonellae are sensitive to chloramphenicol, penicillin, tetracycline, erythromycin, clindamycin, lincomycin and metronidazole, but are resistant to the aminoglycosides and vancomycin.

PEPTOCOCCUS

Members of this genus are Gram-positive cocci, usually about 0.5–1 μm in diameter, and occurring singly, in pairs and in irregular masses. Cells of *Peptococcus magnus* may be as large as 2 μm in diameter. Although Holdeman and Moore (1975) described seven species of *Peptococcus*, only two – *Pc. asaccharolyticus* and *Pc. magnus* – appear to occur at all commonly in clinical material. *Pc. prevottii* and *Pc. variabilis* are similar to, if not identical with *Pc. asaccharolyticus* and *Pc. magnus* respectively. Colonies of peptococci on horse blood agar are circular, entire, convex, opaque or translucent, and 1–2 mm in diameter after 48 h incubation; they are non-haemolytic. Growth of peptococci is usually enhanced by Tween 80.

Pc. asaccharolyticus forms part of the normal bacterial flora of the nasopharynx, and of the gastrointestinal and female genital tracts; it is not uncommonly present in clinical specimens. It is entirely non-saccharolytic and non-proteolytic, but unlike *Pc. magnus* it produces indole. The major products of metabolism are acetic and butyric acids.

Pc. magnus is also encountered in pathological material from various parts of the body, although its normal distribution is sparse. Characterized mainly by its large cell size, the organism is non-saccharolytic, non-proteolytic and does not produce indole. Acetic acid is the principal product of metabolism.

The peptococci are sensitive to penicillin, tetracycline, erythromycin, chloramphenicol, clindamycin, lincomycin and metronidazole.

PEPTOSTREPTOCOCCUS

Members of this genus are Gram-positive cocci, usually about 0.5–1 μm in diameter, although the cells of some species are much smaller (0.3–0.5 μm). Cells are arranged in pairs and in short or long chains. Colonies on horse blood agar are 1–2 mm in diameter after 2–4 days' incubation, circular, entire, convex and translucent. Of the five species described by Holdeman and Moore (1975) only *Pst. anaerobius* and *Pst. intermedius* are encountered at all frequently in clinical material; *Pst. productus* is an uncommon isolate. Quite the commonest isolate is *Pst. anaerobius* (*Streptococcus putridus* of Schottmüller, 1911a, b). Like the peptococci, the growth of peptostreptococci is usually stimulated by Tween 80.

While peptostreptococci are essentially obligate anaerobes, it is a not uncommon experience for clinical laboratories to encounter strains of 'anaerobic streptococci' that become 'aerotolerant' after one or two transfers. Most of these organisms are, in fact, carbon dioxide-dependent facultatively anaerobic streptococci of the *Streptococcus mutans* and *Streptococcus milleri* type (*see for example* Bateman, Eykyn and Phillips, 1975); these organisms grow as well in air plus 10% carbon dioxide as they do under anaerobic conditions, and are always resistant to metronidazole. Valid anaerobic streptococci will not grow at all under ordinary atmospheric conditions, and are universally metronidazole-sensitive.

These organisms are associated with a variety of anaerobic infections in man, commonly in association with bacteroides and fusobacteria. They also play an aetiological role with facultative anaerobes in synergistic infections such as Meleney's gangrene and anaerobic streptococcal myositis. The peptostreptococci have a similar sensitivity pattern to that of the peptococci.

Peptostreptococcus anaerobius

Pst. anaerobius forms part of the normal bacterial flora of the oropharynx, the gastrointestinal and the female genital tracts. It is an obligate anaerobe, all cultures of which develop a foul odour.

It is a Gram-positive coccus about 0.8 μm in diameter, typically arranged in long chains. On horse blood agar colonies are about 1.5–2 mm in diameter after 3–4 days' incubation, circular, entire, convex, smooth and pearly grey in colour. Some strains may produce dark or even black colonies, but this is not a constant feature. The organism is non-haemolytic. It grows abundantly in relatively simple fluid media such as peptone water and cooked meat broth with the production of gas. Most strains weakly ferment glucose and fructose, hydrolysis of these carbohydrates being stimulated by the presence of sulphur compounds such as thioglycollate. *Pst. anaerobius* is entirely non-proteolytic, indole is not formed but hydrogen sulphide is produced.

The major product of metabolism is acetic acid with small amounts of propionic, isobutyric, butyric and isocaproic acids.

Peptostreptococcus productus

Pst. productus forms part of the normal bacterial flora of the gastro-intestinal tract, but it is seldom encountered in clinical material, and is probably non-pathogenic.

It is a Gram-positive coccus, about 0.7–1.2 μm in diameter, most frequently arranged in short chains. Elongated cells are not uncommonly seen and may be mistaken for Gram-positive bacilli. The organism ferments a range of substances including glucose, maltose, lactose, sucrose, fructose, galactose and xylose, but it is entirely non-proteolytic. The major products of metabolism are acetic acid with a smaller amount of succinic acid.

Peptostreptococcus intermedius

Pst. intermedius has the same bodily distribution as *Pst. anaerobius*, but it is less frequently encountered in pathological material. *Pst. intermedius* is composed of organisms with a variety of characteristics, and so represents a convenient specific repository for 'anaerobic streptococci' whose characters do not accord with those of *Pst. anaerobius* or *Pst. productus*. Most strains ferment glucose, maltose, lactose, sucrose and fructose, but the organism is non-proteolytic. The major product of metabolism is lactic acid; many of these strains fall into the 'aerotolerant' or carbon dioxide-dependent catagories and are resistant to metronidazole.

REFERENCES

Aalbaek, B. (1971). '*Sphaerophorus necrophorus*. A study of 23 strains.' *Acta Veterinaria Scandinavica,* **12,** 344

Bailit, H. L., Baldwin, D. C. and Hunt, E. E. (1964). 'The increasing prevalence of gingival *Bacteroides melaninogenicus* with age in children.' *Archives of oral Biology,* **9,** 435

Bateman, N. T., Eykyn, S. J. and Phillips, I. (1975). 'Pyogenic liver abscess caused by *Streptococcus milleri.*' *Lancet,* **1,** 657

Batty, I. (1958). '*Actinomyces odontolyticus*, a new species of actinomycete regularly isolated from deep carious dentine.' *Journal of Pathology and Bacteriology,* **75,** 455

Boone, C. J. and Pine, L. (1968). 'Rapid method for characterization of actinomycetes by cell wall composition.' *Applied Microbiology,* **16,** 279

Bowden, G. H., and Hardie, J. M. (1971). 'Anaerobic organisms from the human mouth.' In *Isolation of Anaerobes*, p. 177. Ed by D. A. Shapton and R. G. Board. London; Academic Press

Bowden, G. H. and Hardie, J. M. (1973). 'Commensal and pathogenic *Actinomyces* species in man.' In *Actinomycetales. Characteristics and Practical Importance,* p. 277. Ed by G. Sykes and F. A. Skinner, London; Academic Press

Brock, D. W. and Georg, L. K. (1969a). 'Determination and analysis of *Actinomyces israelii* serotypes by fluorescent-antibody procedures.' *Journal of Bacteriology,* **97,** 581

Brock, D. W. and Georg, L. K. (1969b). 'Characterization of *Actinomyces israelii* serotypes 1 and 2.' *Journal of Bacteriology,* **97,** 589

Brock, D. W., Georg, L. K., Brown, J. M. and Hicklin, M. D. (1973). 'Actinomycosis caused by *Arachnia propionica.*' *American Journal of clinical Pathology,* **59,** 66

Buchanan, R. E. and Gibbons, N. E. (1974). *Bergey's Manual of Determinative Bacteriology.* 8th edn. Baltimore; Williams and Wilkins

Buchanan, B. B. and Pine, L. (1962). 'Characterization of a propionic acid producing actinomycete, *Actinomyces propionicus,* sp. nov.' *Journal of general Microbiology,* **28,** 305

Chow, A. W., Patten, V. and Guze, L. B. (1975). 'Rapid screening of *Veillinella* by ultraviolet fluorescence.' *Journal of clinical Microbiology,* **2,** 546

Coleman, R. M. and Georg, L. K. (1969). 'Comparative pathogenicity of *Actinomyces naeslundii* and *Actinomyces israelii.*' *Applied Microbiology,* **18,** 427

Cross, T. and Goodfellow, M. (1973). 'Taxonomy and classification of the *Actinomycetes.*' In *Actinomycetales. Characteristics and Practical Importance.* p. 11. Ed. by G. Sykes and F. A. Skinner. London; Academic Press

Cunliffe, W. J. and Cotterill, J. A. (1973). 'Clindamycin as an alternative to tetracycline in severe acne vulgaris.' *Practitioner,* **210,** 698

Cunliffe, W. J., Forster, R. A., Greenwood, N. D., Hetherington, C., Holland, K. T., Holmes, R. L., Khan, S., Roberts, C. D., Williams, M. and Williamson, B. (1973). 'Tetracycline and acne vulgaris: a clinical and laboratory investigation.' *British medical Journal,* **4,** 332

Dormer, B. J. and Babett, J. A. (1972). 'Orofacial infection due to bacteroides a neglected pathogen.' *Journal of oral Surgery,* **30,** 658

Dowell, V. R. and Hawkins, T. M. (1974). *Laboratory Methods in Anaerobic Bacteriology*. Atlanta; US Department of Health, Education, and Welfare

Eiken, M. (1958). 'Studies on an anaerobic, rod-shaped, gram-negative micro-organism: *Bacteroides corrodens* n. sp.' *Acta Pathologica et Microbiologica Scandinavica,* **43,** 404

Finegold, S. M., Harada, N. E. and Miller, L. G. (1967). 'Antibiotic susceptibility patterns as aids in classification and characterization of Gram-negative anaerobic bacilli.' *Journal of Bacteriology,* **94,** 1443

Finegold, S. M. and Miller, L. G. (1968). 'Susceptibility to antibiotics as an aid in classification of Gram-negative anaerobic bacilli.' In *The Anaerobic Bacteria.* p. 139. Ed. by V. Fredette, Montreal; Montreal University

Georg, L. K. (1970). 'Diagnostic procedures for the isolation and identification of the etiologic agents of actinomycosis.' In *Proceedings of the International Symposium on Mycoses, Washington.* p. 71. Scientific Publication of the Pan-American Health Organization NO. 205

Georg, L. K. (1974). 'The agents of human actinomycosis.' In *Anaerobic Bacteria: Role in Disease*. p. 237. Ed. by A. Balows, R. M. DeHaan, V. R. Dowell and L. B. Guze. Springfield; Thomas

Georg, L. K., Robertstad, G. W., Brinkman, S. A. and Hicklin, M. D. (1965). 'A new pathogenic anaerobic *Actinomyces* species.' *Journal of infectious Diseases,* **115,** 88

Gibbons, R. J. (1974). 'Aspects of the pathogenicity and ecology of the indigenous oral flora of man.' In *Anaerobic Bacteria: Role in Disease.* p. 267. Ed. by A. Balows, R. M. DeHaan, V. R. Dowell and L. B. Guze. Springfield; Thomas

Gibbons, R. J., and Engle, L. P. (1964). 'Vitamin K compounds in bacteria that are obligate anaerobes.' *Science,* **146,** 1307

Gibbons, R. J. and Macdonald, J. B. (1960). 'Hemin and vitamin K compounds as required factors for the cultivation of certain strains of *Bacteroides melaninogenicus.' Journal of Bacteriology,* **80,** 164

Hardie, J. (1974). 'Anaerobes in the mouth.' In *Infection with Non-sporing Anaerobic Bacteria*. p. 99. Ed. by I. Phillips and M. Sussman. Edinburgh; Churchill Livingstone

Hare, R. (1967). 'The anaerobic cocci.' In *Recent advances in Medical Microbiology*. p. 284. Ed. by A. P. Waterson. London; Churchill

Hart, P. (1969). 'Bacteroides with special reference to *Bacteroides melaninogenicus.' Journal of medical Laboratory Technology,* **26,** 144

Hassing, G. S. (1971). 'Partial purification and some properties of a lipase from *Corynebacterium acnes.' Biochimica et Biophysica Acta,* **242,** 381

Hofstad, T. and Kristoffersen, T. (1971). 'Preparation and chemical characteristics of endotoxic lipopolysaccharide from three strains of *Sphaerophorus necrophorus.' Acta Pathologica et Microbiologica Scandinavica,* **79,** 385

Holdeman, L. V. and Moore, W. E. C. (1975). *Anaerobe Laboratory Manual*. Blacksburg; Virgina Polytechnic Institute Anaerobe Laboratory

Holmberg, K. and Forsum, U. (1973). 'Identification of *Actinomyces, Arachnia, Bacterionema, Rothia,* and *Propionibacterium* species by defined immunofluorescence.' *Applied Microbiology,* **25,** 834

Jackson, F. L., Goodman, Y. F., Bel, F. R., Wong, P. C. and Whitehouse, R. L. S. (1971). 'Taxonomic status of facultative and strictly anaerobic "Corroding Bacilli" that have been classified as *Bacteroides corrodens.*' *Journal of medical Microbiology,* **4,** 171

James, A. L. and Robinson, J. V. A. (1974). 'A comparison of the biochemical activities of *Bacteroides corrodens* and *Eikenella corrodens* with those of certain other Gram-negative bacteria.' *Journal of medical Microbiology,* **8,** 59

Johnson, J. L. (1973). 'Use of nucleic-acid homologies in the taxonomy of anaerobic bacteria.' *International Journal of systematic Bacteriology,* **23,** 308

Journal of antimicrobial Chemotherapy. (1975). 'An evaluation of metronidazole in the prophylaxis and treatment of anaerobic infections in surgical patients.' **1,** 393 (Study Group)*

Kelstrup, J. (1966). 'The incidence of *Bacteroides melaninogenicus* in human gingival sulci, and its prevalence in the oral cavity at different ages.' *Periodontics,* **4,** 14

Lambe, D. W., Vroon, D. H. and Rietz, C. W. (1974). 'Infections due to anaerobic cocci.' In *Anaerobic Bacteria: Role in Disease.* p. 585. Ed. by A. Balows, R. M. DeHaan, V. R. Dowell and L. B. Guze. Springfield; Thomas

Leyden, J. J., Marples, R. R., Mills, O. H. and Kligman, A. M. (1974). 'Tretinoin and antibiotic therapy in acne vulgaris.' *Southern medical Journal,* **67,** 20

Macdonald, J. B. (1962). 'On the pathogenesis of mixed anaerobic infections of mucous membranes.' *Annals of the Royal College of Surgeons of England,* **31,** 361

Macdonald, J. B. and Gibbons, R. J. (1962). 'The relationship of indigenous bacteria to periodontal disease.' *Journal of dental Research,* **41,** 320

Macdonald, J. B., Socransky, S. S. and Gibbons, R. J. (1963). 'Aspects of the pathogenesis of mixed anaerobic infections of mucous membranes.' *Journal of dental Research,* **42,** 529

Marples, R. R., Downing, D. T. and Kligman, A. M. (1971). 'Control of free fatty acids in human surface lipids by *Corynebacterium acnes.*' *Journal of Investigative Dermatology,* **56,** 127

Marples, R. R., Leyden, J. J., Stewart, R. N., Mills. O. H. and Kligman, A. M. (1974). 'The skin microflora in acne vulgaris.' *Journal of investigative Dermatology,* **62,** 57

Marples, R. R. and McGinley, K. J. (1974). '*Corynebacterium acnes* and other anaerobic diphtheroids from human skin.' *Journal of medical Microbiology,* **7,** 349

Marples, R. R., McGinley, K. J. and Mills, O. H. (1973). 'Microbiology of comedones in acne vulgaris.' *Journal of investigative Dermatology,* **60,** 80

Pien, F. D., Thompson, R. L. and Martin, W. J. (1972). 'Clinical and bacteriologic studies of anaerobic Gram-positive cocci.' *Mayo Clinic proceedings,* **47,** 251

Pine, L. and Georg, L. K. (1969). 'Reclassification of *Actinomyces propionicus.*' *International Journal of systematic Bacteriology,* **19,** 267

* 'Study Group'refers to the Luton and Dunstable Hospital Study Group which consisted of; A. T. Willis, C. L. Bullen, I. R. Ferguson, P. H. Jones, K. D. Phillips, P. V. Tearle, K. Williams and S. E. J. Young, (Department of Clinical Microbiology); G. H. Bancroft-Livingstone, G. C. Brentnall and S. A. Seligman (Department of Obstetrics and Gynaecology); R. B. Berry, R. V. Fiddian, D. F. Graham, D. H. C. Harland, D. F. R. Hughes, D. B. Innes, D. Knight, W. M. Mee, N. Pashby, R. L. Rothwell-Jackson, and A. K. Sachdeva (Department of Surgery); D. Edwards, C. Kilbey and I. Sutch (Pharmacy).

Prévot, A. R. (1957). *Manual for the Classification and Determination of the Anaerobic Bacteria*. 3rd Edn. Paris; Masson (French)

Prévot, A. R. and Fredette, V. (1966). *Manual for the Classification and Determination of the Anaerobic Bacteria*. 1st American edn. Philadelphia; Lea and Febiger

Robinson, J. V. A. and James, A. L. (1974). 'Some observations on the colony morphology of "Corroding bacilli".' *Journal of applied Bacteriology*, **37**, 101

Rogosa, M. (1964). 'The genus *Veillonella*. I. General cultural, ecological and biochemical considerations.' *Journal of Bacteriology*, **87**, 162

Rogosa, M. and Bishop, F. S. (1964a). 'The genus *Veillonella*. II. Nutritional studies.' *Journal of Bacteriology*, **87**, 574

Rogosa, M. and Bishop, F. S. (1964b). 'The genus *Veillonella*. III. Hydrogen sulphide production by growing cultures.' *Journal of Bacteriology*, **88**, 37

Sawyer, S. J., Macdonald, J. B. and Gibbons, R. J. (1962). 'Biochemical characteristics of *Bacteroides melaninogenicus*. A study of thirty-one strains.' *Archives of oral Biology*, **7**, 685

Schottmüller, H. (1911a). 'The aetiology of puerperal fever.' *Münchener Medizinische Wochenschrift*, **1**, 557 (German)

Schottmüller, H. (1911b). 'On bacteriological investigations, and their methods in puerperal fever.' *Münchener Medizinische Wochenschrift*, **1**, 787 (German)

Schwabacher, H., Lucas, D. R. and Rimington, C. (1947). '*Bacterium melaninogenicum* – a misnomer.' *Journal of general Microbiology*, **1**, 109

Simon, P. C. and Stovell, P. L. (1969). 'Diseases of animals associated with *Sphaerophorus necrophorus*. Characteristics of the organism.' *Veterinary Bulletin*, **39**, 311

Slack, J. M. and Gerencser, M. A. (1975). *Actinomyces, Filamentous Bacteria: Biology and Pathogenicity*. Minneapolis; Burgess

Slack, J. M., Landfried, S. and Gerencser, M. A. (1969). 'Morphological, biochemical and serological studies of 64 strains of *Actinomyces israelii*.' *Journal of Bacteriology*, **97**, 873

Sonnenwirth, A. C. (1974). 'Incidence of intestinal anaerobes in blood cultures.' In *Anaerobic Bacteria: Role in Disease*. p. 157. Ed. by A. Balows, R. M. DeHaan, V. R. Dowell and L. B. Guze. Springfield; Thomas

Sutter, V. L. and Finegold, S. M. (1971). 'Antibiotic disc susceptibility tests for rapid presumptive identification of Gram-negative anaerobic bacilli.' *Applied Microbiology*, **21**, 13

Warner, J. F., Fales, W. H., Sutherland, R. C. and Teresa, G. W. (1975). 'Endotoxin from *Fusobacterium necrophorum* of bovine hepatic abscess origin.' *American Journal of veterinary Research*, **36**, 1015

Weinberg, A. N. (1974). 'Infections due to anaerobic cocci.' In *Anaerobic Bacteria: Role in Disease*. p. 257. Ed. by A. Balows, R. M. DeHaan, V. R. Dowell, and L. B. Guze. Springfield; Thomas

Werner, H. (1974). 'Differentiation and medical importance of saccharolytic intestinal bacteroides.' *Arzneimittel-Forschung*, **24**, 340

Werner, H. and Pulverer, G. (1971). 'Incidence and medical significance of *Bacteroides* and *Spherophorus* strains.' *Deutsche Medizinische Wochenschrift*, **96**, 1325 (German)

Williams, R. A. D., Bowden, G. H., Hardie, J. M. and Shah, H. (1975). 'Biochemical properties of *Bacteroides melaninogenicus* subspecies.' *International Journal of systematic Bacteriology*, **25**, 298

Zabransky, R. J. (1970). 'Isolation of anaerobic bacteria from clinical specimens'. *Mayo Clinic Proceedings*, **45**, 256

6

Infections Due to Non-Clostridial Anaerobes

Organisms that are most frequently implicated in non-clostridial anaerobic infections in man include various species of *Bacteroides, Fusobacteria* and anaerobic cocci (*Peptococci, Peptostreptococci* and *Veillonella*), and *Actinomyces israelii.*

In contrast to the clostridial diseases, infections due to the non-sporing anaerobes are of endogenous origin, are less acute in their onset and prone to chronicity, and do not commonly produce a severe toxaemia. In general, these organisms are lowly pathogens, not given to producing primary infections, but they commonly cause disease at sites that have been debilitated or are the seat of some preceding pathological change. Together, the Gram-negative non-sporing anaerobic bacilli and the anaerobic cocci are by far the commonest cause of anaerobic bacterial disease, whose frequent occurrence as significant pathogens is often overlooked by the clinician and missed by the microbiologist. 'Infections due to them are at least ten times as common as those due to clostridia and undoubtedly represent the most frequently undiagnosed or misdiagnosed of the bacterial infections' (Finegold, 1970).

Infections due to the non-clostridial anaerobes are usually endogenous because most of these organisms are obligate parasites and form part of the normal bacterial flora of the mucosal surfaces of the oropharynx, and of the alimentary and female genital tracts. While strains of bacteroides are normally present in all these situations, fusobacteria are most commonly encountered in the oropharynx; and anaerobic cocci are prevalent in the upper respiratory and genital tracts, but are sparse in the intestinal tract.

From these situations the non-sporing anaerobes may invade adjacent tissues that are debilitated or are the seat of some other pathological change. Since the normal faecal flora is dominated by bacteroides, many serious infections due to them are derived from this reservoir. In the same way, the common occurrence of anaerobic cocci and bacteroides in the normal vagina explains why these organisms are the commonest cause of puerperal pyrexia, septic abortion and post-operative pelvic infections in gynaecological patients. The main portal of entry for the fusobacteria is the respiratory tract.

Although the development of anaerobiosis in the tissues must clearly pave the way for their invasion by these anaerobic species, many of which are intolerant of relatively small amounts of oxygen, the conditions necessary for the development of many of these endogenous infections are not understood. When the local resistance of the tissues is lowered by inadequate diet or by unrelated infections such as measles and pertussis, rapidly progressive synergistic fusobacterial infections such as Vincent's stomatitis, gangrenous stomatitis, periodontal infections, and pulmonary abscess and gangrene may result. Infection of the female genital tract by bacteroides and anaerobic cocci is greatly favoured by blood loss and tissue damage, as may occur in septic abortion, in prolonged labour and in postpartum haemorrhage. The trauma of abdominal surgery, especially bowel and gynaecological surgery, provides conditions favourable to the development of bacteroides infections. Malignant diseases and diabetes appear to be general predisposing causes. The use of neomycin for the preoperative 'sterilization' of the intestine may also be a predisposing factor in bowel surgery, since anaerobes are insensitive to this drug. Both the anaerobic cocci and the bacteroides induce septic thrombophlebitis much more readily than aerobic species, a factor which explains some common characters of these infections – local extension, bacteraemia and metastatic abscess formation.

The bacteroides and anaerobic cocci cause a variety of infections of virtually all types which are most commonly initiated in the vicinity of their normal habitats. In addition, the anaerobic cocci play an important part in two synergistic wound infections – anaerobic streptococcal myositis and Meleney's synergistic gangrene. In these infections, anaerobic cocci are associated with pyogenic aerobic organisms such as *Staphylococcus aureus.* These conditions are distinguished from gas gangrene by the absence of severe toxaemia, and on the microscopic appearances of the wound exudate (*see* pp. 223 and 234).

The march of events in infections due to bacteroides and anaerobic cocci may be illustrated by reference to the female genital tract. Almost any condition that causes bleeding and tissue damage in the female genital tract predisposes to invasion of the pelvic viscera by non-sporing anaerobes. Important among these are malignant disease, gynaecological surgery including curettage and cervical cauterization, childbirth and incomplete abortion. A variety of localized vulvovaginal infection may be caused by these organisms – incisional abscess, periurethral abscess, Bartholinitis, perirectal abscess and labial abscess. Much the most dangerous obstetrical and gynaecological infections, however, are those that commence in the uterus. Septic thrombophlebitis originating in the uterine sinusoids may extend via venous and lymphatic routes into the broad ligaments are retroperitoneal spaces in the pelvic floor. Adnexal abscesses thus formed may rupture into the peritoneal cavity producing general peritonitis or multiple intra-abdominal abscesses. Septic thrombophlebitis of the pelvic veins results in a continuous bacteraemia, rarely endocarditis may develop, and there may be metastatic abscess formation in the lungs, brain, liver, bones and joints, and other organs.

Non-clostridial anaerobic infections commonly declare themselves as a bacteraemia, patients showing the classic symptoms of hectic fever, rigors and profuse sweating; some patients with bacteroides bacteraemia develop the syndrome of disseminated intravascular coagulation (Yoshikawa, Tanaka and Guze, 1971). Confirmation by blood culture of a non-clostridial anaerobic bacteraemia should initiate a search for a primary focus of infection related to one of the three main portals of entry; in the female, the genital tract is of particular importance. 'Localized' infections due to these organisms, whether deep seated or superficial, may be suspected clinically from their anatomical positions, by the presence of large, necrotic, foul-smelling abscesses, by the development of clinical 'Gram-negative' bacteraemia, and by relevant history such as recent intestinal or gynaecological surgery, abdominal malignancy, and childbirth. Although it is still widely believed that foul smelling pus is a product of infection by coliform and related bacilli, pure *E. coli* pus is, in fact, odourless. Foul smelling suppurative inflammation is almost always due to obligate anaerobes (Altemeier, 1938b; Hoffman and Gierhake, 1969; Bodner, Koenig and Goodman, 1970; Gorbach and Bartlett, 1974a, b).

Like the staphylococcus, the non-clostridial anaerobes are able to produce a great diversity of infections (*Table 6.1*), a brief review of which forms the text of this chapter. Useful general reviews have been published by Bornstein *et al.* (1964), Saksena *et al.* (1968), Finegold (1970), Finegold *et al.* (1972), Finegold and Rosenblatt (1973), Levison

(1973), Goldsand (1974), Gorbach and Bartlett (1974a, b), Phillips and Sussman (1974), Miraglia (1974), Balows *et al.* (1974), DeHaan, Schellenberg and Pfeifer (1974), Willis (1974), Geddes (1975), Ledger (1975), Poindexter and Washington (1975), and Dunkle, Brotherton and Feigin (1976).

Table 6.1 A summary of some endogenous non-clostridial anaerobic infections

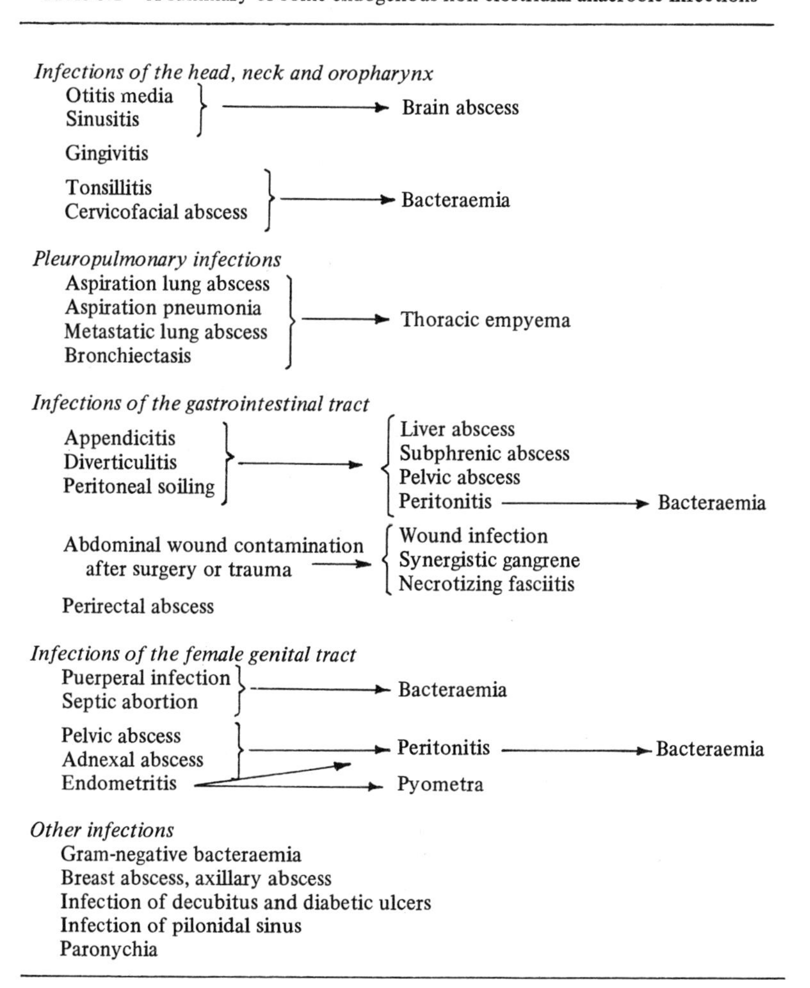

OBSTETRICAL AND GYNAECOLOGICAL INFECTIONS

Non-sporing anaerobic bacteria make up a large portion of the normal vaginal flora (Hare and Polunin, 1960; Hite, Hesseltine and Goldstein, 1947; Bollinger, 1964; Ansbacher, Boyson and Morris, 1967; Neary *et al.*, 1973; Gupta, Oumachigui and Hingorani, 1973; Gorbach *et al.*, 1973; Bullen, Willis and Williams, 1973; Swenson, 1974; Bartizal *et al.*, 1974; Hurley *et al.*, 1974; *Lancet* 1974; *Journal of antimicrobial Chemotherapy,* 1975). Bacteroides and anaerobic cocci are the predominant anaerobes, various species of which are encountered in about 70% of high vaginal and cervical swabs. The commonest species are *B. fragilis* and *B. melaninogenicus* (Burdon, 1928; *Journal of antimicrobial Chemotherapy,* 1975).

There is some evidence to suggest that the incidence of non-sporing anaerobes in the vaginal flora is subject to variations that depend on normal physiological influences. Thus, Hite, Hesseltine and Goldstein (1947) did not find *Bacteroides* species in the vaginas of any of 61 normal antenatal patients, and Louvois and his colleagues (1975) isolated bacteroides from the vaginas of only 5.2% of 265 women during the first half of pregnancy (*see also* Hurley *et al.*, 1974). Similar studies carried out in my laboratory have shown that anaerobes are usually absent from the vaginal flora of pregnant women at term, but appear in large numbers in the lochia soon after delivery. In a study of the preoperative vaginal flora in gynaecological patients, Neary *et al.*, (1973) found that although bacterial colonization of the vagina was common in the first week of the menstrual cycle, there was a steady decrease in carriage rate, especially of *Bacteroides* species, during the following three weeks; they, too, noted a low incidence of bacteroides in the vaginas of pregnant women. It was suggested that the presence of blood in the vagina might account for the menstrual cyclical variations in the anaerobic bacterial flora referred to above, a factor that may also explain the rapid appearance of vaginal anaerobes during the puerperium. It seems likely, however, that the cyclical effects of oestrogens on the vaginas of non-pregnant women, and their continuous effects in pregnant women, are also of importance in determining the bacterial flora.

The anaerobic bacteria present in the introitus and lower vagina are representatives of the faecal flora and are more properly regarded as faecal contaminants rather than normal inhabitants.

In view of this common occurrence of non-clostridial anaerobes in the vagina, it is not surprising that these organisms are frequently involved in infections of the female genital tract. From the great majority of obstetrical and gynaecological infections multiple bacterial species, both anaerobic and facultatively anaerobic, may be isolated. Modern experience leaves little room for doubt, however, that obligate anaerobes are the predominant pathogens in most serious, and in many of the more trifling, infections of the female genital tract (Clark and Wiersma, 1952; Carter *et al.*, 1953; Jones *et al.*, 1959; Ledger, Sweet and Headington, 1971; Ledger, 1972; Sweet, 1975; Chow, Marshall and Guze, 1975; Chow *et al.*, 1975). In any patient in whom a pelvic infection of unexplained origin is found or in whom the response to therapy is not prompt, non-clostridial anaerobic infections should be considered.

It must be said at once that the isolation of facultative anaerobes such as *E. coli* and faecal streptococci from infected lesions of the female genital tract cannot be ignored; the anaerobic 'bandwagon' does not prevent these species from causing infections! Their likely significance, however, cannot be assessed in the absence of information about the anaerobic flora. Indeed, aerobic cultures *alone* from infections of these sorts are more often than not misleading because the information they provide is always incomplete, and often irrelevant. In a recent survey of postoperative sepsis amont 100 consecutive hysterectomies, there were 22 cases of pelvic infection; in 20 of these the predominant bacterial flora was composed of non-clostridial anaerobes, in one the only organism isolated was *Staphylococcus aureus*, and in the other the isolates were *E. coli*, a *Proteus* species and a faecal streptococcus (*Journal of antimicrobial Chemotherapy*, 1975).

The anaerobic pathogens most commonly implicated in female pelvic infections are *B. melaninogenicus*, *B. fragilis*, *B. capillosus*, fusobacteria and anaerobic cocci (Thadepalli, Gorbach and Keith, 1973; Swenson, 1974; *Journal of antimicrobial Chemotherapy*, 1975).

Vulvovaginal infections

Vulvovaginal infections due to anaerobic bacteria have been reported upon by Parker and Jones (1966), Pearson and Anderson (1970a), Roberts and Hester (1972), Thadepalli, Gorbach and Keith (1973), Swenson *et al.* (1973), Gorbach and Bartlett (1974a) and Swenson (1974).

Bartholinitis

The majority of non-venereal Bartholin's abscesses are caused by anaerobic bacteria (Carter, 1963). Parker and Jones reported that 34 of 45 (69%) of their patients were infected with anaerobes, 12 of them with anaerobes alone. The most frequently cultured organisms were anaerobic cocci in 33 patients; 5 patients were infected with bacteroides. Ten of 15 cases of Bartholin's abscesses reported upon by Swenson and his colleagues yielded anaerobic bacteria, in 4 of which only anaerobes were present. Pearson and Anderson reported that bacteroides were associated with 26 of 173 patients with Bartholin's abscesses.

Labial abscess

In their report of five years' experience of vulval and vaginal infections, Parker and Jones encountered 10 patients with labial abscesses. Anaerobic cocci were isolated from 7 of these, in 3 of which bacteroides were also present.

Other infections

Anaerobic bacteria are a common cause of a variety of other vulvovaginal abscesses such as hidradenitis, periurethral abscess, perianal abscess and infected sebaceous cyst.

Sequelae of vulvovaginal infections

Most vulvovaginal abscesses remain localized, and respond to surgical drainage.

In some patients, especially those with diabetes, a form of progressive synergistic gangrene may develop, in which a rapidly progressive necrotizing fasciitis extends away from the primary abscess, and is especially prone to advance along deep fascial planes. As a result, there may be a very extensive involvement of tissue, the infection spreading up over the abdominal wall, into the flanks, or into the muscles of the thighs. In one fatal case reported by Roberts and Hester (1972), infection had encircled the abdomen and had reached as far as both axillae. Unlike Meleney's synergistic gangrene (*see* p. 223), necrotizing fasciitis is a rapidly progressive, highly toxic synergistic infection, in which there is marked tissue destruction of deeper structures often without overt cutaneous manifestations except for cellulitis and evidence of gas formation. The bacteriology of this condition is poorly documented. From reports available it seems likely that it may be caused both by mixtures of facultatively anaerobic bacteria such as *Staphylococcus aureus* with group A *Streptococcus pyogenes,* or *Escherichia coli* with faecal streptococci, and by mixtures of anaerobic bacteria such as peptostreptococci with *B. fragilis* (Rea and Wyrick, 1970; Roberts and Hester, 1972; Gorbach and Bartlett, 1974a). In a recent case of necrotizing fasciitis of the scrotum and perineum, which developed at my hospital in a 70-year-old man following prostatectomy, the responsible pathogens were *B. fragilis* and an anaerobic streptococcus.

Pelvic infections

During the past decade it has become clear that the great majority of non-venereal pelvic abscesses of obstetrical and gynaecological origin contain a variety of non-sporing anaerobic bacteria. Although anaerobes are commonly present as the exclusive pathogens, they may be associated with facultative anaerobes (Parker and Jones, 1966; Swenson *et al.,* 1973; Thadepalli, Gorbach and Keith, 1973; Swenson, 1974; *Journal of antimicrobial Chemotherapy,* 1975). In the latter event, modern microbiological experience suggests that the anaerobic component of the bacterial flora is usually the significant one. Similarly, pelvic cellulitis, which may be regarded as an infection that has failed to localize as an abscess, is most frequently due to non-sporing anaerobic bacteria.

The indigenous anaerobic bacteria of the vagina are likely to invade the pelvic viscera following local tissue damage and bleeding. Consequently, surgical interference (obstetrical or gynaecological) is a pre-eminent predisposing cause of infection. Other factors that predispose to postoperative or 'postmanipulative' infection include poor personal hygiene, prolonged uterine bleeding, recent parturition especially where labour was prolonged or difficult, retained products of conception, recent postabortal states and criminal abortion, malignant disease of the vagina, uterus and cervix, obstructed uterine drainage, and unresolved salpingo-oophoritis.

Endometritis and pyometra

Non-clostridial anaerobes are commonly implicated in endometritis and pyometra, which may develop as a sequel to puerperal infection of the uterus (especially following caesarean section), or in association with malignant disease of the uterus in older patients. These infections probably represent direct invasion of the uterine cavity from the vagina (Carter *et al.,* 1951, 1953).

Posthysterectomy vaginal cuff infections

The development of pelvic cellulitis or pelvic abscess following hysterectomy is a not uncommon experience. Most of these infections are due to non-sporing anaerobic bacteria derived from the normal vaginal flora. Of 64 such infections reported upon by Hall *et al.* (1967), 63 were associated with anaerobic organisms, usually mixed with facultative anaerobes. Swenson *et al.* (1973) and Swenson (1974) reported that 16 of 21 vaginal cuff abscesses contained anaerobes, in 11 of which obligate anaerobes were present as the sole pathogens. In a recent prospective study of postoperative infection following hysterectomy (*Lancet,* 1974; *Journal of antimicrobial Chemotherapy,* 1975), 21 patients developed clinical abdominal or pelvic infections, 18 of which were diagnosed microbiologically as due to non-sporing anaerobes. These vaginal cuff infections usually localize as abscesses, from which uneventful recovery may be expected once drainage has been established. In a small proportion of cases, however, pelvic cellulitis and fever persist, and these patients are likely to become bacteraemic unless appropriate antimicrobial therapy is initiated (*see also* Okubadejo, Green and Payne, 1973).

Pelvic abscess

Abscesses, other than posthysterectomy vaginal cuff infections, may occur elsewhere in the pelvis either as a result of postoperative or puerperal sepsis, or secondary to pelvic or intestinal inflammatory disease. Postoperative pelvic abscesses most commonly follow hysterectomy, induced abortion and tubal ligation, and develop as adnexal infections.

In a study of the bacterial flora of tubo-ovarian abscesses, Altemeier (1940) recovered non-sporing anaerobes from 23 of 25 consecutive cases; anaerobic cocci were found in 22 patients, and bacteroides in 10. Subsequently, Willson and Black (1964) recovered anaerobes from 6 of 28 ovarian abscesses; the true incidence of anaerobic infections in their series may have been much higher than this, however, since 15 of their specimens were ‘sterile’. Similarly, in a study of 104 patients with pelvic abscess, Parker and Jones (1966) encountered anaerobic bacteria in 33 cases, in 23 of whom anaerobes were present as the sole pathogens; but 40 of their specimens were ‘sterile’. The significance of non-sporing anaerobes as causal agents of non-specific adnexitis is not easy to assess, because obvious inflammatory exudate is frequently not available from these patients. However, in a study of 108 patients with non-gonococcal adnexitis, Viberg (1964) isolated anaerobes from the cervical canal on 34 occasions (31%), but 30 samples failed to yield significant bacterial growth. More recent studies, which are in accord with my own experience, have shown that the great majority of adnexal infections are associated with bacteroides and anaerobic cocci, although facultative anaerobes, such as *E. coli* and faecal streptococci may also be present (Ledger, Campbell and Willson, 1968; Ledger, Sweet and Headington, 1971; Pearson and Anderson, 1970b; Swenson *et al.*, 1973; Swenson, 1974; Thadepalli, Gorbach and Keith, 1973; Craft, Ghandi and Hardy, 1974).

Acute pelvic inflammatory disease

Spontaneously developing acute inflammation of the female pelvis is still most commonly due to gonococcal infections. Chow *et al.* (1975) expressed the view that although the gonococcus was important in initiating acute pelvic inflammation, its primary role was to pave the way for secondary invaders from the normal vaginal flora. The evidence that non-sporing anaerobes play a part in this sequence of events is not strong.

Puerperal sepsis

The great majority (more than 70%) of puerperal infections are due to non-sporing anaerobes. As with other forms of anaerobic pelvic infection in the female, most earlier workers considered that anaerobic cocci were the predominant pathogen, although the observations of Harris and Brown (1927) adumbrated the importance of the bacteroides (Schwarz and Dieckman, 1927; Schwarz and Brown, 1936; Douglas and Davis, 1946; Colebrook, 1930; Hill, 1963). During the last 20 years, however, it has become increasingly clear that bacteroides species, often in association with anaerobic cocci, are frequently the significant pathogens in these infections. The non-sporing anaerobes of the vaginal flora easily gain access to the uterus during even normal deliveries (Hite, Hesseltine and Goldstein, 1947). Factors that favour the subsequent development of infection include anaemia and tissue damage, premature rupture of the membranes, prolonged and difficult labour, instrumentation, excessive blood loss and retained products of conception (Lewis, Johnson and Miller, 1976).

The clinical picture is variable. Commonly there is merely a mild fever and a heavy or offensive lochia; extension of localized infections such as endometritis may lead to the development of pyometra, or adnexal or pelvic abscess. In more severe infections a swinging temperature with peritonitis, pelvic thrombophlebitis or bacteraemia may develop. Thrombosis of the pelvic veins may be followed by dissemination of infected thrombi causing metastatic abscesses in other organs, especially the lungs. The more severe forms of bacteroides puerperal sepsis are associated particularly with caesarean section (Pearson and Anderson, 1970b; Bosio and Taylor, 1973).

Parturient infections

Bacteroides infections of the female genital tract that occur during pregnancy may first declare themselves as a bacteraemia, and although the treated infection is seldom fatal for the mother, the prognosis for the infant may be poor (Pearson and Anderson, 1967). Perinatal infections due to anaerobes usually present as bacteraemic states (Robinow and Simonelli, 1965; Chow *et al.*, 1974; Harrod and Stevens, 1974).

Septic abortion

Non-clostridial anaerobes are an important cause of postabortal sepsis (Clark and Wiersma, 1952; Rotheram and Schick, 1969; Pearson and Anderson, 1970b; Thadepalli, Gorbach and Keith, 1973; Rotheram, 1974). Pearson and Anderson referred to 111 women who had incomplete septic abortion due to mixed infections with bacteroides, half of whom became bacteraemic. They estimated that 10–15% of all cases of septic abortion are associated with these organisms. My own experience, and the observations of Thadepalli and his colleagues, and of Rotheram, suggest that a much higher proportion of these infections is due to anaerobes (over 80%), although Rotheram inappropriately included carbon dioxide-dependent organisms in his anaerobic group.

Bacteraemia complicating obstetrical and gynaecological infections

The likely development of bacteraemia and septic thromboembolic disease from foci of anaerobic infection in the female pelvis was noted earlier. Bacteroides bacteraemias of obstetrical and gynaecological origin have been reported upon by Tynes and Frommeyer (1962), Pearson and Anderson (1967, 1970b), Bodner, Koenig and Goodman

(1970), Gelb and Seligman (1970), Felner and Dowell (1971), Bosio and Taylor (1973), Chow and Guze (1974) and Ledger *et al.* (1975). Amongst obstetrical and gynaecological patients with anaerobic infections, those with septic abortion are much more likely to become bacteraemic than others.

OTHER GENITOURINARY INFECTIONS

Urinary tract infections

Infections of the urinary tract are rarely caused by anaerobic bacteria, and it seems likely that many alleged cases of anaerobic bacterial cystitis represent contamination of the urine either with normal bacterial flora, or with pus from an adjacent anaerobic infection such as periurethral abscess, prostatic abscess or Cowperitis (Finegold *et al.*, 1965; Headington and Beyerlein, 1966; Segura *et al.*, 1972; Kumazawa *et al.*, 1974).

Prostatic infections and urethritis

Fishbach and Finegold (1973) reported upon a single case of prostatic abscess with bacteraemia in which the offending organism was *F. gonidiaformans.* There is no published evidence to suggest that obligate anaerobes play any common role in chronic prostatitis, or in chronic or non-specific urethritis (Justesen, Nielsen and Hattel, 1973).

INFECTIONS RELATED TO THE GASTROINTESTINAL TRACT

Pioneer studies on the incidence of *Bacteroides* species in faeces were made by Eggerth and Gagnon (1933), Weiss and Rettger (1937), Misra (1938) and Lewis and Rettger (1940), all of whom agreed that human adult faeces contains a predominance of non-sporing obligately anaerobic bacteria. There have been numerous more recent studies of the normal bacterial flora of the intestinal tract of man, in which attention has been paid not only to the incidence and relative porportions of different bacterial species in different parts of the gastrointestinal tract, but also to the effect on these bacterial populations of such factors as race, diet and antibiotic therapy. Although some reports may differ in detail, there is general agreement that the largest concentration of anaerobes is in the colon. There, the non-sporing anaerobic bacilli are the overwhelmingly predominant bacteria, accounting for around 99% of the cultivable flora. The two commonest anaerobic species present are *B. fragilis* and bifidobacteria, both being present at levels of the order of 10^{10}–10^{11} organisms per g of faeces. Enterobacteria (*E. coli,* coliform bacilli and so on) are the next most prominent group of organisms in the colon, being present at levels of about 10^6–10^7 organisms per g of faeces; 'coliforms' thus account for less than 0.1% of the total bacterial population of faeces.

The stomach and proximal part of the small intestine are much less densely populated than the colon. There, the bacterial flora is made up largely of Gram-positive facultative anaerobes such as yeasts, green-forming streptococci and lactobacilli, although small numbers of anaerobic bacilli may also be present, especially in the jejunum. Total counts in the stomach and jejunal contents are of the order of 10^2–10^3/ml and 10^3–10^5/ml respectively, and it seems likely that most of these organisms are contaminants transported from the mouth and respiratory tract rather than permanent residents. As one progresses from the jejunum to the ileum, the bacterial flora increases both in number and variety, so that in the terminal ileum about equal numbers of aerobes and anaerobes are present in a total bacterial population of around 10^8 organisms/ml.

Enumerations of the microbiological floras of different parts of the gastrointestinal tract have been made by numerous workers (Rosebury, 1962; Houte and Gibbons, 1966; Donaldson, 1964; Finegold, 1969; Aries *et al.*, 1969; Drasar, Shiner and Mcleod, 1969; Moore, Cato and Holdeman, 1969; Finegold *et al.*, 1970; Finegold, Attebery and Sutter, 1974; Haenel, 1970; Gorbach, 1971; Peach *et al.*, 1974; Attebery, Sutter and Finegold, 1974; Drasar, 1974; Ueno *et al.*, 1974; Gossling and Slack, 1974; Holdeman, Good and Moore, 1976; Sykes, Boulter and Schofield, 1976; Gorbach and Tabaqchali, 1969).

Although clostridia, notably *Cl. perfringens,* are well known to occur commonly in cases of biliary tract disease (*see* Willis, 1969), little attention seems to have been paid to the incidence of non-sporing anaerobes in this situation. In a recent study, Sapala, Ponka and Neblett (1975) isolated non-clostridial anaerobes from only 7 of 178 samples of bile from patients with inflammatory biliary tract disease; *B. fragilis* and anaerobic cocci were each isolated on two occasions. According to Shimada (1976), however, there is a high incidence of *B. fragilis* in the bile of patients with obstructive biliary tract disease. Although the organism usually occurs mixed with facultative anaerobes, Shimada regards *B. fragilis* as a significant pathogen in this clinical setting.

Intra-abdominal sepsis

Most cases of localized and generalized peritonitis are caused by organisms derived from the intestinal tract, and develop as a result of contamination by bowel content. Although mixed infections with obligate anaerobes and facultatively anaerobic organisms are the rule in intra-abdominal sepsis, in the great majority of cases the overwhelmingly predominant bacteria present are anaerobic species – usually bacteroides; and modern experience leaves little room for doubt that the anaerobic component of the bacterial flora is the clinically significant one (Gillespie and Guy, 1956; Moore, Cato and Holdeman, 1969; Gillespie, 1970; Altemeier *et al.*, 1973; Gorbach and Bartlett, 1974a; Gorbach, Thadepalli and Norsen, 1974; Lorber and Swenson, 1975; *British medical Journal*, 1976).

Virtually any conditions that interfere with the integrity of the gut wall may predispose to intra-abdominal anaerobic infection; these include appendicitis, diverticulitis, inflammatory bowel disease, malignant disease, biliary tract disease, perforated ulcer, accidental trauma and gastrointestinal surgery (Hoffman and Gierhake, 1969; Thadepalli *et al.*, 1972, 1973; Herter, 1972; Baird, 1973; Nichols *et al.*, 1973; McLaughlin, Meban and Thompson, 1973; Okubadejo, Green and Payne, 1973; Leigh, Simmons and Norman, 1974; Leigh, 1975). Initially there is localized or generalized peritonitis; subsequently, localization of the infection leads to the formation of abscesses, of which the most commonly encountered are appendical, pelvic, subphrenic and liver abscess (Butler and McCarthy, 1969; Harley, 1955; Altemeier *et al.*, 1973; Altemeier, 1974; Bateman, Eykyn and Phillips, 1975; Bonfils-Roberts, Barone and Nealon, 1975). The anaerobes most frequently implicated in intra-abdominal sepsis derived from the gastrointestinal tract are *B. fragilis* and anaerobic cocci, especially peptostreptococci; in contrast to infections derived from the female genital tract, *B. melaninogenicus* is uncommonly isolated (Gillespie and Guy, 1956; Moore, Cato and Holdeman, 1969; Gorbach, Thadepalli and Norsen, 1974; *Journal of antimicrobial Chemotherapy*, 1975; Block *et al.*, 1964).

Appendicitis

Although it seems likely that the non-clostridial anaerobes play an aetiological role in some cases of appendicitis, there is no doubt that these organisms are important causes of postappendicectomy sepsis. The commonest complication of appendicectomy is undoubtedly intra-abdominal or wound infection, the incidence of which varies from 4% for normal appendices to 77% for gangrenous or perforated appendices. The average frequency of postoperative infection is probably of the order of 30% (Barnes *et al.*, 1962; *Lancet*, 1970, 1971; Magarey *et al.*, 1971; Airan, Levine and Sice, 1973). With the exception of Meleney's synergistic gangrene (*see below*), the less serious infections involve only the abdominal wound, and are commonly due to facultative anaerobes such as *E. coli* and enterococci. In infected abdominal wounds in which there is dehiscence, abscess formation or other marked disturbance of wound healing, the significant pathogens are usually obligate anaerobes, although they are frequently accompanied by smaller numbers of facultative anaerobes (Hoffmann and Gierhake, 1969). Of more serious import is postoperative intra-abdominal sepsis, which may be caused primarily by facultatively anaerobic Gram-negative enteric bacilli, but is much more commonly due to non-sporing anaerobes, notably *B. fragilis* in association with anaerobic cocci. Although it is still widely believed that postsurgical intra-abdominal infections are due to *Enterobacteriaceae* and enterococci, modern experience confirms the early observation of Veillon and Zuber (1898) and of Altemeier (1938a) that the significant pathogens are usually non-clostridial anaerobes (Baird, 1973; Leigh, Simmons and Norman, 1974; Gorbach and Bartlett, 1974a; Douglas and Vesey, 1975; *British medical Journal,* 1976). The presence of an accompanying facultatively anaerobic flora is the rule, but aerobic species are present in smaller numbers. Recent experimental work by Bartlett *et al.* (1975) who used the rat as an animal model suggests that *E. coli* and coliform bacilli are primarily responsible for acute peritonitis, while non-sporing anaerobes are the principal cause of intra-abdominal abscess formation.

Other causes of intra-abdominal anaerobic sepsis

It was noted that a variety of inflammatory and other pathological states of the gastrointestinal tract may predispose to intra-abdominal anaerobic infection. In general, what has been said of anaerobic infection in relation to appendicitis is equally applicable to them. Peritoneal soiling that originates from the upper reaches of the gastrointestinal tract, as in perforated duodenal ulcer or elective gastrectomy, is less likely to be complicated by anaerobic sepsis than peritoneal contamination by large bowel contents (*see* Nichols, Miller and Smith, 1975). This is doubtless a reflection of the varying bacterial 'population density' at different levels of the intestinal tract. In this connection, Thadepalli *et al.* (1972, 1973) noted that perforating injuries of the upper intestine have a lower sepsis rate and morbidity than similar lesions of the colon. Anaerobic sepsis is less likely to develop as a complication of simple or elective gastrointestinal surgical procedures than of acute or difficult operations largely because the risks of peritoneal soiling are less. In order to reduce the risks of postoperative infection in patients who are to be submitted to elective colonic surgery, it is customary to employ a regimen of preoperative bowel preparation which aims at removing or suppressing the resident bacterial flora. Many of these preparative procedures fail to take account of the major presence of anaerobes, and utilize antibacterial agents that are active only against facultatively anaerobic bacteria. Clearly, unless an 'anti-anaerobic' agent is also used in these circumstances, selective elimination of the facultatively anaerobic flora may actively predispose to the subsequent development of anaerobic sepsis (Nichols *et al.*, 1973; Goldring *et al.*, 1975).

Anorectal suppurative infections

Because of their proximity to the large bowel, infections such as gluteal abscess, perianal and perirectal abscess, and ischiorectal abscess are commonly associated with a mixed bacterial flora (Brightmore, 1972; Lindell, Fletcher and Krippaehne, 1973). Non-clostridial anaerobes, especially *B. fragilis* and peptostreptococci, and sometimes also *B. melaninogenicus,* can be isolated from most of these infections, in which they are frequently the predominant and pathogenically significant organisms.

Meleney's progressive synergistic gangrene

Brewer and Meleney (1926a, b) and Meleney (1931) recorded cases of progressive gangrene of the abdominal wall (Meleney's progressive synergistic gangrene) following drainage of appendical abscesses. In these cases *Staphylococcus aureus* and an anaerobic coccus were isolated from the gangrenous margin of the lesion, and it was shown that a similar gangrenous lesion could be produced in dogs and guinea pigs only when the two organisms were injected together; pure cultures of the individual strains had little or no effect. Subsequent work by Mergenhagen, Thonard and Scherp (1958) showed that staphylococcal hyaluronidase together with an unidentified growth factor enabled the anaerobic coccus to invade the tissues.

A good deal of experimental evidence has since accumulated which supports the conception of a synergistic bacterial aetiology for a number of anaerobic infections. Willard (1936) confirmed Meleney's clinical and experimental findings in a patient with chronic progressive post-operative gangrene of the abdominal wall, and similar clinical conditions have been described by subsequent workers. The 'progressive strepto-coccal ulceration' of Langston (1938), the 'spreading subcutaneous or cutaneous gangrene' of Mitchiner and Cowell (1939), and the 'wound phagedaena' of Callam and Duff (1941), are probably all essentially the same as Meleney's synergistic gangrene. The condition is an uncommon instance of chronic, superficial, progressive gangrene which is differen-tiated clinically from other types of superficial gangrene by its slow and relentless progression, its severe local symptoms and the absence of severe systemic symptoms. It most commonly involves the abdominal wall postoperatively. Bacteriologically it is typically a synergistic infection due to anaerobic cocci mixed with facultative anaerobes such as *Staph. aureus, Strep. pyogenes, Proteus* species or *Ps. aeruginosa,* and it must be distinguished from true postoperative clostridial myone-crosis of the abdominal wall (*see also* Stewart-Wallace, 1934–35; Meleney, Friedman and Harvey, 1945; Smith, 1956; Grainger, MacKenzie and McLachlin, 1967).

Somewhat similar conditions to Meleney's gangrene are *synergistic necrotizing cellulitis* and *necrotizing fasciitis* (Rea and Wyrick, 1970; Stone and Martin, 1972; *Lancet,* 1973a; Puhvel and Reisner, 1974; Ledingham and Tehrani, 1975). These conditions are much more acute than Meleney's gangrene, are associated with a high degree of systemic toxicity and a not insignificant mortality rate. Synergistic necrotizing cellulitis, which is most commonly caused by *Enterobacteriaceae* mixed with anaerobic cocci or bacteroides, occurs in the perineum or thighs, usually in debilitated patients who have diabetes or other arterial insufficiency. There is rapidly spreading necrosis of the skin, subcutaneous tissues and connective tissues of adjoining fascial planes, and an associated serosanguineous discharge.

Necrotizing fasciitis is briefly considered on p. 214. A case of bac-teroides necrotizing fasciitis of the upper extremity in a drug addict was reported upon by Rein and Cosman (1971).

Bacteraemia complicating abdominal infections

Bacteraemic states are a not uncommon consequence of abdominal infections due to non-sporing anaerobes. Thus, gangrenous appendi-citis may be complicated by mesenteric vein thrombosis or pylephle-bitis, leading to liver abscess, subphrenic abscess or bacteraemia. In the same way, anaerobic infections originating in the region of the rectum may progress to a bacteraemia following thrombophlebitis of the rectal, inferior mesenteric and portal veins. Patients who are particularly prone to develop anaerobic bacteraemia are those with gastrointestinal malignant disease and those submitted to gastrointestinal surgery. Bacteroides bacteraemias derived from the gastrointestinal tract have been reported by Lemierre (1936), Gunn (1956), Tynes and Frommeyer (1962), Bodner, Koenig and Goodman (1970), Gelb and Seligman (1970), Felner and Dowell (1971), Ellner and Wasilauskas (1971), Wilson *et al.* (1972), Mitchell and Simpson (1973), MacKenzie and Litton (1974) and Chow and Guze (1974).

Transient bacteraemia may be associated with manipulative procedures such as endoscopy of the upper and lower gastrointestinal tract and with barium enema (Frock *et al.*, 1973, 1975; Shull *et al.*, 1975).

INFECTIONS OF THE RESPIRATORY TRACT

The diverse complexity in structure and function of different parts of the mouth and upper respiratory tract accounts, at least in part, for the great variety of indigenous anaerobic bacteria that make up a substantial part of the normal bacterial flora of the oropharynx. The nose, the pharynx and different parts of the mouth (tongue, tooth surfaces, gingiva and so on) each has an associated microflora which is more or less peculiar to the site. A detailed consideration of this oral microbiological ecology is beyond the scope of the present work; excellent review articles have been published by Bisset and Davis (1960), Bowden and Hardie (1971), Socransky and Manganiello (1971), Davies (1972) and Hardie and Bowden (1974).

The saliva may be regarded as containing the pooled micro-organisms from these various sources, although clearly the salivary flora cannot represent very closely that of any particular site; indeed, according to Gibbons and van Houte (1975) the salivary flora appears to be proportionally representative of that of the dorsum of the tongue. The principal anaerobic inhabitants of the saliva are *Veillonella* species, peptostreptococci, peptococci, *B. melaninogenicus, B. oralis, F. nucleatum,* spirochaetes, *Actinomyces* species, and species of *Leptotrichia* and *Bifidobacterium* (Rosebury, 1962; Gordon and Jong, 1968; Drasar, Shiner and McLeod, 1969; Bowden and Hardie, 1971; Loesche, 1974; Hardie, 1974). In the nose, propionibacteria appear to be the predominant anaerobes, while in the throat veillonellae predominate (Watson *et al.*, 1962; Smith, 1975). In conditions of health the respiratory tract below the larynx is regarded as sterile (Pecora and Yegian, 1958; Lees and McNaught, 1959).

Oropharyngeal fusospirochaetal infections

Acute ulcerative gingivitis (Vincent's gingivitis; trench mouth), Vincent's stomatitis, Ludwig's angina and cancrum oris are often referred to as fusospirochaetal infections because they share a common microbial aetiology; they are all synergistic anaerobic bacterial infections involving *F. nucleatum* and *Borrelia vincentii* (*see* Blake, 1968).

Acute ulcerative gingivitis

This acute ulcerative disorder affects mainly the gingival margin. It occurs most frequently in young adult males during autumn and winter. The single most important predisposing factor to infection is pre-existing gingivitis due to poor oral hygiene, although a variety of systemic factors such as stress, overwork and exposure to cold and damp are common and important antecedents. The simultaneous occurrence of a number of these predisposing factors explains the high incidence of 'trench mouth' during the First World War; 'epidemics' of this endogenous infection are due to a predisposing debility of the population, not to contagion.

Clinically, the prominent symptoms of acute ulcerative gingivitis are bleeding gums, soreness of the mouth and a bad taste. There is a typical ulceration spreading around the gingival margin of adjacent teeth, always more prominent on the labial and buccal side. The gingiva of a few teeth, or of the whole mouth may be involved. The infection may also occur as a complication of agranulocytosis, acute leukaemia and glandular fever.

Vincent's gingivitis responds rapidly to treatment with oral metronidazole (Shinn, 1962; Shinn, Squires and McFadzean, 1965). Subsequent dental care and instruction in oral hygiene is important, for with inadequate treatment or with recurrences, chronic periodontal disease may follow (Manson and Rand, 1961).

Vincent's angina

This is an ulcerating tonsillar and pharyngeal infection associated with an adherent pseudomembrane and a foul discharge. The onset is acute with sore throat, cervical lymphadenopathy, malaise and fever. Vincent's angina must be distinguished clinically from streptococcal infection, diphtheria and infectious mononucleosis. In cases of doubt the patient should be treated for a fusospirochaetal infection as aspiration of infected material may lead to fulminating infection of the lungs.

Ludwig's angina

This is a rare but serious form of fusospirochaetal infection (Williams and Guralnick, 1943) in which there is an indurated cellulitis of both the sublingual and submaxillary spaces. The infection spreads rapidly along fascial planes into the neck, leading to elevation of the tongue and to laryngeal oedema with stridor and dysphagia. Early and adequate antimicrobial therapy is mandatory, and surgical intervention may be necessary to ensure an adequate airway.

Cancrum oris

Cancrum oris is a rare disease that occurs in conditions of severe malnutrition and debility, especially in children (*see* Tempest, 1966). It is a rapidly destructive lesion of the mouth and cheeks, commonly preceded by an acute ulcerative gingivitis (Emslie, 1963), and is doubtless due to a fusospirochaetal invasion of the soft tissues of the mouth. The treatment includes antimicrobial therapy, penicillin and metronidazole being the drugs of choice, and attention to nutritional requirements and to underlying causes of debility.

Pericoronitis

Acute pericoronitis is an infection of the tissues over the crown of an unerrupted tooth, and is most commonly seen around the lower third molar. Anaerobic bacteria, especially anaerobic cocci, are frequently implicated in the infection which, however, may present as a typical acute ulcerative gingivitis, i.e. as a fusopirochaetal infection. Typically the onset is acute with pain, swelling, fever and general malaise; submandibular lymphadenopathy is common and trismus may be present. The immediate treatment is antimicrobial therapy, penicillin being the drug of choice; alternatively tetracycline or erythromycin may be used. In the presence of acute ulcerative gingivitis, metronidazole is also indicated. Subsequent dental treatment is commonly necessary to remove the source of trauma to the lower gum (usually the upper third molar) which is the essential predisposing cause of the infection.

Dento-alveolar abscesses

Periapical abscesses commonly have an anaerobic bacterial aetiology. This is not surprising since the intact pulp chambers of non-vital teeth commonly harbour a variety of anaerobic organisms (MacDonald, Hare and Wood, 1957; Crawford and Shankle, 1961; Kantz and Henry, 1974). In a study of the flora of eight alveolar abscesses, Sabiston and Gold (1974) found that *F. nucleatum* was the commonest significant pathogen, although *Bacteroides* species were also common isolates. The onset of the infection is acute with pain and swelling; the particular clinical signs clearly depend on the tooth involved. The treatment of first importance is drainage of the pus, which may be effected either by widely opening the tooth to drain the root-canal, or by tooth extraction. Anaerobic abscesses of this sort characteristically contain a large amount of foul smelling pus (*see* Quayle, 1974).

Other upper respiratory tract infections

Apart from periodontal infections, the anaerobic infections most commonly encountered in the upper respiratory passages are peritonsillar and pharyngeal abscess (Rubinstein, Onderdonk and Rahal, 1974), chronic sinusitis (Urdal and Berdal, 1949; Fredette, Auger and Forget, 1961; Heineman and Braude, 1963; Bornstein *et al.*, 1964; Frederick and Braude, 1974, and chronic otitis media and mastoiditis (Smith, McCall and Blake, 1944; Vos *et al.*, 1975). Orofacial and oropharyngeal infections due to non-sporing anaerobes may also occur as a complication of trauma (Leake, 1972; Sharp, Meador and Martin, 1974; Sims, 1974).

Prior to the introduction of antibiotics, the oropharynx was a common site from which anaerobic bacteraemic states were initiated (Alston, 1955). The most usual initial cause was a tonsillar or peritonsillar abscess, opened surgically either inadequately or too late in the course of the disease, from which septic thrombophlebitis of the tonsillar and peritonsillar veins developed, with subsequent involvement of the internal jugular vein or even of the facial vein. Bacteraemic states due to anaerobes also developed as a result of periodontal disease, otitis media, mastoiditis and cervical adenitis (Lemierre, 1936). In a review of 148 cases of bacteroides bacteraemia Gunn (1956) reported that over 30% originated from nasopharyngeal infections (*see also* Tynes and Utz, 1960). As is clear from the reports of Tynes and Frommeyer (1962), Felner and Dowell (1971) and Chow and Guze (1974), infections of the oropharynx these days are a much less frequent cause of bacteroides bacteraemia than are infections at other sites, although the oropharynx remains a common portal of entry for anaerobic bacterial endocarditis (*see below*). This is doubtless attributable to a real reduction in the incidence of serious oropharyngeal infections, due to early and effective treatment – surgical, dental and antimicrobial – and to improved oral hygiene. Nevertheless, anaerobic bacteraemias derived from the upper respiratory tract are encountered from time to time; one such case was described recently by Mitre and Rotheram (1974) in a patient who developed internal jugular vein septic thrombophlebitis with anaerobic bacteraemia and metastatic abscess formation following dental treatment.

In view of the number and variety of non-sporing anaerobes present in the normal mouth, it is not surprising that dental manipulations should lead to transient bacteraemias due to them (Francis *et al.*, 1962; Khairat, 1966a, b; Crawford *et al.*, 1974). Among 100 consecutive patients who had tooth extraction, Khairat demonstrated post-extraction bacteraemia due to non-sporing anaerobes in 32, while in their study of 25 patients having tooth extraction, Crawford and his colleagues isolated non-clostridial anaerobes from post-extraction blood cultures from 19; in many of these a number of anaerobic bacterial species was present.

It is important to note that, as in endocarditis due to facultative anaerobes, the commonest portal of entry for organisms causing anaerobic bacterial endocarditis is the oropharynx. Felner and Dowell (1970) reported upon 33 patients with endocarditis due to anaerobic bacteria; among 25 in whom the infection was due to non-clostridial anaerobes, the oropharynx was the portal of entry in 9. Somewhat similar figures were reported by Nastro and Finegold (1973) in their review of 37 published cases of endocarditis due to Gram-negative anaerobic bacilli; among 25 patients in whom a probable portal of entry was established, 11 had pre-existing oropharyngeal sepsis.

Pulmonary infections

Anaerobic pleuropulmonary infections have been the subject of recent excellent reviews by Bartlett and Finegold (1970, 1972, 1974), Bartlett, Sutter and Finegold (1974), Finegold (1974b) and Gorbach and Bartlett (1974a); *see also Lancet* (1976a). Broadly speaking three clinical types of intrathoracic anaerobic infections may be recognized – simple pneumonia, necrotizing pneumonia and pulmonary abscess – any of which may occur with or without empyema; empyema in the absence of parenchymal infection is rare (Bartlett and Finegold, 1974).

Pathogenesis of pulmonary infections

Although there is a variety of general and local conditions that are known to predispose to anaerobic chest infections, it is important to note that commonly no predisposing conditions are detected; that apparently primary anaerobic pulmonary infections not infrequently occur in young, healthy individuals. General underlying conditions that predispose to anaerobic pleuropulmonary infections include malnutrition, diabetes mellitus, and corticosteroid and cytotoxic therapy.

Aspiration is undoubtedly the most common single event that leads to the development of pulmonary infections, and may occur under circumstances such as unconsciousness (anaesthesia, alcoholism, drug addiction and so on), dental and other oral surgery, oesophageal dysfunction, and vomiting. Pulmonary infections that result from aspiration occur most frequently in dependent parts of the lung (posterior segments of the upper lobes and superior segments of the lower lobes) and there is commonly some pre-existing periodontal or other oropharyngeal anaerobic infection (Maier and Grace, 1942; Alston, 1955; Smith and Fekety, 1968; Bartlett and Finegold, 1972; Sullivan *et al.*, 1973).

Various local conditions may predispose to anaerobic infection. Secondary infection by oropharyngeal anaerobes is likely to occur where there is stasis or tissue necrosis, and so may complicate inhaled foreign body, bronchiectasis, bronchogenic carcinoma and pulmonary infarction (Tillotson and Lerner, 1968; Sullivan *et al.*, 1973). Indeed, anaerobic bacterial infection may be the first indication of underlying malignant disease.

Anaerobic infections in other parts of the body may lead to secondary infection of the lungs. Reference is made elsewhere to the facility with which non-sporing anaerobic bacteria induce septic thrombophlebitis, with subsequent intravascular dissemination of the infection from a primary site. Although metastatic abscess formation may occur in a variety of organs and tissues, dissemination of septic emboli to the thorax is especially common, and may occur in primary infections involving the oropharynx, the gastrointestinal tract and the female genital tract. Moreover, bacteroides bacteraemia (as distinct from septic thromboembolism) may cause multiple foci of secondary infection in the lungs (Marcoux *et al.*, 1970; Felner and Dowell, 1971).

Occasionally, lower lobe consolidation with or without empyema may develop adjacent to a subphrenic abscess, suggesting a transdiaphragmatic route of spread of infection (Maier and Grace, 1942).

Bacterial aetiology

The common participation of 'fusospirochaetal' organisms in bronchiectasis, necrotizing pneumonia and lung abscess was recognized over

50 years ago (Varney, 1920; Pilot and Davis, 1924; Smith, 1927–28, 1932; Kline and Berger, 1935), and is due to the fact that the majority of pulmonary infections caused by anaerobic bacteria are initiated by aspiration of oropharyngeal secretions (Gonzalez and Calia, 1975). Surgical excision of lung abscesses, which was practised during the pre-antibiotic era, facilitated definitive studies of their bacterial flora. In one such study of 16 excised abscesses, Cohen (1932) noted that the predominant flora was an anaerobic one, and that the common specific isolates were anaerobic streptococci, *B. fragilis, B. melaninogenicus* and fusiform bacilli. Similar, more recent observations were made by Swenson and Lorber (1974) and Bartlett, Gorbach and Finegold (1974). Bartlett and his co-workers reported upon 54 patients with aspiration pneumonia, in 50 of whom anaerobes were found, and in 25 of whom anaerobes were present as the exclusive pathogens; the predominant isolates were anaerobic cocci, *B. melaninogenicus* and *F. nucleatum.* In a somewhat smaller study of patients with pulmonary infections, Gonzalez and Calia (1975) compared the incidence of anaerobes in specimens from patients with aspiration-induced infections and from those in whom aspiration was an unlikely antecedent event. Of 17 patients in the aspiration group, anaerobes were isolated from 11, in 6 of whom anaerobes were present as the sole pathogens. Among 17 patients in the non-aspiration group, on the other hand, anaerobes were isolated on only three occasions. Since the bacteriology of empyema mirrors to some extent the bacteriology of underlying parenchymal infection, it is pertinent to note here that Bartlett *et al.* (1974) recovered anaerobic bacteria from 76% of 83 empyema fluids studied, that anaerobes were the exclusive isolates in 29 of these, and that the predominant isolates were anaerobic cocci, *F. nucleatum, B. melaninogenicus* and *B. fragilis.*

Bacteriological findings of these sorts indicate that the anaerobic bacteria that predominate in aspiration-induced infections generally reflect the oropharyngeal flora. In much the same way, the anaerobic bacteria present in metastatic infections of the lungs are likely to be representative of the flora of the region from which the septic embolism was derived – most frequently the female genital tract, less commonly the intestinal tract, and occasionally the oropharynx. Thus, while pulmonary infections derived from the upper respiratory tract (both aspiration and metastatic) are likely to be caused by penicillin-sensitive bacteria (*B. melaninogenicus, Fusobacterium* species and anaerobic cocci), metastatic infections of the lungs derived from other parts of the body are much more likely to contain a penicillin-resistant component in the form of *B. fragilis* (*see for example* Beazley, Polakavetz and Miller, 1972).

Despite these convenient generalizations, it will be recognized that, just as with anaerobic infections in other parts of the body, the recovery of multiple organisms, both facultatively anaerobic and anaerobic, from pulmonary infections is the rule, and that in this event a microbiological 'value judgment' is required to confirm a clinical diagnosis, and to indicate appropriate antimicrobial chemotherapy.

Clinical considerations

As with non-clostridial anaerobic infections in other parts of the body, pulmonary infections are characterized by the production of large amounts of foul-smelling pus; thus, foul halitosis in the absence of obvious oropharyngeal infection, and production of copious putrid sputum or empyema fluid are virtually diagnostic hallmarks of anaerobic bacterial pleuropulmonary disease. The absence of a putrid discharge, however, does not exclude anaerobic infection (Bartlett and Finegold, 1974). Although the development of anaerobic pulmonary infections is very commonly subacute or chronic, an acute onset is by no means unknown (Willis, Young and Ferguson, 1974), and these are always the most difficult to diagnose. In the absence of appropriate treatment, progressive pulmonary infection leads to destruction of lung parenchyma with abscess formation, necrotizing pneumonia and empyema, sometimes with bronchopleural fistula. Secondary brain abscess may result from haematogenous dissemination of septic emboli (Heineman and Braude, 1963; Sandler, 1965; Swartz and Karchmer, 1974). Bacteraemia, however, is an uncommon consequence of pleuropulmonary infections.

The treatment of anaerobic infections that are restricted to the lung parenchyma is essentially medical, antimicrobial therapy being of major importance. For the management of anaerobic empyema, surgical drainage is mandatory.

INTRACRANIAL SUPPURATIVE DISEASE

Anaerobic infections of the central nervous system have been excellently reviewed by Swartz and Karchmer (1974).

Apart from infections that occur following accidental trauma, e.g. penetrating wounds of the head, and fractured base of the skull, intracranial suppuration is most commonly due to non-sporing anaerobic bacteria, in particular to *Bacteroides* species and to anaerobic cocci (Heineman and Braude, 1963; Heineman, Braude and Osterholm, 1971; Schoolman, Lun and Rodecker, 1966; Yoshikawa, Chow and Guze,

1975). It usually develops as a complication of suppuration elsewhere in the body. The commonest primary foci are chronic otitis and mastoiditis, sinusitis, and pleuropulmonary infections such as bronchiectasis, lung abscess and empyema (Smith, McCall and Blake, 1944; Heineman, Braude and Osterholm, 1971; Ingham *et al.*, 1975a, b). Brain abscesses complicating cyanotic heart disease may also be due to anaerobic bacteria (Newton, 1956).

The majority of brain abscesses are sited in the temporal or frontal lobes, a reflection of the common pre-existence of chronic ear and sinus infections (Swartz and Karchmer, 1974).

Anaerobic bacteria have rarely been implicated in cases of pyogenic meningitis. Occasionally bacteroides meningitis may occur secondary to bacteraemia (Lifshitz, Liu and Thurn, 1963; Cooke, 1975), or secondary to spread of infection from chronic mastoiditis (Smith, McCall and Blake, 1944). More commonly, meningitis due to bacteroides or to anaerobic cocci may result from intraventricular leakage or rupture of a cerebral abscess. Indeed, the development of anaerobic bacterial meningitis should always raise the suspicion of brain abscess (Swartz and Karchmer, 1974).

BONE AND JOINT INFECTIONS

Anaerobes are infrequently involved in osteomyelitis. Useful reviews on these uncommon infections have been published by Ziment and his colleagues (1968, 1969). Among 17 cases of anaerobic osteomyelitis studied by these workers, 14 represented local extension into bone from adjacent soft tissue infections – eight of these were patients with osteomyelitis of the foot (seven with diabetes) – and five had developed osteomyelitis of skull bones as a complication of chronic otitis or sinusitis; two others were infections of the femur resulting from haematogenous dissemination from distant sites. Multiple bacterial species, both aerobic and anaerobic, were involved in all but one of these infections, the predominant anaerobic isolates being *B. melaninogenicus*, *B. fragilis*, anaerobic cocci and *Fusobacterium* species.

Bacteroides osteomyelitis of the mandible has been discussed by Leake (1972).

As with osteomyelitis, suppurative arthritis is rarely due to anaerobic bacteria. In a review of the world literature, Ziment, Davis and Finegold (1969) summarized the data on 47 cases. They concluded that the sternoclavicular and sacroiliac joints are particularly susceptible to bacteroides infection. In addition to single cases of *B. fragilis* arthritis of the knee reported by these workers and by Brorson, Edmar and

Holm (1975), Ament and Gaal (1967) described a similar infection in a child, Sanders and Stevenson (1968) described bacteroides pyogenic arthritis in two brothers with agammaglobulinaemia, and Kamme *et al.* (1974) reported the presence of anaerobic bacteria (anaerobic cocci and *P. acnes*) in late infections that developed after hip arthroplasty.

SUPERFICIAL AND SOFT TISSUE INFECTIONS

Meleney's synergistic gangrene and necrotizing fasciitis are considered on pp. 214 and 223.

Anaerobic streptococcal wound infection

Anaerobic streptococcal myositis is a massive infection of muscle by anaerobic cocci in association with aerobic pyogenic cocci, the most important of which are *Staph. aureus* and haemolytic streptococci

Table 6.2 The clinical differences between clostridial and anaerobic streptococcal myositis (adapted from Lowry and Curtis, 1947)

	Clostridial myositis	*Anaerobic streptococcal myositis*
Temperature	100°–102°F	High
Pulse	Rapid, poor quality	Rapid, proportionate to temperature
Blood pressure	Low	Normal
Toxaemia	Constant	Mild, varies with temperature
Anaemia	Usually present	Not characteristic
Pain	Often present, severe	Often present, severe
Wound		
1. Discharge	Slightly water to profuse brown	Wet, oedematous, profuse, blood-stained, seropurulent
2. Odour	Inconstant	Foul
3. Crepitation	Inconstant	Slight, late
4. Muscle	Always involved, diffuse	Usually involved, focal

Reproduced by courtesy of Lowry, K. F., and Curtis, G. M., and the Editor of the *American Journal of Surgery*

(MacLennan, 1943). Neglected cases progress to true gangrene of muscle. This type of infection, which is relatively uncommon, may be distinguished from clostridial myonecrosis both clinically (*Table 6.2*) and bacteriologically (MacLennan, 1943, 1962; Lowry and Curtis, 1947). Distinction between the two conditions is important, since radical surgery may be contraindicated in cases of anaerobic streptococcal myositis.

Unlike clostridial infections, a bacteriological diagnosis may be made before the condition can be distinguished clinically from true gas gangrene. In streptococcal myositis, Gram-stained muscle smears show a constant and typical appearance. No Gram-positive bacilli are seen, but instead, vast numbers of streptococci are present along with many pus cells. This is in sharp contrast with the diverse bacterial flora and small number of pus cells in smears from clostridial myonecrosis. The bacteriological diagnosis is subsequently confirmed by isolation of the anaerobic and aerobic cocci. The treatment of streptococcal myositis consists of prompt conservative surgery and appropriate antibiotic therapy. Workers disagree about the extent of surgical interference indicated. MacLennan (1943) considered that only the most conservative surgery should be contemplated, deep relaxing incisions being made through fascia and muscles. Hayward and Pilcher (1945) and Anderson, Marr and Jaffe (1972), however, stressed the importance of evacuating collections of pus and of excising any necrotic tissue present in the wound.

Decubitus ulcers

Because of their common close anatomical association with the anus, most sacral decubitus ulcers become infected with organisms derived from the faecal flora. Microbiological studies of material from these infected ulcers commonly show the presence in varying proportions, but always in impressive numbers, of a variety of facultative anaerobes such as *E. coli,* enterococci, staphylococci, *Proteus* species and *Ps. aeruginosa.* It is not generally appreciated, either by clinicians or laboratory workers, that despite these plentiful facultatively anaerobic flora, the predominant and significant pathogens in these lesions are almost always obligate anaerobes (Ellner and Wasilauskas, 1971; Kagnoff, Armstrong and Blerins, 1971; Rissing *et al.,* 1974). The organisms most commonly implicated in these infections are *B. fragilis* and *B. melaninogenicus.*

Other superficial infections

In my experience a variety of other soft tissue and superficial infections may be caused commonly by the non-sporing anaerobes. These include infections of pilonidal sinuses, of varicose ulcers, and of diabetic perforating ulcers of the feet, breast abscess, axillary abscess and paronchia (*see* Pearson, 1967; Pearson and Smiley, 1968; Sinniah, Sandiford and Dugdale, 1972; Hale, Perinpanayagam and Smith, 1976). Human and animal bites are particularly likely to produce severely infected wounds in which bacteroides and fusobacteria may play a predominant part (Linscheid and Dobyns, 1975). There is a single recorded case of acute suppurative thyroiditis due to *B. melaninogenicus* (Sharma and Rapkin, 1974).

BACTERAEMIA

Although reference has already been made to the development of bacteraemic states from non-clostridial anaerobic infections at various anatomical sites, it is appropriate to briefly refer to 'bacteroides' bacteraemia here. It will be recalled that clinically significant bacteraemia is almost invariably a complication of some pre-existing anaerobic sepsis, which itself is commonly a complication of surgical interference. Useful general reviews have been published by many workers (Gunn, 1956; Tynes and Utz, 1960; Bodner, Koenig and Goodman, 1970; Marcoux *et al.,* 1970; Sinkovics and Smith, 1970; Felner and Dowell, 1971; Wilson *et al.,* 1972; Washington, 1973; Chow and Guze, 1974; and *see also British medical Journal,* 1973; *Lancet,* 1973b).

Neonatal bacteroides bacteraemia may occur as a transient and often self-limiting phenomenon following premature rupture of membranes and maternal amnionitis, or as a result of postoperative sepsis. The literature on the subject has been reviewed by Chow *et al.* (1974); *see also* Harrod and Stevens (1974).

The predominant anaerobic isolates in bacteraemia are Gram-negative bacilli, of which much the commonest is *B. fragilis* (Wilson *et al.,* 1972; Sonnenwirth, 1974), although in the past *Fusobacterium* species were not uncommonly encountered (Tynes and Utz, 1960; Robinow and Simonelli, 1965; Felner and Dowell, 1971); anaerobic cocci are infrequent causes of bacteraemia (Wilson *et al.,* 1972). The reported incidence of polymicrobial bacteraemia varies; Wilson *et al.* (1972) encountered other obligate anaerobes or facultative anaerobes in 31% of patients with bacteroides bacteraemia, and Marcoux *et al.* (1970) reported an incidence of polymicrobial bacteraemia of 24% among 123 patients with bacteroides bacteraemia.

In common with bacteraemia due to many other organisms, that due to anaerobic bacteria may be transient, symptomless and benign, such as may occur following dental procedures (Khairat, 1966a, b), or it may be a severe and life-threatening event. The clinical features of bacteraemia due to anaerobic Gram-negative bacilli vary little from those due to sepsis caused by any Gram-negative bacillus. There is a sudden onset of hectic fever with rigors and profuse sweating, and endotoxic shock may develop. The portal of entry of bacteroides is most commonly the gastrointestinal tract and the female genital tract; that of fusobacteria is usually the respiratory tract. Unlike bacteraemias due to facultatively anaerobic Gram-negative bacilli, the urinary tract is rarely a site of entry for anaerobes. Other important distinguishing features of anaerobic bacteraemia include a high incidence of jaundice, septic thrombophlebitis and metastatic abscess formation. Gunn (1956) summarized the common clinical sequence of events in bacteroides bacteraemia very succinctly: '(1) Symptoms and signs of the primary lesion, e.g. appendicitis, (2) a period of apparent recovery, (3) symptoms and signs of a spreading infection at the site of the original lesion, (4) abrupt onset of septicaemia with rigors, profuse sweating, anaemia and icterus, and (5) symptoms and signs of metastatic infective lesions'.

Although the development of a bacteraemia may be the first declaration of the presence of an underlying 'silent' bacteroides infection, e.g. liver abscess, bacteroides bacteraemias most commonly follow from a failure to recognize the likely anaerobic aetiology of obvious pre-existing sepsis. Not uncommonly the preceding infection is mistakenly attributed to facultatively anaerobic bacteria. There can be little doubt that most anaerobic bacteraemias may be prevented, and that established bacteraemias may be recognized by fostering a high index of suspicion and a knowledge of the natural history and pathogenesis of bacteroides infections.

When appropriate antimicrobial agents are used for the treatment of anaerobic bacterial bacteraemia, a fatality rate a good deal lower than 10% may be expected. In the absence of treatment, or with the use of inappropriate antibiotics, the mortality rate is high – of the order of 60–80% (Gunn, 1956; Felner and Dowell, 1971; Nobles, 1973; Okubadejo, Green and Payne, 1973; Chow and Guze, 1974; MacKenzie and Litton, 1974; Ledingham, 1975).

ENDOCARDITIS

Non-clostridial anaerobes appear to be an uncommon but important cause of endocarditis. Felner and Dowell (1970) referred to an aggregate of 889 cases of blood culture-positive endocarditis reported in the earlier literature, of which only 14 (1.5%) had yielded anaerobic bacteria. Of these 14 isolates, 12 were anaerobic streptococci, one was a 'bacteroides', and one a diphtheroid. This preponderance of anaerobic streptococci in anaerobic bacterial endocarditis is not evident in subsequent reports of the condition, and it seems likely that at least some, probably many, of the earlier isolates were carbon dioxide-dependent cocci rather than obligate anaerobes. Among 39 cases of non-clostridial anaerobic endocarditis recently reviewed by Felner (1974), 15 were due to bacteroides (14 *B. fragilis,* 1 *B. melaninogenicus*), 7 to fusobacteria, 15 to *P. acnes* and only 2 to anaerobic cocci.

Infective endocarditis due to anaerobic bacteria has been reviewed by Felner and Dowell (1970), Nastro and Finegold (1973) and Felner (1974). The clinical features of endocarditis due to anaerobes do not differ appreciably from those described for endocarditis due to facultative anaerobes. Important differences include the common occurrence of major embolic phenomena in anaerobic infections, and a greater susceptibility to anaerobic infections in the absence of pre-existing heart disease.

TREATMENT OF BACTEROIDES AND RELATED INFECTIONS

Many of the publications referred to in the preceding text deal with general or specific aspects of the management of non-clostridial anaerobic infections. The subject as a whole has been reviewed recently by Finegold *et al.* (1975).

Surgical treatment

Copious pus, often with the formation of large abscesses, is a common denominator of almost all non-clostridial anaerobic infections, for which the treatment of first importance is surgical drainage. It is clear that in many instances establishment of drainage alone may be sufficient to effect a cure, as for example in tooth extraction for periapical abscess, and in surgical incision of superficial abscesses such as ischiorectal abscess, Bartholin's cyst abscess and axillary abscess. In more serious infections such as thoracic empyema, subphrenic abscess, septic abortion and brain abscess, it is a common practice to institute antibiotic therapy in addition to surgical drainage. In accordance with accepted surgical principles, however, it is clear that antimicrobial therapy can in no way replace surgical drainage when pus has accumulated in the tissues.

Limited experience indicates that hyperbaric oxygen therapy is not of value in the treatment of serious non-clostridial anaerobic infections (Schreiner, Tönjum and Digranes, 1974; Slack, 1976).

Table 6.3 Some reports upon the activity of antimicrobial agents against anaerobes

Antimicrobial agent	*Authors*
Aminoglycosides	Gillespie and Guy (1956); Finegold *et al.* (1971); Bodner *et al.* (1972); Kislak (1972); Mitchell (1973)
Colistin and polymyxin	Finegold *et al.* (1967); Bodner *et al.* (1970); Kislak (1972); Mitchell (1973)
Co-trimoxazole	Bushby (1973); Okubadejo *et al.* (1973); Phillips and Warren (1974); Rosenblatt and Stewart (1974)
Beta-lactam agents	Gillespie and Guy (1956); Zabransky *et al.* (1973); Blazevic and Matsen (1974); DeHaan *et al.* (1974b); Dornbusch *et al.* (1974); Schoutens and Yourassowsky (1974); Staneck and Washington (1974); Fiedelman and Webb (1975); Sutter and Finegold (1975); Tally *et al.* (1975b)
Tetracyclines	Kislak (1972); Martin *et al.* (1972); Nastro and Finegold (1972); Mitchell (1973); Zabransky *et al.* (1973); DeHaan *et al.* (1974b); Dornbusch *et al.* (1974); Leigh (1974); Monif and Baer (1974); Staneck and Washington (1974); Chow *et al.* (1975c); Leigh and Simmons (1975)
Erythromycin	Bodner *et al.* (1972); Kislak (1972); Martin *et al.* (1972); Mitchell (1973); Sutter *et al.* (1973); Zabransky *et al.* (1973); Dornbusch *et al.* (1974); Leigh (1974)
Chloramphenicol	Gillespie and Guy (1956); Kislak (1972); Martin *et al.* (1972); Mitchell (1973); Sutter *et al.* (1973); Nobles (1973); Zabransky *et al.* (1973); Chow and Guze (1974); DeHaan *et al.* (1974b); Dornbusch *et al.* (1974); Monif and Baer (1974)
Lincomycin and clindamycin	Finegold *et al.* (1965); Ingham *et al.* (1968); Bartlett *et al.* (1972); Kislak (1972); Martin *et al.* (1972); Mitchell (1973); Zabransky *et al.* (1973); Nobles (1973); DeHaan *et al.* (1974b); Dornbusch *et al.* (1974); Leigh (1974); Savage (1974); Staneck and Washington (1974); Finch *et al.* (1975); Okubadejo and Allen (1975)
Rifamycins	Finegold *et al.* (1971); Kislak (1972); Martin *et al.* (1972); Nastro and Finegold (1972); Werner (1972); Mitchell (1973); Zabransky *et al.* (1973); Chow and Guze (1974); Leigh (1974)
Metronidazole	Shinn *et al.* (1965); Freeman *et al.* (1968); Ueno *et al.* (1971); Nastro and Finegold (1972); Tally *et al.* (1972); Mitchell (1973); Leigh (1974); Staneck and Washington (1974); Sutter and Finegold (1975)

Antimicrobial therapy

Antimicrobial agents active against anaerobes

There is developing a fairly considerable literature concerning the activity of antimicrobial agents against the non-clostridial anaerobes, and it is proposed only to briefly summarize the subject here. *Table 6.3* lists some of the *in vitro* studies that have been made with particular antimicrobial agents. Antimicrobial agents that act against anaerobes have been excellently reviewed by Hamilton-Miller (1975).

Of the various drugs listed in *Table 6.3,* chloramphenicol, lincomycin, clindamycin, metronidazole and rifampicin are the most widely active against the non-clostridial anaerobes, and for the empirical treatment of anaerobic infections the choice must lie with one or other of these drugs. From this choice, however, we may for the time being exclude rifampicin, the use of which is quite properly restricted to the treatment of tuberculosis (*see Lancet,* 1976b).

BENZYLPENICILLIN AND OTHER β-LACTAM ANTIBIOTICS

These drugs are active against most anaerobes, *B. fragilis* being an important exception. Thus, for those anaerobic infections that are known to be due to organisms other than *B. fragilis,* benzylpenicillin may be regarded as a drug of choice. In this connection it is important to note that although many of the anaerobic infections that occur 'above the diaphragm' are due to penicillin-sensitive organisms derived from the mouth and upper respiratory tract, this is by no means always the case; a significant proportion is due to *B. fragilis* (Bartlett, Gorbach and Finegold, 1974).

Of the other β-lactam agents, cefoxitin appears to be highly active against most anaerobes including *B. fragilis,* while the other cephalosporins show lower orders of activity; cephalexin appears to be the least active. Carbenicillin also shows good activity against most anaerobes including many strains of *B. fragilis* (Sutter and Finegold, 1975; Fiedelman and Webb, 1975).

TETRACYCLINE AND ERYTHROMYCIN

Tetracycline and erythromycin are erratically active against a variety of non-clostridial anaerobes (Gillespie and Guy, 1956; Martin, Gardner and Washington, 1972; DeHaan, Pfeiffer and Schellenberg, 1974). At one time tetracycline was regarded as the drug of choice for anaerobic infections, including those caused by *B. fragilis.* Over the last few years, however, there has been a notable increase in the number of tetracycline-resistant isolates among both Gram-positive and Gram-negative groups of anaerobes. Newer tetracyclines, such as minocycline and doxycycline are more active than tetracycline (Kislak, 1972; Phillips, 1974; Leigh and Simmons, 1975; Marks, Rogers and Moses, 1975). It should be noted that carbon dioxide in the atmosphere of the anaerobic jar tends to reduce the activity of erythromycin in sensitivity testing (Ingham *et al.*, 1970).

COLISTIN AND POLYMYXIN

These drugs show very restricted activity against anaerobic bacteria. All Gram-positive organisms are resistant, and among the Gram-negative anaerobes *B. fragilis* is insusceptible. Fusobacteria and *B. melaninogenicus,* however, are moderately sensitive.

AMINOGLYCOSIDES

Aminoglycosides such as gentamicin, kanamycin, streptomycin and neomycin have little activity against anaerobic bacteria, even in high concentrations. They are consequently widely used as selective agents for the isolation of anaerobes in direct plating media (Finegold, Harada and Miller, 1967; Finegold, Sugihara and Sutter, 1971).

SPECTINOMYCIN

Spectinomycin is active against *B. fragilis* (Phillips and Warren, 1975; Ferguson and Smith, 1975, 1976a). There appear to be no reports of the action of the drug against other anaerobes.

CO-TRIMOXAZOLE

The few reports upon the action of co-trimoxazole against the anaerobes are conflicting. Okubadejo, Green and Payne (1973) and Phillips and Warren (1974, 1976) found that strains of *B. fragilis* were sensitive. The latter workers reported that the sulphonamide component was usually more active against their strains than was trimethoprim, although there was significant synergy between the components. Bushby (1973), on the other hand, found no synergy in tests against three (only) strains of *B. fragilis,* while Rosenblatt and Stewart (1974) reported that trimethoprim and sulphamethoxazole both independently and in combination were inactive against most of 98 strains of anaerobes tested.

CHLORAMPHENICOL

Chloramphenicol is active against virtually all anaerobes and is regarded by some as a first line drug for patients who are seriously ill with anaerobic infections (Finegold, 1974a). Some workers recommend it as the drug of choice for infections due to *Bacteroides* species (Bodner *et al.,* 1972; Levison, 1973; Finegold, 1976). The great disadvantage of chloramphenicol is its potential activity as a bone marrow depressant.

LINCOMYCIN AND CLINDAMYCIN

Lincomycin and clindamycin are active against the whole range of non-clostridial anaerobes, and have for some years been regarded as the drugs of choice for infections due to these organisms. Lincomycin is available for oral and parenteral use; clindamycin, which is a more active derivative of lincomycin, is available for oral use only, although recent successful clinical trials have been conducted with a parenteral form of the drug – clindamycin phosphate (Finch, Phillips and Geddes, 1975; Chow, Montgomerie and Guze, 1974).

It should be noted that carbon dioxide in the atmosphere of the anaerobic jar tends to reduce the activity of lincomycin (but not of clindamycin) in sensitivity testing (Ingham *et al.,* 1970).

Therapy with these drugs is commonly associated with a marked disturbance of the gut flora, and with looseness of the bowels or frank diarrhoea (Finegold, Harada and Miller, 1966; Sutter and Finegold, 1974). Pseudomembranous colitis is a much more serious side effect that has been attributed to lincomycin and clindamycin therapy (*see British medical Journal,* 1974; Smart *et al.,* 1976; Dane and King, 1976), although there are conflicting reports about its frequency. Tedesco, Barton and Alpers (1974) reported an incidence of 21% diarrhoea and 10% pseudomembranous colitis among 200 consecutive patients receiving clindamycin. Wilson (1974), on the other hand, encountered no cases of pseudomembranous colitis and only one case of severe diarrhoea among 22 000 patients who received lincomycin or clindamycin.

METRONIDAZOLE

For some years metronidazole has been the drug of choice for the treatment of trichomoniasis, giardiasis and amoebiasis. The first indication that this nitroimidazole compound might be of value in the treatment of anaerobic infections followed from the chance observation of Shinn (1962) that the drug was highly effective in the treatment of Vincent's stomatitis. A patient of Shinn's, who was being treated with metronidazole for vaginal trichomoniases, who also had marginal gingivitis, revealed that at the end of one week's treatment she had undergone a 'double cure'. Subsequently the *in vitro* susceptibility to metronidazole of a wide range of clinically important anaerobes has been determined (Davies, McFadzean and Squires, 1964; Freeman, McFadzean and Whelan, 1968; Prince *et al.*, 1969; Ueno *et al.*, 1971; Tally, Sutter and Finegold, 1972, 1975a; Nastro and Finegold, 1972; Whelan and Hale, 1973; Sutter and Finegold, 1975; Ralph and Kirby, 1975). From these studies it is clear that metronidazole has a universally bactericidal effect on anaerobic organisms, that aerobic and facultatively anaerobic bacteria are universally resistant to it, and that the *in vitro* minimum inhibitory and minimum bactericidal concentrations of the drug are equivalent. The antimicrobial activity of metronidazole has been reviewed by Ingham, Selkon and Hale (1975).

Although the potential of metronidazole in the treatment of anaerobic infections has been recognized for some time (Tally, Sutter and Finegold, 1972), it is only recently that serious attention has been paid to its use in the management of anaerobic infections in man. Recent studies have shown that oral metronidazole is highly effective in the treatment and prophylaxis of a wide variety of non-clostridial anaerobic infections (Tally, Sutter and Finegold, 1975; *Lancet,* 1974; *Journal of antimicrobial Chemotherapy,* 1975; *British medical Journal,* 1976), and that the drug can also be administered effectively and safely by the intravenous and rectal routes (Ingham *et al.*, 1975a, b; *Journal of antimicrobial Chemotherapy,* 1975; *British medical Journal,* 1976; Goldring *et al.*, 1975; *Drug and Therapeutics Bulletin,* 1976). A recent study by Salem, Jackson and McFadzean (1975) showed that metronidazole is compatible with most other antimicrobial agents in common use.

Several other nitroimidazoles, such as tinidazole and nimorazole, also show activity against non-clostridial anaerobes (Lozdan *et al.*,

1971; Dornbusch and Nord, 1974; Reynolds, Hamilton-Miller and Brumfitt, 1975; Ferguson and Smith, 1976b), and may find a place in the antimicrobial therapy of anaerobic bacterial infections in the future.

Antimicrobial agents in treatment

For many non-clostridial anaerobic infections there is a clear requirement for antimicrobial therapy. Absolute indications include bacteraemia, endocarditis, meningitis, pneumonia, pulmonary abscess, septic thrombophlebitis and embolic phenomena, unresolving cellulitis, acute ulcerative gingivitis and stomatitis, anaerobic streptococcal myositis, and Meleney's synergistic gangrene and related infections. In addition, it is a common and acceptable practice to combine antibiotic therapy with surgical evacuation of pus from localized infections at sites that are known to carry a special risk of bacteraemia or of thromboembolic complications; of particular importance in this respect are intra-abdominal infections, and pelvic sepsis associated with the female genital tract.

Because the response of the non-clostridial anaerobes to the various first line drugs is fairly predictable, there should rarely be any difficulty in choosing the most appropriate antimicrobial agent in particular circumstances. Chloramphenicol, lincomycin, clindamycin and metronidazole are all active against the whole range of non-clostridial anaerobes, so that these drugs may be used empirically, either when an anaerobic infection is suspected clinically or in cases of known anaerobic sepsis before the results of sensitivity tests become available. Under these circumstances, the choice of drug will be influenced by such factors as its toxicity, the required route of administration, the severity of the infection, and the presence of concomitant facultative infection.

For patients with severe Gram-negative sepsis, in particular those with bacteraemia, in whom antimicrobial therapy must begin before a bacteriological diagnosis is made, it is necessary to cover the possibilities that the infection may be due either to anaerobic or to facultatively anaerobic bacteria. Under these circumstances it is a common practice to initiate combined antimicrobial therapy, which may then be modified in the light of the bacteriological findings. One such, much used and very successful, regimen is lincomycin or clindamycin combined with gentamicin. Another, which has proved equally successful, is metronidazole combined with gentamicin. Although many non-clostridial anaerobic infections contain facultatively anaerobic bacteria in addition to the predominant anaerobic flora, it is rarely necessary or desirable to use antimicrobial therapy against the facultative components. Treatment of the facultatively anaerobic flora only, invariably fails; but antimicrobial therapy directed against the anaerobic bacteria is consistently successful.

Since non-clostridial anaerobic infections of the head-and-neck and thoracic regions are usually due to organisms that are sensitive to penicillin, high cure rates with benzylpenicillin may be expected in many pleuropulmonary and oropharyngeal infections. It will be recalled, however, that in a proportion of these infections, the presence of *B. fragilis* necessitates the use of a drug other than penicillin.

As in many other types of bacterial infection it is helpful to control antimicrobial therapy by *in vitro* sensitivity testing; this is especially important when drugs such as erythromycin and tetracycline are in use, whose activity against different strains of anaerobic bacteria is likely to be erratic. When first line drugs are used, however, *in vitro* sensitivity testing usually provides essential confirmatory information – that the empirically selected drug, which has led to an improvement in the patient's clinical condition, is, indeed, active against the incriminated anaerobic bacteria.

Antimicrobial agents in prophylaxis

The common occurrence of sepsis as a complication of gastrointestinal and gynaecological surgery has been a continuing problem for many years. The incidence of postappendicectomy sepsis, for example, varies from 4% for normal appendices to 77% for gangrenous or perforated appendices, with an average frequency of postoperative infection of around 30%. As with appendicectomy, infection remains the principal complication of colonic surgery, and it is also an important cause of morbidity following hysterectomy. Over the years, efforts to reduce the incidence of postappendicectomy sepsis have led surgeons to employ a variety of topical and systemic prophylactic antibacterial agents (*see British medical Journal,* 1976); and various prophylactic oral antibiotic regimens have been used in the preoperative preparation of patients for elective colonic surgery (*see* Goldring *et al.,* 1975). Many of the studies on the chemoprophylaxis of these postoperative infections have been concerned solely with clinical aspects of infection, and have not considered the nature of the infecting agents. This is unfortunate, because in these clinical settings, where a great variety of bacteria is likely to be present, the effectiveness of any prophylactic antibiotic must clearly depend on its spectrum of antibacterial activity. Despite an increasing awareness of the importance of the non-clostridial anaerobes as common causes of postoperative infection, it is still widely believed that postsurgical abdominal infections are usually caused by the *Enterobacteriaceae* (e.g. *E. coli*) and enterococci. Indeed, conventional procedures for the preoperative 'sterilization' of the gut utilize antimicrobial agents such as neomycin and kanamycin, which act only upon the facultative inhabitants of the colon; obligate anaerobes, which outnumber the facultatively anaerobic faecal flora by 1000 to 1, are universally resistant to these drugs. Under these circumstances, it may not be an overstatement to suggest that this selective suppression of the aerobic flora may actually increase the risk of infection by anaerobes. In any event, it is clear that any prophylactic regimen aimed at preventing infection following surgery upon the gastro-intestinal and female genital tracts, must take account of both the aerobic and anaerobic components of the indigenous bacterial flora. There is no doubt that carefully chosen short term antibiotic prophylaxis in surgery significantly reduces the incidence of postoperative sepsis (Stokes *et al.*, 1974). The prophylactic use of metronidazole in gastrointestinal and female genital tract surgery has been conspicuously successful in the prevention of postoperative anaerobic sepsis (*Lancet*, 1974; *Journal of antimicrobial Chemotherapy*, 1975; *British medical Journal*, 1976; Goldring *et al.*, 1975).

Although anaerobic bacterial endocarditis is not common, patients with valvular defects require appropriate antimicrobial prophylaxis during dental surgery. Fortunately, penicillin, which is commonly used for the prophylaxis of facultatively anaerobic endocarditis, is also highly active against the majority of oral anaerobes (*B. melaninogenicus*, *Fusobacterium* species and anaerobic cocci).

DIAGNOSIS OF NON-CLOSTRIDIAL ANAEROBIC INFECTIONS

Clinical considerations

The diagnosis of the various synergistic infections is usually not difficult; acute ulcerative gingivitis and stomatitis (p. 226), Meleney's synergistic gangrene (p. 223). necrotizing fasciitis (p. 214) and anaerobic strepto-coccal myositis (p. 234) each presents its own fairly typical clinical picture; and in all of these conditions the appearances of direct Gram-stained films are characteristic.

The more general forms of non-clostridial anaerobic infection (*Table 6.2*) share many of the clinical features that are common to most types of pyogenic sepsis. There are, however, a number of features that may help to distinguish non-clostridial anaerobic sepsis from other types of bacterial infection.

1. The proximity of the infection to a mucosal surface – oropharyngeal, gastrointestinal or vaginal – reflects the portal of entry of endogenously derived organisms.
2. Pre-existing states that compromise the integrity of the mucosal surfaces, such as accidental trauma, surgery, appendicitis, perforation of ulcer, prolonged or difficult labour, and malignant disease.
3. Infections that are related to the use of aminoglycosides; and those that follow human and animal bites.
4. Copious foul-smelling discharge is the hallmark of non-clostridial anaerobic infections. It may take the form of frank pus from superficial or deep seated abscesses (including brain abscess), and from such infections as thoracic empyema, chronic otitis media, sinusitis and mastoiditis. Alternatively, the discharge may be an offensive lochia, a purulent sputum or a serosanguinous exudate from infections such as peritonitis, abdominal wound cellulitis and decubitus ulcer. Contrary to popular belief *Enterobacteriaceae* pus is never foul smelling and is usually odourless. On the other hand the discharge from some non-clostridial anaerobic infections is also odourless.
5. Untreated non-clostridial anaerobic infections are prone to progress to a bacteraemic state, and may be complicated by septic thromboembolic phenomena. Consequently, bacteraemia, especially when associated with icterus, septic thrombophlebitis and metastatic abscess formation, are important pointers to an underlying anaerobic infection.

Bacteriological diagnosis

The specimen

Non-clostridial anaerobes are prevalent throughout the body, especially in association with mucosal surfaces, and this 'normal bacterial presence' must be borne in mind when considering the microbiology of pathological specimens. It is this problem that has led some workers to take the view that clinical specimens for anaerobic culture must be collected by methods that avoid contamination with the normal bacterial flora (Sutter and Finegold, 1973; Finegold and Rosenblatt, 1973; Gorbach and Bartlett, 1974a). Ellner, Granato and May (1973) have gone so far as to recommend that 'only specimens from normally sterile areas should be cultured anaerobically'. Thus, sputum collected by expectoration or at bronchoscopy is regarded by some as unsuitable since it is contaminated with the indigenous anaerobes of the oropharynx, and it is recommended that this specimen should be obtained by percutaneous transtracheal aspiration. Similarly, some quite elaborate sampling methods have been devised for cases of uterine infection in order to avoid contamination of the specimen with the normal vaginal flora.

Clearly, contamination of specimens with the normal flora of the colon (faeces) is likely to render them useless because pathologically significant organisms will be overwhelmed by the sheer diversity and weight of numbers of the faecal flora. In all other cases, however, including pulmonary and uterine infections, contamination of samples by the normal bacterial flora causes much less technical difficulty than is generally supposed. There is usually little difficulty in interpreting direct plate cultures of expectorated sputum provided that the inoculum is carefully chosen from the sample, and that multiple cultures are incubated in parallel so that direct comparison can be made between the anaerobic, facultatively anaerobic and carbon dioxide-dependent flora. It is unlikely that clinicians will spontaneously submit transtracheal aspirations, and it seems unreasonable for microbiologists to request them. The same sorts of considerations apply to specimens collected from the female genital tract; although a 'perineal wipe' is no substitute for a high vaginal swab, samples collected carefully under vision from the cervix or the vaginal vault are entirely satisfactory clinical specimens for anaerobic study.

Because many non-sporing anaerobes are intolerant of oxygen it is important that specimens in which they are to be sought are exposed to the atmosphere as little as possible. Close liaison between the clinician and the microbiologist helps to ensure that the appropriate specimen, freshly collected, is transmitted to the laboratory without delay for immediate culture. Many workers consider that special precautions should be taken to protect 'anaerobic' bacteriological specimens from the toxic effects of oxygen, and a variety of specimen transport systems have been designed for this purpose. These include double stoppered tubes containing oxygen-free carbon dioxide or nitrogen (gassed-out tubes) into which fluid specimens are injected at the bedside, pre-reduced and anaerobically sterilized (PRAS) semisolid transport medium for the transport of swabs, and a disposable 'mini' anaerobic jar for the transport of samples of tissue (Attebery and Finegold, 1969, 1970; Sutter *et al.,* 1972; Holdeman and Moore, 1975; Wilkins and Jimenez-Ulate, 1975). There is, however, some evidence to suggest that many exacting anaerobes will tolerate at least short periods of exposure to atmospheric oxygen, and in my own experience there is rarely much difficulty in recovering exacting species from pathological material on swabs provided that the interval between collection of the specimen and its culture is short – say 20–30 min (Collee *et al.*, 1974; Tally *et al.*, 1975a; Yrios *et al.*, 1975; Justesen and Nielsen, 1976). Alternatively, swabs may be sent to the laboratory in a non-nutrient transport medium, such as Stuart's medium, provided that the swab is not held longer than about 4 h. A successful transport system implies that specimens at culture contain the same viable micro-organisms in the same proportions as in the infected lesion; thus, the viability of oxygen-intolerant species must be maintained, and their overgrowth by more robust anaerobes, or by facultative anaerobes must be prevented. Under no circumstances should specimens be transported in nutrient media.

Quite the best specimens for anaerobic bacteriological examination are volumes of pus or other exudate, and pieces of tissue. Exudates are conveniently collected and transported in disposable plastic syringes; soiled dressings from infected wounds may be sealed in plastic bags; samples of excised tissue should be placed in small sterile screw-capped bottles. Early culture of these materials is not so critical as for swabs (Bartlett *et al.*, 1976), but should, nevertheless, be effected without undue delay.

Specimens that must be examined for non-clostridial anaerobes include material from most infective lesions of the gastrointestinal, female genital and lower respiratory tracts; blood for culture; pus from any deep-seated abscess; peritoneal and pleural exudate; joint effusions; middle ear infections; chronic sinus and mastoid infections; and cerebrospinal fluid in cases of meningitis, especially those associated with middle-ear infections, brain abscess and fractured base of skull. A surprising number of superficial abscesses and other infections commonly contain anaerobes – breast abscess, axillary abscess, ischiorectal abscess, infected pilonidal sinus, paronychia, infected abdominal and episiotomy wounds and so on. In addition, any specimen that is sterile or shows only a minimal aerobic flora without a satisfactory explanation is suspect.

Direct microscopy

Bacteroides and related organisms are often easy to recognize in Gram-stained films of pathological material. They may present as small faintly stained Gram-negative bacilli, not dissimilar in appearance from *Haemophilus influenzae*, as fusiform-shaped rods or spheroids. Marked pleomorphism and irregular staining are characteristic of the Gram-negative anaerobes. It is worth remembering, however, that Gram-negative organisms may sometimes not be seen in direct films of specimens, even when present in large numbers.

Reference has been made elsewhere to the characteristic microscopic appearances of Vincent's stomatitis, anaerobic streptococcal myositis and Meleney's synergistic gangrene.

Ultraviolet light fluorescence

Pathological specimens that contain *B. melaninogenicus* commonly show a brick-red fluorescence under ultraviolet light (Myers *et al.*, 1969). All suspect swabs, exudates, samples of pus, wound dressings and pieces of tissue should be 'screened' under a Wood's lamp.

Direct gas liquid chromatography

Gas liquid chromatographic analysis of samples of pus provides a rapid (30 min) and reliable means for the presumptive differentiation of anaerobic from aerobic infections (Phillips, Tearle and Willis, 1976; Gorbach *et al.*, 1976). An aliquot of the specimen is processed in the same way as a bacterial culture: 1 ml of the sample is acidified with a few drops of 50% sulphuric acid and the volatile acids extracted in 1 ml of diethyl ether. One μl of the ether layer is withdrawn for injection into the chromatograph. The detection of volatile fatty acids other than acetic acid is strong presumptive evidence of the presence of significant numbers of anaerobes in the specimen. Exudates from sepsis due to facultatively anaerobic bacteria only, contain no volatile acids or acetic acid only.

Culture

Although non-sporing anaerobes are not infrequently mixed with facultative anaerobes in clinical material, there is rarely any difficulty in determining their likely significance. This is accomplished by conventional methods of direct plating of the specimen on to media for anaerobic and aerobic incubation, and for incubation in air plus 10% carbon dioxide. These cultures are subsequently examined for the identity of isolates and are compared with one another for the relative proportions of aerobic, facultatively anaerobic, anaerobic and carbon dioxide-dependent growth. For the purposes of making this 'value judgment', selective agents should not be added to plating media, and enrichment cultures should not be used. Selective and enrichment techniques may be used subsequently for purification of cultures prior to specific identification. As with the clostridia, the mere presence of non-sporing anaerobes in a lesion does not necessarily imply that they are causing an infection.

Any one of a number of cultural methods may be used successfully for the isolation of non-clostridial anaerobes from clinical material (*see for example* Sutter *et al.,* 1972; Ellner, Granato and May, 1973; Sutter and Finegold, 1973; Dowell and Hawkins, 1974; Goldsand, 1974; Spaulding, 1974; Martin, 1974). These vary in their complexity, and it is important that the method chosen by any particular laboratory is one that can be fitted into the normal daily routine. Sophisticated methods are not necessarily the best; and the best methods may not take account of other laboratory commitments.

Standard horse blood agar prepared from a good quality basal medium is appropriate for the primary isolation of non-clostridial anaerobes. Plates should be freshly prepared, or prereduced by storage in an anaerobic jar until required (*see also* Martin, 1971). Since the growth of many anaerobes is improved by the addition to the medium of a reducing agent such as cysteine hydrochloride, it is a sound practice to do this routinely. Although some strains of anaerobes, notably *B. melaninogenicus,* require vitamin K and/or haemin as growth factors, these organisms can usually be cultured from pathological material without difficulty on the routine horse blood agar medium. However, it is usually necessary to add these growth factors to plating media for the subsequent isolation of these organisms in pure culture.

Each specimen is inoculated onto three horse blood agar plates, for incubation in the three different atmospheres – aerobic, anaerobic with 10% carbon dioxide and aerobic with 10% carbon dioxide. It is convenient to include a fourth culture on a blood agar plate containing 100 μg/ml neomycin sulphate for anaerobic incubation; this culture facilitates the subsequent isolation of anaerobes in pure culture by suppressing the growth of aerobic organisms.

As a guide to the differentiation of the growth of facultative anaerobes from that of obligate anaerobes, metronidazole and gentamicin discs may be placed on the non-selective blood agar plate culture for anaerobic incubation. After incubation, a zone of inhibition about the metronidazole disc indicates the presence of obligate anaerobes; inhibition of growth about the gentamicin disc indicates the presence of facultatively anaerobic bacteria.

All cultures are examined at the end of 24 and 48 h incubation; the incubation of primary anaerobic plates is continued for 96 h. Recognition of the growth of obligate anaerobes is facilitated by direct comparison of the aerobic and anaerobic cultures, and by noting the effects of gentamicin and metronidazole on the anaerobic plate. Carbon dioxide-dependent facultative organisms (sometimes incorrectly referred to as 'microaerophilic'), show growth on plates incubated anaerobically and in air plus carbon dioxide, but not on those incubated in the absence of added carbon dioxide. Carbon dioxide-dependent facultatively anaerobic species are uniformly resistant to metronidazole, but show variable sensitivity to gentamicin.

The presence of *B. melaninogenicus* is often overlooked if cultures are discarded too early, since pigmentation of colonies may not become obvious until the third day or later. Not infrequently young colonies of *B. melaninogenicus* (24 h) fluoresce brick-red in ultraviolet light; some colonies that fail to fluoresce *in situ* produce a red fluorescent solution when extracted with methanol. Fluorescent colonies of *B. melaninogenicus* may induce fluorescence in adjacent colonies of other bacterial species.

The extent to which specific identification of anaerobic isolates is undertaken is governed not only by the requirements of the particular clinical case, but also by the facilities of the investigating laboratory. Every clinical laboratory should be capable of broadly interpreting primary anaerobic plate cultures, and most should have little difficulty in identifying such common isolates as *B. fragilis* and *B. melaninogenicus.* The systematics of the non-clostridial anaerobes are considered in Chapter 5.

Laboratory workers should not lose sight of the fact that the pathological specimen represents an ill patient, whose clinical management and progress may be determined, not only by the results of the bacteriological investigations, but also by the speed with which these findings are transmitted to the clinician. Delays in reaching a bacteriological 'diagnosis in detail' are especially likely to occur with some anaerobic bacteria, because of the special difficulties of purification and identification. It is essential, therefore, that interim reports be issued, and these should contain both the factual bacteriological findings, and the bacteriologist's interpretation of their meaning in the particular clinical setting. Thus, a tentative *clinical* diagnosis of 'bacteroides sepsis' may be supported by the appearance of a Gram-stained film, or confirmed by gas liquid chromatographic analysis of pus, and direct films may be diagnostic in such conditions as acute ulcerative stomatitis and anaerobic streptococcal myositis. Most anaerobic cultures can be interpreted within 24 h, although it may take some days to attach specific names to isolated strains; the early 'interpretation' is of much greater clinical value than a definitive taxonomic report five days later.

ACTINOMYCOSIS

Human actinomycosis is an uncommon condition. It is caused most commonly by *Actinomyces israelii*, one of a number of *Actinomyces* species whose major natural habitat is the human mouth (Emmons, 1937; Sullivan and Goldsworthy, 1940; Slack, 1942; Batty, 1958; Howell *et al.*, 1959; Howell, Stephan and Paul, 1962; Snyder, Bullock and Parker, 1967; Baboolal, 1968). Other oral species include *A. naeslundii* and *A. odontolyticus.* The infection is almost always endogenously derived since the organism is an obligate parasite of man. Direct infection from person to person is very rare, but has been reported to occur in wounds caused by human bites (Cope, 1938).

About 60% of actinomycotic infections in man occur in the cervicofacial region, about 20% in the abdomen, 15% in the thorax and the remainder in the skin and other parts (Cope, 1938, 1952; Bronner and Bronner, 1971; Brown, 1973; Weese and Smith, 1975). Although *A. israelii* is by far the commonest cause of actinomycosis, other species have occasionally been implicated in human infections (*see* Georg, 1974). Thus, *A. eriksonii* was isolated from a brain abscess by Georg *et al.* (1965), and *A. naeslundii* was isolated by Coleman, Georg and Rozzell (1969) from a variety of lesions in man. *Arachnia propionica* (Buchanan and Pine, 1962), an organism closely related to the *Actinomyces,* is also the cause of actinomycotic-type infections in man (Brock *et al.*, 1973).

The pathogenisis of actinomycosis is incompletely understood. It seems probable that poor oral hygiene and necrotic foci about the teeth and gums would favour infection, and it is clear that minor injury to the oral mucosa is an important predisposing factor; many cases have developed following tooth extraction and other injuries to the mouth and throat – the disease has on occasion developed directly after tonsillectomy (Cope, 1938). *A. israelii* is rarely present as the sole organism in actinomycotic lesions, although it seems unlikely that accompanying species such as *B. melaninogenicus,* anaerobic cocci and facultative oral organisms are necessarily pathogenetically related to the infection. It is clear, however, that actinomycotic lesions may become secondarily infected or may develop concomitantly with pyogenic sepsis. 'Through a small superficial abrasion of the mucous membrane of the mouth there may enter both streptococci and actinomycetes; the immediate reaction may resemble . . . an acute cervical cellulitis . . . When the observer thinks that the condition is cured he may be surprised to find the appearance of a subacute or chronic inflammatory lesion which proves to be actinomycosis' (Cope, 1938).

Actinomycosis is characterized by the development of indurated swellings arising chiefly in connective tissue, by limited suppuration and by the development of ramifying sinuses which tend to heal and re-form elsewhere. Typically, the pus which is discharged from the lesions contains small 'sulphur granules' which are colonies of the infecting organism. Local extension of the infection is the rule;

occasionally it is disseminated by the blood stream, but lymphatic spread practically never occurs.

As already noted, cervicofacial actinomycosis is quite the commonest form of the disease, and has been extensively reported upon in the literature (*see for example* Bramley and Orton, 1960; Mitchell, 1966; Rud, 1967; Bronner and Bronner, 1971; Goldstein, Sciubba and Laskin, 1972). Osborn and Roydhouse (1976) consider that *A. israelii* is a dominant aetiological factor in most cases of chronic 'irreversible' tonsillitis. Among the patients they reviewed, no less than 90% had more than 10 colonies of *A. israelii* in one or both tonsils; and colony counts in excess of 100 were not uncommon.

In abdominal actinomycosis the organism is presumed to gain access from the lumen of the intestine although there appears to be no published information on the presence of actinomyces in the intestine. Ileo-caecal actinomycosis may develop as a sequel to perforation of the appendix, and abdominal lesions may follow surgery, accidental trauma or perforation of a peptic ulcer. Rectal and gastric actinomycosis are rare (Manheim *et al.*, 1969; Eastridge *et al.*, 1972; Brown, 1973; Weese and Smith, 1975).

Thoracic actinomycosis appears to result from the inhalation or aspiration of infected material; there may be involvement, not only of the lung parenchyma, but also of the pleural space, chest wall, mediastinum and pericardium (Kay, 1947; McQuarrie and Hall, 1968; Prather *et al.*, 1970; Eastridge *et al.*, 1972; Slade, Slesser and Southgate, 1973; Datta and Raff, 1974).

Actinomycosis may occur in virtually any part of the body. Recent publications include reports of actinomycosis of the tympanomastoid (Leek, 1974), infections of the female genital tract, notably in association with intrauterine contraceptice devices (Henderson, 1973; Dische *et al.*, 1974; Schiffer *et al.*, 1975; Wagman, 1975), and infections of the heart and mediastinum (Brown, 1973; Datta and Raff, 1974; Weese and Smith, 1975). Disseminated actinomycosis has been reported upon by Butas *et al.* (1970) and Smith and Lockwood (1975).

As with other types of non-clostridial anaerobic infection, deep seated actinomycosis may pass unrecognized simply because it is forgotten both by clinicians and microbiologists. Clearly, the symptoms and signs of actinomycosis vary according to the anatomical position of the infection, although in all situations the lesions themselves are essentially similar, and are most obviously displayed in cervicofacial infections. Primary actinomycotic appendicitis usually escapes detection at first. The common sequence of events is appendicectomy for chronic or interval appendicitis, after which temporary recovery ensues. This is followed by the development of a chronic abscess or discharging sinus, and later by metastatic liver abscess. Primary pulmonary actinomycosis is commonly mistaken at first for tuberculosis or malignant disease.

Bacteriological diagnosis is readily made by direct microscopy of actinomycotic pus, and may be confirmed later by culture (*see* Georg, 1970). Pus is examined for 'sulphur granules' by mixing it with sterile distilled water and shaking gently in a screw-capped bottle; on standing the granules settle to the bottom. Granules are collected, crushed between two microscope slides and fixed. In Gram-stained films the crushed granule shows a central felted mass of Gram-positive filaments usually surrounded by Gram-negative clubs arranged in a radial fashion. In Ziehl–Neelsen-stained films the peripheral clubs are acid-fast, while the central mycelial mass takes the counter-stain.

Sulphur granules for culture are crushed between two *sterile* slides, emulsified in a little sterile glucose broth and inoculated onto glucose blood agar plates and into VL or thiol broth for incubation in an anaerobic atmosphere containing 10% carbon dioxide. Surface cultures are examined under the plate microscope at 2 days for the presence of 'spider' microcolonies – young heavily fringed colonies that appear as thin masses of tangled filaments, and again at 5–6 days for the older 'bread crumb' or 'molar tooth' mature colonies. Excellent illustrations of these two colonial forms are published by Georg (1974) and Slack and Gerencser (1975). When growth has developed in the enrichment cultures, Gram films are prepared to check for the presence of branching or filamentous growth, and subcultures are made to solid media.

The treatment of actinomycosis consists of a combination of surgical drainage or excision, and prolonged antibiotic therapy. Penicillin is the drug of choice, although tetracycline, erythromycin and lincomycin have all been used successfully (Garrod, 1952; Herrell, Balows and Dailey, 1955; Blake, 1964; Mohr, Rhoades and Muchmore, 1970; Lerner, 1974). A conventional course of therapy consists of procaine penicillin, 600 000 units intramuscularly twice daily for one month, followed for one month by 250 mg penicillin V four times daily. This is an inconvenient regimen for ambulant patients, and it is reasonable to substitute penicillin V 500 mg four times daily for the procaine penicillin. For patients who are allergic to penicillin, oral tetracycline may be given in a dose of 250 mg four times daily for 1–2 months.

REFERENCES

Airan, M. C., Levine, H. D. and Sice, J. (1973). 'Prevention of wound infection.' *Lancet,* **1,** 1058

Alston, J. M. (1955). 'Necrobacillosus in Great Britain.' *British medical Journal,* **2,** 1524

Altemeier, W. A. (1938a). 'The bacterial flora of acute perforated appendicitis with peritonitis.' *Annals of Surgery,* **107,** 517

Altemeier, W. A. (1938b). 'The cause of the putrid odour of perforated appendicitis with peritonitis.' *Annals of Surgery,* **107,** 634

Altemeier, W. A. (1940). 'The anaerobic streptococci in tubo-ovarian abscess.' *American Journal of Obstetrics and Gynecology,* **39,** 1038

Altemeier, W. A. (1974). 'Liver abscess: The etiologic role of anaerobic bacteria.' In *Anaerobic Bacteria: Role in Disease.* p. 387. Ed. by A. Balows, R. M. DeHaan, V. R. Dowell and L. B. Guze. Springfield; Thomas

Altemeier, W. A., Culbertson, W. R., Fullen, W. D. and Shook, C. D. (1973). 'Intra-abdominal abscesses.' *American Journal of Surgery,* **125,** 70

Ament, M. E. and Gaal, S. A. (1967). '*Bacteroides* arthritis.' *American Journal of Diseases of Children,* **114,** 427

Anderson, C. B., Marr, J. M. and Jaffe, B. M. (1972). 'Anaerobic streptococcal infections simulating gas gangrene.' *Archives of Surgery,* **104,** 186

Ansbacher, R., Boyson, W. A. and Morris, J. A. (1967). 'Sterility of the uterine cavity.' *American Journal of Obstetrics and Gynecology,* **99,** 394

Aries, V., Crowther, J. S., Drasar, B. S., Hill, M. J. and Williams, R. E. O. (1969). 'Bacteria and the aetiology of cancer of the large bowel.' *Gut,* **10,** 334

Attebery, H. R. and Finegold, S. M. (1969). 'Combined screw-cap and rubber-stopper closure for Hungate tubes (pre-reduced anaerobically sterilized roll tubes and liquid media).' *Applied Microbiology,* **18,** 558

Attebery, H. R. and Finegold, S. M. (1970). 'A miniature anaerobic jar for tissue transport or for cultivation of anaerobes.' *American Journal of clinical Pathology,* **53,** 383

Attebery, H. R., Sutter, V. L. and Finegold, S. M. (1974). 'Normal human intestinal flora.' In *Anaerobic Bacteria: Role in Disease.* p. 81. Ed. by A. Balows, R. M. DeHaan, V. R. Dowell and L. B. Guze. Springfield; Thomas

Baboolal, R. (1968). 'Identification of filamentous micro-organisms of human dental plaque by immunofluorescence.' *Caries Research,* **2,** 273

Baird, R. M. (1973). 'Postoperative infections from bacteroides.' *American Surgeon,* **39,** 459

Balows, A., DeHaan, R. M., Dowell, V. R. and Guze, L. B. (1974). *Anaerobic Bacteria: Role in Disease.* Springfield; Thomas

Barnes, B. A., Behringer, G. E., Wheelock, F. C. and Wilkins, E. W. (1962). 'Surgical sepsis: Analysis of factors associated with sepsis following appendicectomy (1937–1959).' *Annals of Surgery,* **156,** 703

Bartizal, F. J., Pacheco, J. C., Malkasian, G. D. and Washington, J. A. (1974). 'Microbial flora found in the products of conception in spontaneous abortions.' *Obstetrics and Gynecology,* **43,** 109

Bartlett, J. G. and Finegold, S. M. (1970). 'Clinical features and diagnosis of anaerobic pleuropulmonary infections.' In *Proceedings of the 10th Interscience Conference on Antimicrobial Agents and Chemotherapy, Chicago, 1970.* p. 78. Ed. by G. L. Hobby. Maryland; Bethesda

Bartlett, J. G. and Finegold, S. M. (1972). 'Anaerobic pleuropulmonary infections.' *Medicine, Baltimore,* **51,** 413

Bartlett, J. G. and Finegold, S. M. (1974). 'Anaerobic infections of the lung and pleural space.' *American Review of Respiratory Disease,* **110,** 56

Bartlett, J. G., Gorbach, S. L. and Finegold, S. M. (1974). 'The bacteriology of aspiration pneumonia.' *American Journal of Medicine,* **56,** 202

Bartlett, J. G., Gorbach, S. L., Thadepalli, H. and Finegold, S. M. (1974). 'Bacteriology of empyema.' *Lancet,* **1,** 338

Bartlett, J. G., Louie, T. J., Onderdonk, A. B. and Gorbach, S. L. (1975). 'Experimental intra-abdominal sepsis.' *Gastroenterology*, **68**, 861

Bartlett, J. G., Sullivan-Gigler, N., Louis, T. J. and Gorbach, S. L. (1976). 'Anaerobes survive in clinical specimens despite delayed processing.' *Journal of clinical Microbiology*, **3**, 133

Bartlett, J. G., Sutter, V. L. and Finegold, S. M. (1972). 'Treatment of anaerobic infections with lincomycin and clindamycin.' *New England Journal of Medicine*, **287**, 1006

Bartlett, J. G., Sutter, V. L. and Finegold, S. M. (1974). 'Anaerobic pleuropulmonary disease. Clinical observations and bacteriology in 100 cases.' In *Anaerobic Bacteria: Role in Disease.* p. 327. Ed. by A. Balows, R. M. DeHaan, V. R. Dowell and L. B. Guze. Springfield; Thomas

Bateman, N. T., Eykyn, S. J. and Phillips, I. (1975). 'Pyogenic liver abscess caused by *Streptococcus milleri.*' *Lancet*, **1**, 657

Batty, I. (1958). '*Actinomyces odontolyticus*, a new species of actinomycete regularly isolated from deep carious dentine.' *Journal of Pathology and Bacteriology*, **75**, 455

Beazley, R. M., Polakavetz, S. H. and Miller, R. M. (1972). 'Bacteroides infections on a university surgical service.' *Surgery, Gynecology and Obstetrics*, **135**, 742

Bisset, K. A. and Davis, G. H. G. (1960). *The Microbial Flora of the Mouth.* London; Heywood

Blake, G. C. (1964). 'Sensitivities of colonies and suspensions of *Actinomyces israelii* to penicillins, tetracyclines, and erythromycin.' *British medical Journal*, **1**, 145

Blake, G. C. (1968). 'The microbiology of acute ulcerative gingivitis with reference to the culture of oral trichomonads and spirochaetes.' *Proceedings of the Royal Society of Medicine*, **61**, 131

Blazevic, D. J. and Matsen, J. M. (1974). 'Susceptibility of anaerobic bacteria to carbenicillin.' *Antimicrobial Agents and Chemotherapy*, **5**, 462

Block, M. A., Schuman, B. M., Eyler, W. R., Truant, J. P. and DuSault, L. A. (1964). 'Surgery of liver abscesses.' *Archives of Surgery*, **88**, 602

Bodner, S. J., Koenig, M. G. and Goodman, J. S. (1970). 'Bacteraemic bacteroides infections.' *Annals of internal Medicine*, **73**, 537

Bodner, S. J., Koenig, M. G., Treanor, L. L. and Goodman, J. S. (1972). 'Antibiotic susceptibility testing of *Bacteroides.*' *Antimicrobial Agents and Chemotherapy*, **2**, 57

Bollinger, C. G. (1964). 'Bacterial flora of the non-pregnant uterus: a new culture technique.' *Obstetrics and Gynecology*, **23**, 251

Bonfils-Roberts, E. A., Barone, J. E. and Nealon, T. F. (1975). 'Treatment of subphrenic abscess.' *Surgical Clinics of North America*, **55**, 1361

Bornstein, D. L., Weinberg, A. N., Swartz, M. N. and Kunz, L. J. (1964). 'Anaerobic infections: Review of current experience.' *Medicine, Baltimore*, **43**, 207

Bosio, B. B. and Taylor, E. S. (1973). 'Bacteroides and puerperal infections.' *Obstetrics and Gynecology*, **42**, 271

Bowden, G. H. and Hardie, J. M. (1971). 'Anaerobic organisms from the human mouth.' In *Isolation of Anaerobes.* p. 177. Ed. by D. A. Shapton and R. G. Board. London; Academic Press

Bramley, P. and Orton, H. S. (1960). 'Cervico-facial actinomycosis.' *British dental Journal*, **109**, 235

Brewer, G. E. and Meleney, F. L. (1926a). 'Progressive gangrenous infection of the skin and subcutaneous tissues, following operation for acute perforated appendicitis: a study in symbiosis.' *Annals of Surgery*, **84**, 438

Brewer, G. E. and Meleney, F. L. (1926b). 'Progressive gangrenous infection of the skin and subcutaneous tissues, following operation for acute perforated appendicitis: a study in symbiosis.' *Transactions of the American Surgical Association,* **44,** 389

Brightmore, T. (1972). 'Perianal gas-producing infection of non-clostridial origin.' *British Journal of Surgery,* **59,** 109

British medical Journal (1973). 'Bacteroides bacteraemia.' **1,** 686

British medical Journal (1974). 'Lincomycin and clindamycin colitis.' **4,** 65

British medical Journal, (1976). 'Metronidazole in the prevention and treatment of bacteroides infections following appendicectomy;' **1,** 318 (Study Group)*

Brock, D. W., Georg, L. K., Brown, J. M. and Hicklin, M. D. (1973). 'Actinomycosis caused by *Arachnia propionica:* report of 11 cases.' *American Journal of Clinical Pathology,* **59,** 66

Bronner, M. and Bronner, M. (1971). *Actinomycosis.* 2nd edn. Bristol; Wright

Brorson, J-E., Edmar, G. and Holm, S. E. (1975). 'Some clinical, immunological and bacteriological observations in a case of pyogenic arthritis due to *Bacteroides fragilis.' Scandinavian Journal of infectious Diseases,* **7,** 222

Brown, J. R. (1973). 'Human actinomycosis. A study of 181 subjects.' *Human Pathology,* **4,** 319

Buchanan, B. B. and Pine, L. (1962). 'Characterization of a propionic acid producing actinomycete, *Actinomyces propionicus,* sp. nov.' *Journal of general Microbiology,* **28,** 305

Bullen, C. L., Willis, A. T. and Williams, K. (1973). 'The significance of bifidobacteria in the intestinal tract of infants.' In *Actinomycetales: Characteristics and Practical Importance.* p. 311. Ed. by G. Sykes and F. A. Skinner. London; Academic Press

Burdon, K. L. (1928). '*Bacterium melaninogenicum* from normal and pathologic tissues.' *Journal of infectious Diseases,* **42,** 161

Bushby, S. R. M. (1973). 'Trimethoprim-sulfamethoxazole: *in vitro* microbiological aspects.' *Journal of infectious Diseases,* **128,** S442

Butas, G. A., Read, S. E., Coleman, R. E. and Abramovitch, H. (1970). 'Disseminated actinomycosis.' *Canadian medical Association Journal,* **103,** 1069

Butler, T. J. and McCarthy, C. F. (1969). 'Pyogenic liver abscess.' *Gut,* **10,** 389

Callam, A. and Duff, A. (1941). 'Wound phagedaena. Report of two cases.' *British medical Journal,* **2,** 801

Carter, B. (1963). 'Anaerobic infections in obstetrics and gynaecology.' *Proceedings of the Royal Society of Medicine,* **56,** 1095

Carter, B., Jones, C. P., Alter, R. L., Creadick, R. N. and Thomas, W. L. (1953). '*Bacteroides* infections in obstetrics and gynecology.' *Obstetrics and Gynecology,* **1,** 491

Carter B., Jones, C. P., Ross, R. A. and Thomas, W. L. (1951). 'A bacteriologic and clinical study of pyometrea.' *American Journal of Obstetrics and Gynecology,* **62,** 793

Chow, A. W. and Guze, L. B. (1974). 'Bacteroidaceae bacteremia: Clinical experience with 112 patients.' *Medicine,* **53,** 93

Chow, A. W., Leake, R. D., Yamauchi, T., Anthony, B. F. and Guze, L. B. (1974). 'The significance of anaerobes in neonatal bacteremia: Analysis of 23 cases and review of the literature.' *Pediatrics,* **54,** 736

Chow, A. W., Malkasian, K. L., Marshall, J. R. and Guze, L. B. (1975). 'The bacteriology of acute pelvic inflammatory disease.' *American Journal of Obstetrics and Gynecology,* **122,** 876

* See footnote on p. 206.

Chow, A. W., Marshall, J. R. and Guze, L. B. (1975). 'Anaerobic infections of the female genital tract: prospects and perspectives.' *Obstetrical and Gynecological Survey,* **30,** 477

Chow, A. W., Montgomerie, J. R. and Guze, L. B. (1974). 'Parenteral clindamycin therapy for severe anaerobic infections.' *Archives of internal Medicine,* **134,** 78

Chow, A. W., Patten, V. and Guze, L. B. (1975). 'Comparative susceptibility of anaerobic bacteria to minocycline, doxycycline, and tetracycline.' *Antimicrobial Agents and Chemotherapy,* **7,** 46

Clark, C. E. and Wiersma, A. F. (1952). 'Bacteroides infections of the female genital tract.' *American Journal of Obstetrics and Gynecology,* **63,** 371

Cohen, J. (1932). 'The bacteriology of abscess of the lung and methods for its study.' *Archives of Surgery,* **24,** 171

Colebrook, L. (1930). 'Infection by anaerobic streptococci in puerperal fever.' *British medical Journal,* **2,** 134

Coleman, R. M., Georg, L. K. and Rozzell, A. R. (1969). '*Actinomyces naeslundii* as an agent of human actinomycosis.' *Applied Microbiology,* **18,** 420

Collee, J. G., Watt, B., Brown, R. and Johnstone, S. (1974). 'The recovery of anaerobic bacteria from swabs.' *Journal of Hygiene, Cambridge,* **72,** 339

Cooke, R. W. I. (1975). '*Bacteroides fragilis* septicaemia and meningitis in early infancy.' *Archives of Disease in Childhood,* **50,** 241

Cope, V. Z. (1938). *Actinomycosis.* London; Oxford University Press

Cope, V. Z. (1952). *Human Actinomycosis.* London; Heinemann

Craft, I., Ghandi, F. and Hardy, R. (1974). 'Bacteroides in gynaecological infection.' *Lancet,* **1,** 677

Crawford, J. J., Sconyers, J. R., Moriarty, J. D., King, R. C. and West, J. F. (1974). 'Bacteraemia after tooth extractions studied with the aid of prereduced anaerobically sterilized culture media.' *Applied Microbiology,* **27,** 927

Crawford, J. J. and Shankle, R. J. (1961). 'Application of newer methods to study the importance of root canals and oral microbiota in endodontics.' *Oral Surgery,* **14,** 1109

Dane, T. E. B. and King, E. G. (1976). 'Fatal pseudomembranous enterocolitis following clindamycin therapy.' *British Journal of Surgery,* **63,** 305

Datta, J. S. and Raff, M. J. (1974). 'Actinomycotic pleuropericarditis.' *American Review of Respiratory Disease,* **110,** 338

Davies, A. H., McFadzean, J. A. and Squires, S. (1964). 'Treatment of Vincent's stomatitis with metronidazole.' *British medical Journal,* **1,** 1149

Davies, R. M. (1972). 'General ecology of the commensal microflora of the mouth.' In *Host Resistance to Commensal Bacteria. The Response to Dental Plaque.* Ed. by T. MacPhee. Edinburgh; Churchill-Livingstone

DeHaan, R. M., Pfeifer, R. T. and Schellenberg, D. (1974). '*In vitro* susceptibility of anaerobic isolates with reference to selection of therapeutic agents.' In *Infectious Disease Reviews.* Vol. 3, p. 37. Ed. by W. J. Holloway, New York; Futura

DeHaan, R. M., Schellenberg, D. and Pfeifer, R. T. (1974). 'Bacterial etiology of some common anaerobic infections.' In *Infectious Disease Reviews.* Vol. 3, p. 59. Ed. by W. J. Holloway: New York; Futura

Dische, F. E., Burt, J. M. Davidson, N. J. H. and Puntambekar, S. (1974). 'Tubo-ovarian actinomycosis associated with intrauterine contraceptive devices.' *Journal of Obstetrics and Gynaecology of the British Commonwealth,* **81,** 724

Donaldson, R. M. (1964). 'Normal bacterial population of the intestine and relation to intestinal function.' *New England Journal of Medicine,* **270,** 938, 994, 1050

Dornbusch, K. and Nord, C-E. (1974). '*In vitro* effect of metronidazole and tinidazole on anaerobic bacteria.' *Medical Microbiology and Immunology,* **160,** 265

Dornbusch, K., Nord, C-E. and Wadström, T. (1974). 'Biochemical characterization and *in vitro* determination of antibiotic susceptibility of clinical isolates of *Bacteroides fragilis.' Scandinavian Journal of infectious Diseases,* **6,** 253

Douglas, B. and Vesey, B. (1975). '*Bacteroides*: A cause of residual abscess?' *Journal of Pediatric Surgery,* **10,** 215

Douglas, R. G. and Davis, I. F. (1946). 'Puerperal infection – etiology, prophylactic and therapeutic considerations.' *American Journal of Obstetrics and Gynecology,* **51,** 352

Dowell, V. R. and Hawkins, T. M. (1974). *Laboratory Methods in Anaerobic Bacteriology; C.D.C. Laboratory Manual.* Atlanta; US Department of Health, Education, and Welfare

Drasar, B. S. (1974). 'Some factors associated with geographical variations in the intestinal microflora.' In *The Normal Microbial Flora of Man.* p. 187. Ed. by F. A. Skinner and J. G. Carr. London; Academic Press

Drasar, B. S., Shiner, M. and McLeod, G. M. (1969). 'Studies on the intestinal flora. I. The bacterial flora of the gastrointestinal tract in healthy and achlorhydric persons.' *Gastroenterology,* **56,** 71

Drug and Therapeutics Bulletin (1976). 'Metronidazole in anaerobic infections.' **14,** 25

Dunkle, L. M., Brotherton, T. J. and Feigin, R. D. (1976). 'Anaerobic infections in children: A prospective study.' *Pediatrics,* **57,** 311

Eastridge, C. E., Prather, J. R., Hughes, F. A., Young, J. M. and McCaughan, J. J. (1972). 'Actinomycosis: A 24-year experience.' *Southern medical Journal,* **65,** 839

Eggarth, A. H. and Gagnon, B. H. (1933). 'The bacteroides of human feces.' *Journal of Bacteriology,* **25,** 389

Ellner, P. D., Granato, P. A. and May, C. B. (1973). 'Recovery and identification of anaerobes: a system suitable for the routine clinical laboratory.' *Applied Microbiology,* **26,** 904

Ellner, P. D. and Wasilauskas, B. L. (1971). 'Bacteroides septicemia in older patients.' *Journal of the American Geriatrics Society,* **19,** 296

Emmons, C. W. (1937). 'Microaerophilic strains of *Actinomyces* isolated from tonsils.' *Mycologia,* **29,** 377

Emslie, R. D. (1963). 'Cancrum oris.' *Dental Practitioner,* **13,** 481

Felner, J. M. (1974). 'Infective endocarditis caused by anaerobic bacteria. In *Anaerobic Bacteria: Role in Disease.* p. 345. Ed. by A. Balows, R. M. DeHaan, V. R. Dowell and L. B. Guze. Springfield; Thomas

Felner, J. M. and Dowell, V. R. (1970). 'Anaerobic bacterial endocarditis.' *New England Journal of Medicine,* **283,** 1188

Felner, J. M. and Dowell, V. R. (1971). ' "Bacteroides" bacteremia.' *American Journal of Medicine,* **50,** 787

Ferguson, I. R. and Smith, L. L. (1975). '*Bacteroides fragilis* and spectinomycin.' *Journal of antimicrobial Chemotherapy,* **1,** 245

Ferguson, I. R. and Smith, L. L. (1976a). 'The susceptibility of *Bacteroides fragilis* to spectinomycin.' In *Chemotherapy, Vol. 1. Clinical Aspects of Infections.* p. 247. Ed. by J. D. Williams and A. M. Geddes. New York; Plenum

Ferguson, I. R. and Smith, L. L. (1976b). '*Bacteroides fragilis* and nitroimidazoles.' *Journal of antimicrobial Chemotherapy,* **2,** 220

Fiedelman, W. and Webb, C. D. (1975). 'Clinical evaluation of carbenicillin in the treatment of infection due to anaerobic bacteria.' *Current therapeutic Research,* **18,** 441

Finch, R. G., Phillips, I. and Geddes, A. M. (1975). 'A clinical, microbiological and toxicological assessment of clindamycin phosphate.' *Journal of antimicrobial Chemotherapy,* **1,** 297

Finegold, S. M. (1969). 'Intestinal bacteria. The role they play in normal physiology, pathologic physiology, and infection.' *Calfornia Medicine,* **110,** 455

Finegold, S. M. (1970). 'Infections due to non-sporing anaerobic bacteria.' *Delaware medical Journal,* **42,** 298

Finegold, S. M. (1974a). 'Intra-abdominal, genitourinary, skin and soft tissue infections due to non-sporing anaerobic bacteria.' In *Infections with Non-sporing Anaerobic Bacteria.* p. 160. Ed. by I. Phillips and M. Sussman. Edinburgh; Churchill-Livingstone

Finegold, S. M. (1974b). 'Anaerobic infections of the lung.' In *Anaerobic Infections.* p. 50. Ed. by J. O. Godden and J. B. R. Duncan. Toronto; Upjohn

Finegold, S. M. (1976). 'Therapy of anaerobic infections.' *Proceedings of the V National Congress of Microbiology, Salamanca, 1975.* p. 483. Barcelona; Zambon

Finegold, S. M., Attebery, H. R. and Sutter, V. L. (1974). 'Effect of diet on human faecal flora: Comparison of Japanese and American diets.' *American Journal of clinical Nutrition,* **27,** 1456

Finegold, S. M. Bartlett, J. G., Chow, A. W., Flora, D. J., Gorbach, S. L., Harder, E. J. and Tally, F. P. (1975). 'Management of anaerobic infections.' *Annals of internal Medicine,* **83,** 375

Finegold, S. M. Harada, N. E. and Miller, L. G. (1966). 'Lincomycin: Activity against anaerobes and effect on normal human fecal flora.' *Antimicrobial Agents and Chemotherapy,* **6,** 659

Finegold, S. M., Harada, N. E. and Miller, L. G. (1967). 'Antibiotic susceptibility patterns as aids in classification and characterization of Gram-negative anaerobic bacilli.' *Journal of Bacteriology,* **94,** 1443

Finegold, S. M., Miller, L. G., Merrill, S. L. and Posnick, D. J. (1965). 'Significance of anaerobic and capnophilic bacteria isolated from the urinary tract.' In *Progress in Pyelonephritis.* p. 159; Ed. by E. H. Kass. Philadelphia; Davis

Finegold, S. M. and Rosenblatt, J. E. (1973). 'Practical aspects of anaerobic sepsis.' *Medicine, Baltimore,* **52,** 311

Finegold, S. M. Rosenblatt, J. E. Sutter, V. L. and Attebery, H. R. (1972). *Scope Monograph on Anaerobic Infections.* Kalamazoo; Upjohn

Finegold, S. M., Sugihara, P. T. and Sutter, V. L. (1971). 'Use of selective media for isolation of anaerobes from humans.' In *Isolation of Anaerobes.* p. 99; Ed. by D. A. Shapton and R. G. Board. London; Academic Press

Finegold, S. M., Sutter, V. L., Boyle, J. D. and Shimada, K. (1970). 'The normal flora of ileostomy and transverse colostomy effluents.' *Journal of infectious Diseases,* **122,** 376

Fishbach, R. S. and Finegold, S. M. (1973). 'Anaerobic prostatic abscess with bacteremia.' *American Journal of clinical Pathology,* **59,** 408

Francis, L. E., de Vries, J. A., Soomsawasdi, P. and Platonow, M. (1962). 'Control of post-extraction bacteraemia.' *Journal of the Canadian dental Association,* 28, 683

Frederick, J. and Braude, A. I. (1974). 'Anaerobic infection of the paranasal sinuses.' *New England Journal of Medicine,* **290,** 135

Fredette, V., Auger, A. and Forget, A. (1961). 'Anaerobic flora of chronic nasal sinusitis in adults.' *Canadian medical Association Journal,* **84,** 164

Freeman, W. A., McFadzean, J. A. and Whelan, J. P. F. (1968). 'Activity of metronidazole against experimental tetanus and gas gangrene.' *Journal of applied Bacteriology,* **31,** 443

Frock, J. le, Ellis, C. A., Klainer, A. S. and Weinstein, L. (1975). 'Transient bacteraemia associated with barium enema.' *Archives of internal Medicine,* **135,** 835

Frock, J. le, Ellis, C. A., Turchik, J. B. and Weinstein, L. (1973). 'Transient bacteraemia associated with sigmoidoscopy.' *New England Journal of Medicine,* **289,** 467

Garrod, L. P. (1952). 'The sensitivity of *Actinomyces israelii* to antibiotics.' *British medical Journal,* **1,** 1263

Geddes, A. M. (1975). 'Bacteroides infections.' In *Medicine,* **3,** *Infectious Diseases Part 3.* p. 101. Ed. by H. Smith and R. T. D. Edmond. London; Medical Education (International)

Gelb, A. F. and Seligman, S. J. (1970). 'Bacteroidaceae bacteremia. Effect of age and focus of infection upon clinical course.' *Journal of the American medican Association,* **212,** 1038

Georg, L. K. (1970). 'Diagnostic procedures for the isolation and identification of the etiologic agents of actinomycosis.' *Proceedings, International Symposium on Mycoses. 1970. Washington, D.C.* p. 71. Scientific Publication No. 205. Pan American Health Organization

Georg, L. K. (1974). 'The agents of human actinomycosis.' In *Anaerobic Bacteria: Role in Disease.* p. 237. Ed. by A. Balows, R. M. DeHaan, V. R. Dowell and L. B. Guze. Springfield; Thomas

Georg, L. K., Robertstad, G. W., Brinkman, S. A. and Hicklin, M. D. (1965). 'A new pathogenic anaerobic *Actinomyces* species.' *Journal of infectious Disease,* **115,** 88

Gibbons, R. J. and Houte, J. van (1975). 'Bacterial adherence in oral microbial ecology.' *Annual Review of Microbiology,* **29,** 19

Gillespie, W. A. (1970). 'The bacteriology of perihepatic and intrahepatic abscess.' *Proceedings of the Royal Society of Medicine,* **63,** 322

Gillespie, W. A. and Guy, J. (1956). 'Bacteroides in intra-abdominal sepsis. Their sensitivity to antibiotics.' *Lancet,* **1,** 1039

Goldring, J., Scott, A., McNaught, W. and Gillespie, G. (1975). 'Prophylactic oral antimicrobial agents in elective colonic surgery.' *Lancet,* **2,** 997

Goldsand, G. (1974). 'The clinical spectrum of bacteroides infections and their treatment.' In *Current Concepts in the Management of Gram-negative Bacterial Infections.* p. 36. Ed. by A. R. Ronald. Amsterdam; Excerpta Medica

Goldstein, B. H., Sciubba, J. J. and Laskin, D. M. (1972). 'Actinomycosis of the maxilla: Review of literature and report of case.' *Journal of oral Surgery,* **30,** 362

Gonzalez, C. L. and Calia, F. M. (1975). 'Bacteriologic flora of aspiration-induced pulmonary infections.' *Archives of internal Medicine,* **135,** 711

Gorbach, S. L. (1971). 'Intestinal microflora.' *Gastroenterology,* **60,** 1110

Gorbach, S. L. and Barlett, J. G. (1974a). 'Anaerobic infections.' *New England Journal of Medicine,* **290,** 1177, 1237, 1289

Gorbach, S. L. and Bartlett, J. G. (1974b). 'Anaerobic infections: Old myths and new realities.' *Journal of infectious Diseases,* **130,** 307

Gorbach, S. L., Mayhew, J. W., Bartlett, J. G., Thadepalli, H. and Onderdonk, A. B. (1976). 'Rapid diagnosis of anaerobic infections by direct gas-liquid chromatography of clinical specimens.' *Journal of clinical Investigation,* **57,** 478

Gorbach, S. L., Menda, K. B., Thadepalli, H. and Keith, L. (1973). 'Anaerobic microflora of the cervix in healthy women.' *American Journal of Obsetetric and Gynecology,* **117,** 1053

Gorbach, S. L. and Tabaqchali, S. (1969). 'Bacteria, bile and the small bowel.' *Gut,* **10,** 963

Gorbach, S. L., Thadepalli, H. and Norsen, J. (1974). 'Anaerobic microorganisms in intraabdominal infections.' In *Anaerobic Bacteria:* Role in Disease. p. 399. Ed. by A. Balows, R. M. DeHaan, V. R. Dowell and L. B. Guze. Springfield; Thomas

Gordon, D. F. and Jong, B. B. (1968). 'Indigenous flora from human saliva.' *Applied Microbiology,* **61,** 428

Gossling, J. and Slack, J. M. (1974). 'Predominant Gram-positive bacteria in human feces: Numbers, variety and persistence.' *Infection and Immunity,* **9,** 719

Grainger, R. W., MacKenzie, D. A. and McLachlin, A. D. (1967). 'Progressive bacterial synergistic gangrene: Chronic undermining ulcer of Meleney.' *Canadian Journal of Surgery,* **10,** 439

Gunn, A. A. (1956). 'Bacteroides septicaemia.' *Journal of the Royal College of Surgeons of Edinburgh,* **2,** 41

Gupta, U., Oumachigui, A. and Hingorani, V. (1973). 'Microbial flora of the vagina with special reference to anaerobic bacteria and mycoplasma.' *Indian Journal of medical Research,* **61,** 1600

Haenel, H. (1970). 'Human normal and abnormal gastrointestinal flora.' *American Journal of clinical Nutrition,* **23,** 1433

Hale, J. E., Perinpanayagam, R. M. and Smith, G. (1976). 'Bacteroides: an unusual cause of breast abscess.' *Lancet,* **2,** 70

Hall, W. L., Sobel, A. I., Jones, C. P. and Parker, R. T. (1967). 'Anaerobic postoperative pelvic infections.' *Obstetrics and Gynecology,* **30,** 1

Hamilton-Miller, J. M. T. (1975). 'Antimicrobial agents acting against anaerobes.' *Journal of antimicrobial Chemotherapy,* **1,** 273

Hardie, J. (1974). 'Anaerobes in the mouth.' In *Infection with Non-spore forming Anaerobic Bacteria.* p. 99. Ed. by I. Phillips and M. Sussman. Edinburgh; Churchill-Livingstone

Hardie, J. M. and Bowden, G. H. (1974). 'The normal microbial flora of the mouth.' In *The Normal Microbial Flora of Man.* p. 47. Ed. by F. A. Skinner and J. G. Carr. London; Academic Press

Hare, R. and Polunin, I. (1960). 'Anaerobic cocci in the vagina of native women in British North Borneo.' *Journal of Obstetrics and Gynaecology of the British Empire,* **67,** 985

Harley, H. R. S. (1955). *Subphrenic Abscess.* Oxford; Blackwell

Harris, J. W. and Brown, J. H. (1927). 'Description of a new organism that may be a factor in the causation of puerperal infection.' *Bulletin of the Johns Hopkins Hospital,* **40,** 203

Harrod, J. R. and Stevens, D. A. (1974). 'Anaerobic infections in the new-born infant.' *Journal of Pediatrics,* **85,** 399

Hayward, N. J. and Pilcher, R. (1945). 'Anaerobic streptococcal myositis.' *Lancet,* **2,** 560

Headington, J. T. and Beyerlein, B. (1966). 'Anaerobic bacteria in routine urine culture.' *Journal of clinical Pathology,* **19,** 573

Heineman, H. S. and Braude, A. I. (1963). 'Anaerobic infection of the brain: Observations on eighteen consecutive cases of brain abscess.' *American Journal of Medicine,* **35,** 682

Heineman, H. S., Braude, A. I. and Osterholm, J. L. (1971). 'Intracranial suppurative disease.' *Journal of the American medical Association,* **218,** 1542

Henderson, S. R. (1973). 'Pelvic actinomycosis associated with an intrauterine device.' *Obstetrics and Gynecology,* **41,** 726

Herrell, W. E., Balows, A. and Dailey, J. S. (1955). 'Erythromycin in treatment of actinomycosis.' *Antibiotic Medicine,* **2,** 507

Herter, F. P. (1972). 'Preparation of the bowel for surgery.' *Surgical Clinics of North America,* **52,** 859

Hill, A. M. (1963). 'Puerperal and abortal infections.' In *Chemotherapy with Antibiotics and Allied Drugs.* p. 94. 2nd Edn. Ed. by J. C. Tolhurst, G. Buckle, and S. W. Williams. Canberra; National Health and Medical Research Council

Hite, K. E., Hesseltine, H. C. and Goldstein, L. (1947). 'A study of the bacterial flora of the normal and pathologic vagina and uterus.' *American Journal of Obstetrics and Gynecology,* **53,** 233

Hoffmann, K. and Gierhake, F. W. (1969). 'Postoperative infection of wounds by anaerobes.' *German medical Monthly,* **14,** 31

Holdeman, L. V., Good, I. J. and Moore, W. E. C. (1976). 'Human fecal flora: Variation in bacterial composition within individuals and a possible effect of emotional stress.' *Applied and environmental Microbiology,* **31,** 359

Holdeman, L. V. and Moore, W. E. C. (1975). *Anaerobe Laboratory Manual.* Blacksburg; Virginia Polytechnic Institute Anaerobe Laboratory

Houte, J. van and Gibbons, R. J. (1966). 'Studies of the cultivable flora of normal human feces.' *Antonie van Leeuwenhoek,* **32,** 212

Howell, A., Murphy, W. C., Paul, F. and Stephan, R. M. (1959). 'Oral strains of *Actinomyces.*' *Journal of Bacteriology,* **78,** 82

Howell, A., Stephan, R. M. and Paul, F. (1962). 'Prevalence of *Actinomyces israelii, A. naeslundii, Bacterionema matruchottii* and *Candida albicans* in selected areas of the oral cavity and saliva.' *Journal of dental Research,* **41,** 1050

Hurley, R., Stanley, V. C., Leask, B. G. S. and Louvois, J. de (1974). 'Microflora of the vagina during pregnancy.' In *The Normal Microbial Flora of Man.* p. 155. Ed. by F. A. Skinner and J. G. Carr. London; Academic Press

Ingham, H. R., Rich, G. E., Selkon, J. B., Hale, J. H., Roxby, C. M., Betty, M. J., Johnson, R. W. G. and Uldall, P. R. (1975a). 'Treatment with metronidazole of three patients with serious infections due to *Bacteroides fragilis.*' *Journal of antimicrobial Chemotherapy,* **1,** 235

Ingham, H. R., Selkon, J. B., Codd, A. A. and Hale, J. H. (1968). 'A study *in vitro* of the sensitivity to antibiotics of *Bacteroides fragilis.*' *Journal of clinical Pathology,* **21,** 432

Ingham, H. R., Selkon, J. B., Codd, A. A. and Hale, J. H. (1970). 'The effect of carbon dioxide on the sensitivity of *Bacteroides fragilis* to certain antibiotics *in vitro.*' *Journal of clinical Pathology,* **23,** 254

Ingham, H. R., Selkon, J. B. and Hale, J. H. (1975). 'The antibacterial activity of metronidazole.' *Journal of antimicrobial Chemotherapy,* **1,** 355

Ingham, H. R., Selkon, J. B., So, S. C. and Weiser, R. (1975b). 'Brain abscess.' *British medical Journal,* **4,** 39

Jones, C. P., Carter, F. B., Thomas, W. L., Peate, C. H. and Cherny, W. L. (1959). 'Non-spore forming anaerobic bacteria of the vagina.' *Annals of the New York Academy of Science,* **83,** 259

Journal of antimicrobial Chemotherapy (1975). 'An evaluation of metronidazole in the prophylaxis and treatment of anaerobic infections in surgical patients.' **1**, 393 (Study Group)*

Justesen, T. and Neilsen, M. L. (1976). 'Survival of anaerobic bacteria during transportation. 1. Experimental investigations on the effect of evacuation of atmospheric air by flushing with carbon dioxide and nitrogen.' *Acta Pathologica et Microbiologica Scandinavica,* **84**, 51

Justesen, T., Nielsen, M. L. and Hattel, T. (1973). 'Anaerobic infections in chronic prostatitis and chronic urethritis.' *Medical Microbiology and Immunilogy,* **158**, 237

Kagnoff, M. F., Armstrong, D. and Blevins, A. (1971). 'Bacteroides bacteremia. Experience in a hospital for neoplastic diseases.' *Cancer,* **29**, 245

Kamme, C., Lidgren, L., Lindberg, L. and Mardh, P-A. (1974). 'Anaerobic bacteria in late infections after total hip arthroplasty.' *Scandinavian Journal of infectious Diseases,* **6**, 161

Kantz, W. E. and Henry, C. A. (1974). 'Isolation and classification of anaerobic bacteria from intact pulp chambers of non-vital teeth in man.' *Archives of oral Biology,* **19**, 91

Kay, E. B. (1947). 'Bronchopulmonary actinomycosis.' *Annals of internal Medicine,* **26**, 581

Khairat, O. (1966a). 'The non-aerobes of post-extraction bacteraemia.' *Journal of dental Research,* **45**, 1191

Khairat, O. (1966b). 'An effective antibiotic cover for the prevention of endocarditis following dental and other post-operative bacteraemias.' *Journal of clinical Pathology,* **19**, 561

Kislak, J. W. (1972). 'The susceptibility of *Bacteroides fragilis* to 24 antibiotics.' *Journal of infectious Diseases,* **125**, 295

Kline, B. S. and Berger, S. S. (1935). 'Pulmonary abscess and pulmonary gangrene.' *Archives of internal Medicine,* **56**, 753

Kumazawa, J., Kiyohara, H., Narahashi, K., Hidaka, M. and Momose, S. (1974). 'Significance of anaerobic bacteria isolated from the urinary tract. 1. Clinical studies.' *Journal of Urology,* **112**, 257

Lancet (1970). 'Wound infection after appendicectomy.' **1**, 930

Lancet (1971). 'Sepsis after appendicectomy.' **2**, 195

Lancet (1973a). 'Synergistic bacterial gangrene.' **2**, 3

Lancet (1973b). 'Bacteroides in the blood.' **1**, 27

Lancet (1974). 'Metronidazole in the prevention and treatment of bacteroides infections in gynaecological patients.' **2**, 1540 (Study Group)*

Lancet (1976a). 'Anaerobes in pleuropulmonary infections.' **1**, 289

Lancet (1976b). 'Rifampicin: For tuberculosis only?' **1**, 291

Langston, H. H. (1938). 'Unusual case of progressive streptococcal ulceration.' *British Journal of Surgery,* **26**, 254

Leake, D. L. (1972). 'Bacteroides osteomyelitis of the mandible.' *Oral Surgery,* **34**, 585

Ledger, W. J. (1972). 'Infections in obstetrics and gynecology. New developments in treatment.' *Surgical Clinics of North America,* **52**, 1447

Ledger, W. J. (1975). 'Anaerobic infections.' *American Journal of Obstetrics and Gynecology,* **123**, 111

Ledger, W. J., Campbell, C. C. and Willson, J. R. (1968). 'Postoperative adnexal infections.' *Obstetrics and Gynecology,* **31**, 83

*see footnote on p. 206.

Ledger, W. J., Norman, M., Gee, C. and Lewis, W. (1975). 'Bacteremia on an obstetric-gynecologic service.' *American Journal of Obstetrics and Gynecology,* **121,** 205

Ledger, W. J., Sweet, R. L. and Headington, J. T. (1971). 'Bacteroides species as a cause of severe infections in obstetric and gynecologic patients.' *Surgery, Gynecology and Obstetrics,* **133,** 837

Ledingham, I. McA. (1975). 'Septic shock.' *British Journal of Surgery,* **62,** 777

Ledingham, I. McA. and Tehrani, M. A. (1975). 'Diagnosis, clinical course and treatment of acute dermal gangrene.' *British Journal of Surgery,* **62,** 364

Leek, J. H. (1974). 'Actinomycosis of the tympanomastoid.' *Laryngoscope, St. Louis,* **84,** 290

Lees, A. W. and McNaught, W. (1959). 'Bacteriology of lower-respiratory secretions, sputum, and upper-respiratory tract secretions in "normals" and chronic bronchitics.' *Lancet,* **2,** 1112

Leigh, D. A. (1974). 'Clinical importance of infections due to *Bacteroides fragilis* and role of antibiotic therapy.' *British medical Journal,* **3,** 225

Leigh, D. A. (1975). 'Wound infections due to *Bacteroides fragilis* following intestinal surgery.' *British Journal of Surgery,* **62,** 375

Leigh, D. A. and Simmons, K. (1975). 'Activity of minocycline against *Bacteroides fragilis.*' *Lancet,* **1,** 51

Leigh, D. A., Simmons, K. and Norman, E. (1974). 'Bacterial flora of the appendix fossa in appendicitis and postoperative wound infections.' *Journal of clinical Pathology,* **27,** 997

Lemierre, A. (1936). 'On certain septicaemias due to anaerobic organisms.' *Lancet,* **1,** 701

Lerner, P. I. (1974). 'Susceptibility of pathogenic actinomycetes to antimicrobial compounds.' *Antimicrobial Agents and Chemotherapy,* **5,** 302

Levison, M. E. (1973). 'The importance of anaerobic bacteria in infectious diseases.' *Medical Clinics of North America,* **57,** 1015

Lewis, J. F., Johnson, P. and Miller, P. (1976). 'Evaluation of amniotic fluid for aerobic and anaerobic bacteria.' *American Journal of clinical Pathology,* **65,** 58

Lewis, K. H. and Rettger, L. F. (1940). 'Non-sporulating anaerobic bacteria of the intestinal tract. I. Occurrence and toxonomic relationships.' *Journal of Bacteriology,* **40,** 287

Lifshitz, F., Liu, C. and Thurn, A. N. (1963). 'Bacteroides meningitis.' *American Journal of Diseases of Children,* **105,** 487

Lindell, T. D., Fletcher, W. S. and Krippaehne, W. W. (1973). 'Anorectal suppurative disease.' *American Journal of Surgery,* **125,** 189

Linscheid, R. L. and Dobyns, J. H. (1975). 'Common and uncommon infections of the hand.' *Orthopedic Clinics of North America,* **6,** 1063

Loesche, W. J. (1974). 'Dental infections.' In *Anaerobic Bacteria: Role in Disease.* p. 409. Ed. by A. Balows, R. M. DeHaan, V. R. Dowell and L. B. Guze. Springfield; Thomas

Lorber, B. and Swenson, R. M. (1975). 'The bacteriology of intra-abdominal infections.' *Surgical Clinics of North America,* **55,** 1349

Louvois, J. de, Hurley, R. and Stanley, V. C. (1975). 'Microbial flora of the lower genital tract during pregnancy: relationship to morbidity.' *Journal of clinical Pathology,* **28,** 731

Louvois, J. de, Stanley, V. C., Hurley, R., Jones, J. B. and Foulkes, J. E. B. (1975). 'Microbial ecology of the female lower genital tract during pregnancy.' *Postgraduate medical Journal,* **51,** 156

Lowry, K. F. and Curtis, G. M. (1947). 'Diagnosis of clostridial myositis.' *American Journal of Surgery,* **74**, 752

Lozdan, J., Sheiham, A., Pearlman, B. A., Keiser, B., Rachanis, C. C. and Meyer, R. (1971). 'The use of nitrimidazine in the treatment of acute ulcerative gingivitis. A double-blind controlled trial.' *British dental Journal,* **130**, 294

MacDonald, J. B., Hare, G. C. and Wood, A. W. S. (1957). 'The bacteriologic status of the pulp chambers in intact teeth found to be non-vital following trauma.' *Oral Surgery,* **10**, 318

MacKenzie, I. and Litton, A. (1974). 'Bacteroides bacteraemia in surgical patients.' *British Journal of Surgery,* **61**, 288

McLaughlin, P., Meban, S. and Thompson, W. G. (1973). 'Anaerobic liver abscess complicating radiation enteritis.' *Canadian medical Association Journal,* **108**, 353

MacLennan, J. D. (1943). 'Streptococcal infection of muscle.' *Lancet,* **1**, 582

MacLennan, J. D. (1962). 'The histotoxic clostridial infections of man.' *Bacteriological Reviews,* **26**, 232

McQuarrie, D. G. and Hall, W. H. (1968). 'Actinomycosis of the lung and chest wall.' *Surgery,* **64**, 905

Magarey, C. J., Chant, A. D. B., Rickford, C. R. K. and Magarey, J. R. (1971). 'Peritoneal drainage and systemic antibiotics after appendicectomy.' *Lancet,* **2**, 179

Maier, H. C. and Grace, E. J. (1942). 'Putrid empyema.' *Surgery, Gynecology and Obstetrics,* **74**, 69

Manheim, S. D., Voleti, C., Ludwig, A. and Jacobson, J. H. (1969). 'Hyperbaric oxygen in the treatment of actinomycosis.' *Journal of the American medical Association,* **210**, 552

Manson, J. D. and Rand, H. (1961). 'Recurrent Vincent's disease. A survey of 61 cases.' *British dental Journal,* **110**, 386

Marcoux, J. A., Zabransky, R. J., Washington, J. A., Wellman, W. E. and Martin, W. J. (1970). 'Bacteroides bacteremia.' *Minnesota Medicine,* **53**, 1169

Marks, G., Rogers, J. F. and Moses, M. L. (1975). 'Clinical effectiveness of intravenous doxycycline: Treatment and prevention of surgical infections of the gastrointestinal tract.' *Current therapeutic Research,* **18**, 460

Martin, W. J. (1971). 'Practical method for isolation of anaerobic bacteria in the clinical laboratory.' *Applied Microbiology,* **22**, 1168

Martin, W. J. (1974). 'Isolation and identification of anaerobic bacteria in the clinical laboratory.' *Mayo Clinic Proceedings,* **49**, 300

Martin, W. J., Gardner, M. and Washington, J. A. (1972). '*In vitro* antimicrobial susceptibility of anaerobic bacteria isolated from clinical specimens.' *Antimicrobial Agents and Chemotherapy,* **1**, 148

Meleney, F. L. (1931). 'Bacterial synergism in disease processes with a confirmation of the synergistic bacterial aetiology of a certain type of progressive gangrene of the abdominal wall.' *Annals of Surgery,* **94**, 961

Meleney, F. L., Friedman, S. T. and Harvey, H. D. (1945). 'The treatment of progressive bacterial synergistic gangrene with penicillin.' *Surgery,* **18**, 423

Mergenhagen, S. E., Thonard, J. C. and Scherp, H. W. (1958). 'Studies on synergistic infections. I. Experimental infections with anaerobic streptococci.' *Journal of infectious Diseases,* **103**, 33

Miraglia, G. J. (1974). 'Pathogenic anaerobic bacteria.' *Critical Reviews in Microbiology,* **3**, 161

Misra, S. S. (1938). 'A note on the predominance of the genus *Bacteroides* in human stools.' *Journal of Pathology and Bacteriology,* **46**, 204

Mitchell, A. A. B. (1973). 'Incidence and isolation of *Bacteroides* species from clinical material and their sensitivity to antibiotics.' *Journal of clinical Pathology,* **26**, 738

Mitchell, A. A. B. and Simpson, R. G. (1973). 'Bacteroides septicaemia.' *Current medical Research and Opinion,* **1**, 385

Mitchell, R. G. (1966). 'Actinomycosis and the dental abscess.' *British dental Journal,* **120**, 423

Mitchiner, P. H. and Cowell, E. M. (1939). *Medical Organisation and Surgical Practice in Air Raids.* London; Churchill

Mitre, R. J. and Rotheram, E. B. (1974). 'Anaerobic septicaemia from thrombophlebitis of the internal jugular vein: Successful treatment with metronidazole.' *Journal of the American medical Association,* **230**, 1168

Mohr, J. A., Rhoades, E. R. and Muchmore, H. G. (1970). 'Actinomycosis treated with lincomycin.' *Journal of the American medical Association,* **212**, 2260

Monif, G. R. G. and Baer, H. (1974). '*In vitro* susceptibility of *Bacteroides* strains to tetracycline, chloramphenicol, doxycycline, and clindamycin.' *Obstetrics and Gynecology,* **43**, 211

Moore, W. E. C., Cato, E. P. and Holdeman, L. V. (1969). 'Anaerobic bacteria of the gastrointestinal flora and their occurrence in clinical infections.' *Journal of infectious Diseases,* **119**, 641

Myers, M. B., Cherry, G., Bornside, B. B. and Bornside, G. H. (1969). 'Ultraviolet red fluorescence of *Bacteroides melaninogenicus.*' *Applied Microbiology,* **17**, 760

Nastro, L. J. and Finegold, S. M. (1972). 'Bactericidal activity of five antimicrobial agents against *Bacteroides fragilis.*' *Journal of infectious Disease,* **126**, 104

Nastro, L. J. and Finegold, S. M. (1973). 'Endocarditis due to anaerobic Gram-negative bacilli.' *American Journal of Medicine,* **54**, 482

Neary, M. P., Allen, J., Okubadejo, O. A. and Payne, D. J. H. (1973). 'Preoperative vaginal bacteria and postoperative infections in gynaecological patients.' *Lancet,* **2**, 1291

Newton, E. J. (1956). 'Haematogenous brain abscess in cyanotic congenital heart disease.' *Quarterly Journal of Medicine,* **25**, 201

Nichols, R. L., Broido, P., Condon, R. E., Gorbach, S. L. and Nyhus, L. M. (1973). 'Effect of preoperative neomycin-erythromycin intestinal preparation on the incidence of infectious complications following colon surgery.' *Annals of Surgery,* **178**, 453

Nichols, R. L., Miller, B. and Smith, J. W. (1975). 'Septic complications following gastric surgery. Relationship to the endogenous gastric microflora.' *Surgical Clinics of North America,* **55**, 1367

Nobles, E. R. (1973). 'Bacteroides infections.' *Annals of Surgery,* **177**, 601

Okubadejo, O. A. and Allen, J. (1975). 'Combined activity of clindamycin and gentamicin on *Bacteroides fragilis* and other bacteria.' *Journal of Antimicrobial Chemotherapy,* **1**, 403

Okubadejo, O. A., Green, P. J. and Payne, D. J. H. (1973). '*Bacteroides* infection among hospital patients.' *British medical Journal,* **2**, 212

Osborn, G. R. and Roydhouse, N. (1976). *The Tonsillitis Habit.* Brookvale; Symes Publishers

Parker, R. T. and Jones, C. P. (1966). 'Anaerobic pelvic infections and developments in hyperbaric oxygen therapy.' *American Journal of Obstetrics and Gynecology,* **96**, 645

Peach, S., Fernandez, F., Johnson, K. and Drasar, B. S. (1974). 'The non-sporing anaerobic bacteria in human faeces.' *Journal of medical Microbiology,* **7,** 213

Pearson, H. E. (1967). 'Bacteroides in aerolar breast abscesses.' *Surgery, Gynecology and Obstetrics,* **125,** 800

Pearson, H. E. and Anderson, G. V. (1967). 'Perinatal deaths associated with bacteroides infections.' *Obstetrics and Gynecology,* **30,** 486

Pearson, H. E. and Anderson, G. V. (1970a). 'Genital bacteroidal abscesses in women.' *American Journal of Obstetrics and Gynecology,* **107,** 1264

Pearson, H. E. and Anderson, G. V. (1970b). 'Bacteroides infections and pregnancy.' *Obstetrics and Gynecology,* **35,** 31

Pearson, H. E. and Smiley, D. F. (1968). 'Bacteroides in pilonidal sinuses.' *American Journal of Surgery,* **115,** 336

Pecora, D. V. and Yegian, D. (1958). 'Bacteriology of lower respiratory tract in health and chronic disease.' *New England Journal of Medicine,* **258,** 71

Phillips, I. (1974). 'Antibiotic sensitivity of non-sporing anaerobes.' In *Infection with Non-sporing Anaerobic Bacteria.* p. 37. Ed. by I. Phillips and M. Sussman. Edinburgh; Churchill-Livingstone

Phillips, I. and Sussman, M. (1974). *Infection with Non-sporing Anaerobic Bacteria.* Edinburgh; Churchill-Livingstone

Phillips, I. and Warren, C. (1974). 'Susceptibility of *Bacteroides fragilis* to trimethoprim and sulphamethoxazole.' *Lancet,* **1,** 827

Phillips, I. and Warren, C. (1975). 'Susceptibility of *Bacteroides fragilis* to spectinomycin.' *Journal of antimicrobial Chemotherapy,* **1,** 91

Phillips, I. and Warren, C. (1976). 'Activity of sulfamethoxazole and trimethoprim against *Bacteroides fragilis.' Antimicrobial Agents and Chemotherapy,* **9,** 736

Phillips, K. D., Tearle, P. V. and Willis, A. T. (1976). 'Rapid diagnosis of anaerobic infections by gas–liquid chromatography of clinical material.' *Journal of clinical Pathology,* **29,** 428

Pilot, I. and Davis, D. J. (1924). 'Studies of fusiform bacilli and spirochaetes. IX. Their role in pulmonary abscess, gangrene and bronchiectasis.' *Archives of internal Medicine,* **34,** 313

Poindexter, H. A. and Washington, D. (1975). 'Bacteroidosis.' *Southern medical Journal,* **68,** 995

Prather, J. R., Eastridge, C. E., Hughes, F. A. and McCaughan, J. J. (1970). 'Actinomycosis of the thorax. Diagnosis and treatment.' *Annals of thoracic Surgery,* **9,** 307

Prince, H. N., Grunberg, E., Titsworth, E. and DeLorenzo, W. F. (1969). 'Effects of 1-(2-nitro-1 imidazolyl)-3-methoxy-2-propanol and 2-methyl-5-nitroimidazole-1-ethanol against anaerobic and aerobic bacteria and protozoa.' *Applied Microbiology,* **18,** 728

Puhvel, S. M. and Reisner, R. M. (1974). 'Dermatologic anaerobic infections (including acne).' In *Anaerobic Bacteria: Role in Disease.* p. 435. Ed. by A. Balows, R. M. DeHaan, V. R. Dowell and L. B. Guze, Springfield; Thomas

Quayle, A. A. (1974). 'Bacteroides infections in oral surgery.' *Journal of oral Surgery,* **32,** 91

Ralph, E. D. and Kirby, W. M. M. (1975). 'Unique bactericidial action of metronidazole against *Bacteroides fragilis* and *Clostridium perfringens.' Antimicrobial Agents and Chemotherapy,* **8,** 409

Rea, W. J. and Wyrick, W. J. (1970). 'Necrotizing fasciitis.' *Annals of Surgery,* **172,** 957

Rein, J. M. and Cosman, B. (1971). 'Bacteroides necrotizing fasciitis of the upper extremity.' *Plastic and Reconstruction Surgery,* **48,** 592

Reynolds, A. V., Hamilton-Miller, J. M. T. and Brumfitt, W. (1975). 'A comparison of the *in vitro* activity of metronidazole, tinidazole and nimorazole against Gram-negative anaerobic bacilli.' *Journal of clinical Pathology,* **28,** 775

Rissing, J. P., Crowder, J. G., Dunfee, T. and White, A. (1974). 'Bacteroides bacteremia from decubitus ulcers.' *Southern medical Journal,* **67,** 1179

Roberts, D. B. and Hester, L. L. (1972). 'Progressive synergistic bacterial gangrene arising from abscesses of the vulva and Bartholin's gland duct.' *American Journal of Obstetrics and Gynecology,* **114,** 285

Robinow, M. and Simonelli, F. A. (1965). '*Fusobacterium* bacteremia in the new-born.' *American Journal of Diseases of Children,* **110,** 92

Rosebury, T. (1962). *Micro-organisms Indigenous to Man.* New York; McGraw-Hill

Rosenblatt, J. E. and Stewart, P. R. (1974). 'Lack of activity of sulfamethoxazole and trimethoprim against anaerobic bacteria.' *Antimicrobial Agents and Chemotherapy,* **6,** 93

Rotheram, E. B. (1974). 'Septic abortion and related infections of pregnancy.' In *Anaerobic Bacteria: Role in Disease.* p. 369. Ed. by A. Balows, R. M. De Haan, V. R. Dowell and L. B. Guze. Springfield; Thomas

Rotheram, E. B. and Schick, S. F. (1969). 'Non-clostridial anaerobic bacteria in septic abortion.' *American Journal of Medicine,* **46,** 80

Rubinstein, E., Onderdonk, A. B. and Rahal, J. J. (1974). 'Peritonsillar infection and bacteremia caused by *Fusobacterium gonidiaformans.*' *Journal of Pediatrics,* **85,** 673

Rud, J. (1967). 'Cervicofacial actinomycosis.' *Journal of oral Surgery,* **25,** 229

Sabiston, C. B. and Gold, W. A. (1974). 'Anaerobic bacteria in oral infections.' *Oral Surgery,* **38,** 187

Saksena, D. S., Block, M. A., McHenry, M. C. and Truant, J. P. (1968). '*Bacteroidaceae:* Anaerobic organisms encountered in surgical infections.' *Surgery,* **63,** 261

Salem, A. R., Jackson, D. D. and McFadzean, J. A. (1975). 'An investigation of interactions between metronidazole (Flagyl) and other antibacterial agents.' *Journal of antimicrobial Chemotherapy,* **1,** 387

Sanders, D. Y. and Stevenson, J. (1968). 'Bacteroides infections in children.' *Journal of Pediatrics,* **72,** 673

Sandler, B. P. (1965). 'The prevention of cerebral abscess secondary to pulmonary suppuration.' *Diseases of the Chest,* **48,** 32

Sapala, J. A., Ponka, J. L. and Neblett, T. R. (1975). 'The bacteriology of the biliary tract.' *Henry Ford Hospital medical Journal,* **23,** 81

Savage, G. M. (1974). 'Lincomycin and clindamycin: Their role in chemotherapy of anaerobic and microaerophilic infections.' *Infection,* **2,** 152

Schiffer, M. A., Elguezabal, A., Sultana, M. and Allen, A. C. (1975). 'Actinomyosis infections associated with intrauterine contraceptive devices.' *Obstetrics and Gynecology,* **45,** 67

Schoolman, A., Liu, C. and Rodecker, C. (1966). 'Brain abscess caused by *Bacteroides* infection.' *Archives of internal Medicine,* **118,** 150

Schoutens, E. and Yourassowsky, E. (1974). 'Speed of bactericidal action of penicillin G, ampicillin, and carbenicillin on *Bacteroides fragilis.*' *Antimicrobial Agents and Chemotherapy,* **6,** 227

Schreiner, A. Tönjum, S. and Digranes, A. (1974). 'Hyperbaric oxygen therapy in bacteroides infections.' *Acta chirurgica Scandinavica,* **140,** 73

Schwarz, O. H. and Brown, T. K. (1936). 'Puerperal infection due to anaerobic streptococci.' *American Journal of Obstetrics and Gynecology,* **31,** 379

Schwarz, O. H. and Dieckman, W. J. (1927). 'Puerperal infections due to anaerobic streptococci.' *American Journal of Obstetrics and Gynecology,* **13,** 467

Segura, J. W., Kelalis, P. P., Martin, W. J. and Smith, L. H. (1972). 'Anaerobic bacteria in the urinary tract.' *Mayo Clinic Proceedings,* **47**, 30

Sharma, R. K. and Rapkin, R. H. (1974). 'Acute suppurative thyroiditis caused by *Bacteroides melaninogenicus.*' *Journal of the American medical Association,* **229,** 1470

Sharp, P. M., Meador, R. C. and Martin, R. R. (1974). 'A case of mixed anaerobic infection of the jaw.' *Journal of oral Surgery,* **32,** 457

Shimada, K. (1976). 'Anaerobic bacteria in biliary diseases.' *Proceedings of the Fifth National Congress of Microbiology, Salamanca, 1975.* p. 445. Barcelona; Zambon

Shinn, D. L. S. (1962). 'Metronidazole in acute ulcerative gingivitis.' *Lancet,* **1,** 1191

Shinn, D. L. S., Squires, S. and McFadzean, J. A. (1965). 'The treatment of Vincent's disease with metronidazole.' *Dental Practitioner,* **15,** 275

Shull, H. J., Greene, B. M., Allen, S. A., Dunn, G. D. and Schenker, S. (1975). 'Bacteremia with upper gastrointestinal endoscopy.' *Annals of internal Medicine,* **83,** 212

Sims, W. (1974). 'The clinical bacteriology of purulent oral infections.' *British Journal of oral Surgery,* **12,** 1

Sinkovics, J. G. and Smith, J. P. (1970). 'Septicemia with bacteroides in patients with malignant disease.' *Cancer, Philadelphia,* **25,** 663

Sinniah, D., Sandiford, B. R. and Dugdale, A. E. (1972). 'Subungual infection in the newborn. An institutional outbreak of unknown etiology, possibly due to *Veillonella.*' *Clinical Pediatrics,* **11,** 690

Slack, J. (1942). 'The source of infection in actinomycosis.' *Journal of Bacteriology,* **43,** 193

Slack, J. M. and Gerencser, M. A. (1975). *Actinomyces. Filamentous Bacteria. Biology and Pathogenicity*. Minneapolis; Burgess

Slack, W. K. (1976). 'Hyperbaric oxygen therapy in anaerobic infections: Gas gangrene.' *Proceedings of the Royal Society of Medicine,* **69,** 326

Slade, P. R., Slesser, B. V. and Southgate, J. (1973). 'Thoracic actinomycosis.' *Thorax,* **28,** 73

Smart, R. F., Ramsden, D. A., Gear, M. W. L., Nicol, A. and Lennox, W. M. (1976). 'Severe pseudomembranous colitis after lincomycin and clindamycin.' *British Journal of Surgery,* **63,** 25

Smith, D. D. and Fekety, F. R. (1968). 'Bacteroides empyema.' *Annals of internal Medicine,* **68,** 1176

Smith, D. L. and Lockwood, W. R. (1975). 'Disseminated actinomycosis.' *Chest,* **67,** 242

Smith, D. T. (1927–28). 'Fuso-spirochaetal diseases of the lungs.' *Tubercle,* **9,** 420

Smith, D. T. (1932). *Oral Spirochaetes and Related Organisms in Fuso-spirochaetal Disease.* Baltimore; Williams and Wilkins

Smith, J. G. (1956). 'Progressive bacterial synergistic gangrene: Report of a case treated with chloramphenicol.' *Annals of internal Medicine,* **44,** 1007

Smith, L. DS. (1975). *The Pathogenic Anaerobic Bacteria.* 2nd edn. Springfield; Thomas

Smith, W. E., McCall, R. E. and Blake, T. J. (1944). '*Bacteroides* infections of the central nervous system.' *Annals of internal Medicine,* **20,** 920

Snyder, M. L., Bullock, W. W. and Parker, R. B. (1967). 'Morphology of Gram-positive filamentous bacteria identified in dental plaque by fluorescent antibody technique.' *Archives of oral Biology,* **12,** 1269

Socransky, S. S. and Manganiello, S. D. (1971). 'The oral microbiota of man from birth to senility.' *Journal of Periodontology,* **42,** 485

Sonnonwirth, A. C. (1974). 'Incidence of intestinal anaerobes in blood cultures.' In *Anaerobic Bacteria: Role in Disease.* p. 157. Ed. by A. Balows, R. M. DeHaan, V. R. Dowell and L. B. Guze. Springfield; Thomas

Spaulding, E. H. (1974). 'Laboratory diagnosis of anaerobic infections.' In *Infectious Disease Review.* Vol. 3. p. 77 Ed. by W. J. Holloway. New York; Futura

Staneck, J. L. and Washington, J. A. (1974). 'Antimicrobial susceptibilities of anaerobic bacteria: Recent clinical isolates.' *Antimicrobial Agents and Chemotherapy,* **6,** 311

Stewart-Wallace, A. M. (1934–35). 'Progressive postoperative gangrene of skin.' *British Journal of Surgery,* **22,** 642

Stokes, E. J., Waterworth, P. M., Franks, V., Watson, B. and Clark, C. G. (1974). 'Short term routine antibiotic prophylaxis in surgery.' *British Journal of Surgery,* **61,** 739

Stone, H. H. and Martin, J. D. (1972). 'Synergistic necrotizing cellulitis.' *Annals of Surgery,* **175,** 702

Sullivan, H. R. and Goldsworthy, N. E. (1940). 'A comparative study of anaerobic strains of *Actinomyces* from clinically normal mouths and from actinomycotic lesions.' *Journal of Pathology and Bacteriology,* **51,** 253

Sullivan, K. M., O'Toole, R. D., Fisher, R. H. and Sullivan, K. N. (1973). 'Anaerobic empyema thoracis.' *Archives of internal Medicine,* **131,** 521

Sutter, V. L., Attebery, H. R., Rosenblatt, J. E., Bricknell, K. S. and Finegold, S. M. (1972). *Anaerobic Bacteriology Manual.* Los Angeles; University of California

Sutter, V. L. and Finegold, S. M. (1973). 'Anaerobic bacteria: Their recognition and significance in the clinical laboratory.' *Progress in clinical Pathology,* **5,** 219

Sutter, V. L. and Finegold, S. M. (1974). 'The effect of antimicrobial agents on human faecal flora: studies with cephalexin, cyclacillin and clindamycin.' In *The Normal Microbial Flora of Man.* p. 229. Ed. by F. A. Skinner and J. G. Carr. London; Academic Press

Sutter, V. L. and Finegold, S. M. (1975). 'Susceptibility of anaerobic bacteria to carbenicillin, cefoxitin, and related drugs.' *Journal of infectious Diseases,* **131,** 417

Sutter, V. L., Kwok, Y. Y. and Finegold, S. M. (1973). 'Susceptibility of *Bacteroides fragilis* to six antibiotics determined by standardized antimicrobial disc susceptibility testing.' *Antimicrobial Agents and Chemotherapy,* **3,** 188

Swartz, M. N. and Karchmer, A. W. (1974). 'Infections of the central nervous system.' In *Anaerobic Bacteria: Role in Disease.* p. 309. Ed. by A. Balows, R. M. DeHaan, V. R. Dowell and L. B. Guze. Springfield; Thomas

Sweet, R. L. (1975). 'Anaerobic infections of the female genital tract.' *American Journal of Obstetrics and Gynecology,* **122,** 891

Swenson, R. M. (1974). 'Anaerobic bacteria in infections of the female genital tract.' In *Anaerobic Bacteria: Role in Disease.* p. 379. Ed, by A. Balows, R. M. DeHaan, V. R. Dowell and L. B. Guze, Springfield; Thomas

Swenson, R. M. and Lorber, B. (1974). 'Bacteriology of aspiration pneumonia.' *Clinical Research,* **22,** 38a

Swenson, R. M., Michaelson, T. C., Daly, M. J. and Spaulding, E. H. (1973). 'Anaerobic bacterial infections of the female genital tract.' *Obstetrics and Gynecology,* **42,** 538

Sykes, P. A., Boulter, K. H. and Schofield, P. F. (1976). 'Alteration in small-bowel microflora in acute intestinal obstruction.' *Journal of medical Microbiology*, 9, 13

Tally, F. P., Jacobus, N. V., Bartlett, J. G. and Gorbach, S. L. (1975a). '*In vitro* activity of penicillins against anaerobes.' *Antimicrobial Agents and Chemotherapy*, 7, 413

Tally, F. P., Stewart, P. R., Sutter, V. L. and Rosenblatt, J. E. (1975b). 'Oxygen tolerance of fresh clinical anaerobic bacteria.' *Journal of clinical Microbiology*, **1**, 161

Tally, F. P., Sutter, V. L. and Finegold, S. M. (1972). 'Metronidazole versus anaerobes.' *California Medicine*, **117**, 22

Tally, F. P., Sutter, V. L. and Finegold, S. M. (1975). 'Treatment of anaerobic infections with metronidazole.' *Antimicrobial Agents and Chemotherapy*, 7, 672

Tedesco, F. J., Barton, R. W. and Alpers, D. H. (1974). 'Clindamycin-associated colitis. A prospective study.' *Annals of internal Medicine*, **81**, 429

Tempest, M. N. (1966). 'Cancrum oris.' *British Journal of Surgery*, **53**, 949

Thadepalli, H., Gorbach, S. L., Broido, P. W. and Norsen, J. (1972). 'A prospective study of infections in penetrating abdominal trauma.' *American Journal of clinical Nutrition*, **25**, 1405

Thadepalli, H., Gorbach, S. L., Broido, P. W., Norsen, J. and Nyhus, L. (1973). 'Abdominal trauma, anaerobes and antibiotics.' *Surgery, Gynecology and Obstetrics*, **137**, 270

Thadepalli, H., Gorbach, S. L. and Keith, L. (1973). 'Anaerobic infections of the female genital tract: Bacteriologic and therapeutic aspects.' *American Journal of Obstetrics and Gynecology*, **117**, 1034

Tillotson, J. R. and Lerner, A. M. (1968). 'Bacteroides pneumonias. Characteristics of cases with empyema.' *Annals of internal Medicine*, **68**, 308

Tynes, B. S. and Frommeyer, W. B. (1962). 'Bacteroides septicaemia. Cultural, clinical and therapeutic features in a series of twenty-five patients.' *Annals of internal Medicine*, **56**, 12

Tynes, B. S. and Utz, J. P. (1960). 'Fusobacterium septicaemia.' *American Journal of Medicine*, **29**, 879

Ueno, K., Ninomiya, K. and Suzuki, S. (1971). 'Antibacterial activity of metronidazole against anaerobic bacteria.' *Chemotherapy*, **19**, 111

Ueno, K., Sugihara, P. T., Bricknell, K. S., Attebery, H. R., Sutter, V. L. and Finegold, S. M. (1974). 'Comparison of characteristics of Gram-negative anaerobic bacilli isolated from feces of individuals in Japan and the United States.' In *Anaerobic Bacteria: Role in Disease*. p. 135. Ed. by A. Balows, R. M. DeHaan, V. R. Dowell and L. B. Guze; Springfield; Thomas

Urdal, K. and Berdal, P. (1949). 'The microbial flora in 81 cases of maxillary sinusitis.' *Acta Otolaryngologica*, **37**, 20

Varney, P. L. (1920). 'The bacterial flora of treated and untreated abscesses of the lung.' *Archives of Surgery*, **19**, 1602

Veillon, M. A. and Zuber (1898). 'Studies on some strict anaerobic bacteria and their role in pathology.' *Archive de Medicene Expérimentale et d'Anatomie Pathologique*, **10**, 517 (French)

Viberg, L. (1964). 'Acute imflammatory conditions of the uterine adnexa.' *Acta Obstetrica et Gynecologica Scandinavica*, **43**, Supplement 4

Vos, M. de, Straeten, M. van der, Blaauw, M. and Hombrouck, R. (1975). '*Bacteroides fragilis* septicemia originating in the middle ear and the lungs.' *Infection*, **3**, 19

Wagman, H. (1975). 'Genital actinomycosis.' *Proceedings of the Royal Society of Medicine*, **68**, 228

Washington, J. A. (1973). 'Bacteremia due to anaerobic, unusual and fastidious bacteria.' In *Bacteremia: Laboratory and Clinical Aspects.* p. 47. Ed. by A. C. Sonnenwirth. Springfield; Thomas

Watson, E. D., Hoffmann, N. J., Simmers, R. W. and Rosebury, T. (1962). 'Aerobic and anaerobic bacterial counts of nasal washings: presence of organisms resembling *Corynebacterium acnes.' Journal of Bacteriology,* **83,** 144

Weese, W. C. and Smith, I. M. (1975). 'A study of 57 cases of actinomycosis over a 36-year period.' *Archives of internal Medicine,* **135,** 1562

Weiss, J. E. and Rettger, L. F. (1937). 'The Gram-negative bacteroides of the intestine.' *Journal of Bacteriology,* **33,** 423

Werner, H. (1972). 'The susceptibility of *Bacteroides, Fusobacterium, Leptotrichia* and *Sphaerophorus* strains to rifampicin.' *Arzneimittel-Forschung,* **22,** 1043

Whelan, J. P. S. and Hale, J. H. (1973). 'Bactericidal activity of metronidazole against *Bacteroides fragilis.' Journal of clinical Pathology,* **26,** 393

Wilkins, T. D. and Jimenez-Ulate, F. (1975). 'Anaerobic specimen transport device.' *Journal of clinical Microbiology,* **2,** 441

Willard, H. G. (1936). 'Chronic progressive postoperative gangrene of the abdominal wall.' *Annals of Surgery,* **104,** 227

Williams, A. C. and Guralnick, W. C. (1943). 'The diagnosis and treatment of Ludwig's angina: a report of twenty cases.' *New England Journal of Medicine,* **228,** 443

Willis, A. T. (1969). *Clostridia of Wound Infection.* London; Butterworths

Willis, A. T. (1974). 'Anaerobic bacterial diseases.' In *Progress in Chemotherapy. Proceedings of the 8th International Congress of Chemotherapy, Athens.* Vol. 2, p. 5. Ed. by G. K. Daikos

Willis, A. T., Young, S. E. J. and Ferguson, I. R. (1974). 'Infections due to non-sporing anaerobes: Some illustrative cases.' In *Infection with Non-sporing Anaerobic Bacteria.* p. 189. Ed. by I. Phillips and M. Sussman, London; Churchill-Livingstone

Willson, J. R. and Black, J. R. (1964). 'Ovarian abscess.' *American Journal of Obstetrics and Gynecology,* **90,** 34

Wilson, D. H. (1974). 'Lincomycin and clindamycin colitis.' *British medical Journal,* **4,** 288

Wilson, W. R., Martin, W. J., Wilkowske, C. J. and Washington, J. A. (1972). 'Anaerobic bacteremia.' *Mayo Clinic Proceedings,* **47,** 639

Yoshikawa, T. T., Chow, A. W. and Guze, L. B. (1975). 'Role of anerobic bacteria in subdural empyema.' *American Journal of Medicine,* **58,** 99

Yoshikawa, T. T., Tanaka, K. R. and Guze, L. B. (1971). 'Infection and disseminated intravascular coagulation.' *Medicine,* **50,** 237

Yrios, J. W., Balish, E., Helstad, A., Field, C. and Inhorn, S. (1975). 'Survival of anaerobic and aerobic bacteria on cotton swabs in three transport systems.' *Journal of clinical Microbiology,* **1,** 196

Zabransky, R. J., Johnston, J. A. and Hauser, K. J. (1973). 'Bacteriostatic and bactericidal activities of various antibiotics against *Bacteroides fragilis.' Antimicrobial Agents and Chemotherapy,* **3,** 152

Ziment, I., Davis, A. and Finegold, S. M. (1969). 'Joint infection by anaerobic bacteria: A case report and review of the literature.' *Arthritis and Rheumatism,* **12,** 627

Ziment, I., Miller, L. G. and Finegold, S. M. (1968). 'Nonsporulating anerobic bacteria in osteomyelitis.' In *Antimicrobial Agents and Chemotherapy.* p. 77. Ed. by G. Hobby. Bethesda; American Society for Microbiology

7

Clostridial Infections

WOUND INFECTIONS

Clostridial infections of man are usually of traumatic origin and have been encountered most commonly in war. Their comparative rarity in peace time, however, in no way reduces their importance, nor does it provide an excuse for their neglect in the clinical microbiology laboratory.

Most clinicians are familiar with the classic conditions of anaerobic myonecrosis (gas gangrene) and tetanus. Both of these syndromes may occur following accidental trauma and result from contamination of the wound by the appropriate organism. Tetanus is specifically caused by *Cl. tetani*; anaerobic myonecrosis may be caused by any of a number of pathogenic clostridia of which the commonest is *Cl. perfringens*. These anaerobic spore-forming bacilli are widely distributed in nature, being present in soil, dust, clothing and so on, and in the intestinal tract of animals and man. Their very ubiquity ensures that a large proportion of accidental wounds are exposed to the risk of contamination at the time of injury. However, mere contamination of a wound is not inevitably followed by the development of tetanus or gas gangrene, because the very special conditions of anaerobiosis must prevail in the lesion before the organisms can multiply. In the absence of an anaerobic environment, the spores of toxigenic clostridia are dormant, but potentially dangerous contaminants.

The most important single factor which enables clostridia to flourish in a wound is a low oxygen tension. This may obviously be present in severe and extensive wounds, such as open fractures, and those caused by high velocity missiles. Here, anaerobiosis is initiated not only by the presence of necrotic tissue, blood clot and foreign bodies at the site of wounding, but also by more distant vascular damage which may greatly reduce the blood supply to the part. Although it is this type of injury which is the classic precursor of gas gangrene, early surgical intervention after wounding ensures that serious anaerobic infections rarely develop. The form of anaerobic myonecrosis most commonly encountered today is postoperative gas gangrene, in which the infection is derived, not from the inanimate environment (exogenous), but from the patient's own resident bacterial flora (endogenous).

Three types of clostridial wound infections are distinguished (MacLennan, 1943b):

1. simple contamination in which one or more clostridia are present, but from which subsequent invasion of the underlying tissues does not necessarily occur;
2. clostridial cellulitis, a condition characterized by invasion of fascial planes by the organisms, but without invasion of muscle tissue, and with minimal toxin production; and
3. clostridial myositis, in which there is invasion of healthy muscle tissue, with abundant formation of exotoxins.

Simple contamination of wounds with pathogenic clostridia is not uncommon, and many such wounds heal by first intention without special treatment, and without sequelae (*see* Gorbach and Thadepalli, 1975). This is due to the fact that the conditions in such lesions are unsuitable for the multiplication of the contaminating organisms or for toxin production by them. A high oxidation-reduction potential (Eh) due to healthy surrounding tissues and possibly also to the presence of atmospheric oxygen, and restriction of the organism's nutritional requirements are probably the chief factors which prevent colonization of the tissues.

In the absence of treatment, clostridial cellulitis or myositis may develop from simple contamination, so that the three types of 'infection' may be looked upon as three ascending grades of severity.

The limiting Eh values at which different anaerobic species will grow have been variously estimated by different workers (Fildes, 1929; Knight and Fildes, 1930; Hanke and Bailey, 1945). The available evidence indicates, however, that the Eh of healthy tissues (at pH about 7.5) is well above the levels necessary for the initiation of anaerobic growth. Thus, Hanke and Tuta (1928) found that the Eh of the circulating blood varied between +0.126V and +0.246V; the highest Eh recorded for anaerobic growth is +0.11V, which was determined by

Knight and Fildes (1930) for the germination of spores of *Cl. tetani*. The limiting oxygen tension for the growth of *Cl. perfringens* is reported to be about 5% while that for *Cl. tetani, Cl. septicum* and *Cl. novyi* between 0.2 and 0.3% (Futter and Richardson, 1971). Anaerobes are thus unable to grow or multiply in normal tissues, and are therefore unable to produce disease. This was demonstrated many years ago by Vaillard and Rouget (1892), who showed that the atraumatic inoculation of guinea pigs with toxin-free spores of *Cl. tetani* did not give rise to the disease. Most wounds, however, are not atraumatic, and it is the trauma of wounding which frequently initiates the chain of events leading to colonization of the tissues by contaminating anaerobes.

Vascular damage is probably the most important predisposing factor in clostridial wound infections. North (1947) reported that 72% of cases of gas gangrene had injuries to the vessels supplying the affected muscle, while Lowry and Curtis (1947) recorded that 87% of cases submitted to amputation for gas gangrene showed evidence of vascular damage.

The type of wound produced by a high-velocity missile, such as a bomb fragment or bullet, provides ideal conditions for the proliferation of anaerobes. On entry, the missile produces a relatively small wound in the skin compared with the extensive muscular damage beneath. Shock-wave pressures cause tearing of soft tissues and blood vessels, extensive cavity formation, and fracture of long bones at a distance from the site of wounding. Consequent upon these injuries, extensive haemorrhages occur in the damaged tissues and large areas of muscle are rendered anoxic. The Eh and pH of the damaged tissues fall, and these changes, together with the breakdown of some protein to amino acids, enable anaerobic growth to commence. Growth of the organisms is associated with a further drop in Eh, and with the production of bacterial toxins and other products of metabolism. These factors promote the invasion of uninjured tissues by the organisms. In addition, the natural defences of the body are hampered, for neither phagocytic cells nor antibodies can enter the necrotic area. These questions in relation to gas gangrene have been fully discussed by Smith (1955), Oakley (1954) and Willis (1969). An excellent article on infection in missile wounds was published by Matheson (1968).

While exogenously derived clostridial infections of wounds are most commonly associated with accidental trauma, they sometimes follow elective surgical procedures, usually due to some breakdown in theatre sterility. Sevitt (1949, 1953) recorded two cases of post-operative tetanus (one of them fatal), and one case of gas gangrene, a result of the exhaust system of ventilation which sucked air and dust into the theatre from the corridor and other parts of the hospital.

Another outbreak of postoperative tetanus affecting five patients, two of whom died, which resulted from widespread contamination of an operating theatre and its contents with *Cl. tetani*, is recorded in the *Report of the Ministry of Health for 1957* (1958). Studies of the clostridial flora of hospital and operating theatre air have been made by Lowbury and Lilly (1958a and b) and Gye, Rountree and Loewenthal (1961). The latter workers recorded 16 cases of *Cl. perfringens* infection occurring in clean postoperative wounds, and noted the wide distribution of clostridia in operating theatres and wards. They drew attention to the fact that patients undergoing surgery for treatment of occlusive vascular disease, and for ischaemic complications of diabetes mellitus, are at special risk from anaerobic infections. In retrospect, it seems likely that some of the cases referred to by Gye and his colleagues were examples of endogenous infection (*see below*). Dickinson and Edgar (1963) described a case of exogenous *Cl. perfringens* infection in a patient following periurethral prostatectomy and orchidectomy, and Willis and Jacobs (1964) documented a case of *Cl. perfringens* meningitis which developed in a patient following elective craniotomy for the treatment of trigeminal neuralgia.

In modern civilian practice clostridial wound infections of endogenous origin are these days much commoner than those in which the organisms are derived from exogenous sources (including gas gangrene complicating accidental trauma). Most of these infections occur as complications of surgical procedures and are usually due to strains of *Cl. perfringens*. Spontaneous or metastatic gas gangrenous infections in man are rare (Willis, 1969; Marty and Filler, 1969; Kapusta, Mendelson and Niloff, 1972; Engeset *et al.*, 1973; Mzabi, Himal and MacLean, 1975).

In the case of *Cl. perfringens* infections following clean elective surgical procedures, the infecting organisms commonly occur on the patients' skin as faecal contaminants that are implanted in the tissues during surgery (*see* Mills, 1973). Mid-thigh amputation for vascular disease, and lower limb surgery involving the insertion of a foreign body carry a special risk (Ayliffe and Lowbury, 1969; *British medical Journal*, 1969a, b; Parker, 1969; *British medical Journal*, 1971a; Drewett *et al.*, 1972). Consequently, these procedures call for scrupulous preoperative skin preparation, and for the routine use of penicillin prophylaxis (Taylor, 1960; Lowbury, Lilly and Bull, 1964; Parker, 1969; Deveridge and Unsworth 1973; Shaw, Vellar and Vellar, 1973; McKinnon and McDonald, 1973).

The risks of gas gangrene following the injection of adrenaline are well known (Tonge, 1957; Marshall and Sims, 1960; *British medical Journal*, 1961, 1964; Harvey and Purnell, 1967). It has been shown by

Bishop and Marshall (1960) that experimental *Cl. perfringens* infection in guinea pigs is enhanced one-thousand fold by adrenaline. The injection of irritating agents such as iron compounds into the gluteal muscles also carries a risk of gas gangrene. Here, as with adrenaline, the organisms presumably gain entry at the time of inoculation, since the skin is likely to be contaminated with *Cl. perfringens*, especially about the buttocks.

The risk of endogenous clostridial infections following abdominal surgery is well known. Gas gangrene of the abdominal wall, usually due to *Cl. perfringens*, may result from intestinal contamination of the surgical wound (Elliot-Smith and Ellis, 1957; McNally and Crile, 1964; Morgan, Morain and Eraklis, 1971; Fraser-Moodie, 1973; Phillips, Heimbech and Jones, 1974). Spann and McGill (1957) recorded an infection due to *Cl. septicum* following resection of bowel. *Cl. perfringens* bacteraemia following elective cholecystectomy is a rare but devastating event (Pyrtek and Bartus, 1962; Yudis and Zucker, 1967). In these cases the organism is derived from the biliary tract. Abeyatunge (1969) reported briefly upon a case of tetanus following intussusception in which the infection was thought to be of endogenous origin.

TETANUS

General reviews of tetanus have been published by Adams, Laurence and Smith (1969), Willis (1969, 1972–73, 1975) and Furste and Wheeler (1972).

Tetanus is an infective disease usually resulting from the contamination of a wound or raw surface by *Cl. tetani*. These days the disease is rarely produced by infection following the injection of material (vaccines and so on), or by actual contamination of these materials. In the Mulkowal incident of 1902 in the Punjab (Simpson, 1907), the first 19 of 107 persons inoculated with plague vaccine developed fatal tetanus due to contamination of the vaccine at the time of inoculation. Rare though this mode of infection appears to be today, Patel *et al.* (1960a, b) recorded 20 cases of tetanus in Bombay in which the portal of entry was a smallpox vaccination site, and a further 32 cases in which the tetanus bacillus appeared to have been introduced during intramuscular inoculation of a drug. Although the author has not seen a case of tetanus attributable to this type of injury, other anaerobic infections, for example, gas gangrene following the injection of adrenaline, and the intramuscular injection of irritating agents such as iron compounds in the gluteal region, are not unknown.

Osbourne (1892) recorded a case of fatal tetanus which resulted from infection of the injection site after self-administration of morphia, and a somewhat similar, but more recent case was reported by Martin and McDowell (1954), in which a patient developed tetanus twice within 10 months after self-administration of heroin (*see also* Doane, 1924). Although drug addiction has been regarded as a noteworthy but minor cause of tetanus, more recently emphasis has been placed on it as a cause of the disease in urban populations (Levinson, Marske and Shein, 1955; Cherubin, 1967, 1971; *Journal of the American medical Association*, 1968; Blake and Feldman, 1975). It was reported that between 1955 and 1965, nearly 75% of all known cases of tetanus in New York City occurred in narcotic addicts. There were no fewer than 102 cases of addict tetanus with a mortality rate of almost 90%.

Today, tetanus is more likely to develop from severe, badly soiled wounds, such as are encountered in war, than from clean superficial injuries. That tetanus nowadays more commonly follows mild injuries is due to the fact that protective measures are routinely initiated in cases of severe trauma, and are so frequently omitted in cases of mild wounding. A slight penetrating wound produced by a rusty nail, a splinter of wood or a thorn, or even a dirty abrasion, are the types of lesions which now most commonly precede the development of clinical tetanus (Cole, 1951; Moynihan, 1956; Cox, Knowelden and Sharrard, 1963; Tetanus Surveillance, 1974).

The social environment and habits of a population are sometimes related to the nature of the lesions from which tetanus may develop. Infection of the umbilical cord in neonates in less highly developed communities is noted below. Johnson (1956), Phadke, Godbole and Pande (1958) and Pitt (1971) referred to the high incidence of tetanus amongst those who go barefoot in Queensland, Poona and Nepal. Johnson studied 144 cases of tetanus, in 65 (45%) of which the portal of entry was shown to be puncture wounds in the foot; he drew attention to the fact that a high proportion of the population of Queensland lives in rural districts, and that favourable climatic conditions permit children to go barefoot thoughout the year. In Poona and Nepal, the hazard is greater due to the comparatively low standards of living, public sanitation and hygiene.

In a series of 33 cases of tetanus reported from Sheffield by Cox, Knowelden and Sharrard (1963), 10 followed abrasions, 7 were due to lacerations, 2 to varicose ulcer of the leg, 2 to penetrating wounds of the foot, 1 to a thorn in the finger, 1 to cracks on the hands, and in 6 cases no injury was found. Other portals of entry from which tetanus may develop are plaster sores, boils, paronychia, chronic ulcers of the leg, epistaxis and ear-piercing (Cole, 1951; Riis, 1958; Cormie, 1962;

Rey, Armengaud and Mar, 1967). A study of otogenic tetanus has been made by Shah (1955), and tetanus associated with burn injuries has been discussed by Sherman (1970), Marshall *et al.* (1972), and Larkin and Moylan (1975).

Thus, although infection of extensive wounds by *Cl. tetani* is extremely dangerous and of an overwhelming nature, tetanus more often follows very minor injuries. The organism produces no tissue-destroying enzymes and does not invade normal tissues. In the presence of minimal tissue necrosis, and in the absence of oxygen, the tetanus bacillus can proliferate and produce its neurotoxin in lethal amount from a comparatively small site. The presence of soil greatly favours the growth and multiplication of *Cl. tetani*, and indeed of all clostridia, in the tissues. This is due to the presence in soil (especially cultivated soil) of calcium chloride, which produces the tissue necrosis so necessary for the establishment of anaerobic infections (Bullock and Cramer, 1919).

TETANUS NEONATORUM

This results from contamination of the cut surface of the umbilical cord. It is extremely rare in Great Britain and the United States but is common in many underdeveloped communities. Among the native population in Nigeria the cord may be cut with a sharp splinter of palm wood or a broken bottle, tied with string and then 'dressed' with clay, salt or 'native medicine' which may contain cowdung (Jelliffe, 1950; Johnstone, 1958; Tompkins, 1958; Daramola, 1968). In Sierra Leone, the topical umbilical dressing consists of the juice of young banana shoots (Wilkinson, 1961). According to Stahlie (1960), tetanus appears to be the leading cause of neonatal death in Thailand; in New Guinea, Schofield, Tucker and Westbrook (1961) recorded an incidence among live births of 8%; and in Baghdad, Critchley (1958) reported seeing nearly 10 cases per month over a three year period. Neonatal tetanus has a mortality rate of about 90%.

An interesting historical account of infantile tetanus in the Scottish Western Isles during the latter part of the nineteeth century has been published by Ferguson (1958).

POSTABORTAL AND PUERPERAL TETANUS

Like the neonatal disease, postabortal and puerperal tetanus are rare in the United Kingdom but quite common in some underdeveloped communities (Adams, 1968; Daramola, 1968; Orr and Coffey, 1969; Adadevoh and Akinla, 1970; Pitt, 1971; Afonja, Jaiyeola and Tunwashe,

1973). They result from infection of the genital tract, often from criminal interference in the case of postabortal infections.

Adams and Morton (1955) briefly reviewed the literature and found that only 210 cases of postabortal tetanus and 114 cases of postpartum tetanus had been previously recorded. A more detailed review on postabortal tetanus was published by Weinstein and Beacham (1941), together with brief descriptions of 14 of their own cases. These authors recorded a mortality rate of 57% in their own cases as compared with an average mortality of 90% recorded by other workers.

In modern communities puerperal tetanus is one of the rarest of puerperal infections. Ramsay, France and Dempsey (1956) recorded only one case at the Royal Free Hospital, London, over a period of 31 years, during which time there had been 6000 cases of puerperal infection due to other organisms. However, it is the most dangerous form of puerperal infection, and probably also the most dangerous form of tetanus. The source of *Cl. tetani* in postabortal and puerperal tetanus is unsterile examination, manipulation or instrumentation, or unsterile cotton and cellulose dressings (Pulvertaft, 1937; *Journal of Obstetrics and Gynaecology*, 1941; *British medical Journal*, 1959). In view of the hazardous environmental conditions and the primitive accouchement practices in underdeveloped communities, the relatively high incidence of uterine tetanus among these peoples is not surprising.

Ordinary wound tetanus, as opposed to uterine tetanus, may occur during pregnancy, but this is a rare event (Singleton and Witt, 1956; Holmdahl and Thoren, 1962; Januszkiewicz *et al.*, 1973).

RECURRENT TETANUS

A second attack of clinical tetanus is not unknown. Martin and McDowell (1954) reported a second attack, as did Beare (1953), Aguileiro Moreira, Braneiro and Ghigliazza (1960) and Alhady (1961). Patel *et al.* (1961) recorded no less than 17 cases of recurrent tetanus amongst 2007 cases studied by them over a period of 4 years in Bombay. More recently, Pace and Busuttil (1971) reported 3 cases from Malta; one of these patients suffered the disease three times in 18 months following a penetrating wound in the foot.

Clinical aspects

Tetanus is characterized by muscular spasms which may rarely commence near the site of the infection (local tetanus). The commonest early symptom is trismus, which has led to the popular synonym 'lockjaw'.

This is often combined with pain and stiffness in the neck, back and abdomen. Occasionally dysphagia appears first. These signs increase slowly or rapidly according to the severity of the attack. The whole of the somatic musculature may ultimately be involved in severe reflex spasms. Between the time of wounding and the onset of clinical tetanus there is an incubation period of 3–30 days, the commonest time of onset being 7–10 days after injury. The clinical types, modes of presentation and clinical features are fully discussed by Knott and Cole (1950), Adams, Laurence and Smith (1969), Willis (1969, 1972–73), and Furste and Wheeler (1972).

Incidence

In general terms, the social environment and habits of a population are related to the incidence of tetanus and to the form that the disease is likely to take. For example, it was noted that neonatal and uterine tetanus are common in underdeveloped communities where standards of living, sanitation and hygiene are low. Also, the habit of going barefoot heightens the risks of tetanus, especially in agricultural areas. Physical injury, whether accidental or intentional, is a *sine qua non* for the development of the disease. The incidence of tetanus is therefore influenced by such factors as military commitments, agricultural activity and drug addiction on the one hand, and by the standard and availability of wound treatment and the immune status of the population on the other (Glenn, 1946; Bamford, 1967).

Tetanus affects men more often than women because they are generally at greater risk. For the same reason, rural populations are more prone to infection than urban ones (Young, 1927; Anusz, 1967).

During the First World War, the incidence of tetanus among the wounded on the Western Front was 1.47 per 1000. An instructive series of cases during this period was recorded by Dean (1917). In the European and African Campaigns of the Second World War the incidence of tetanus was only 0.12 per 1000 (Boyd, 1946). This marked improvement may be attributed to the almost universal prophylactic immunization of troops before exposure to risk during the 1939–45 war, and to improved surgical techniques.

In any large community it is difficult to determine accurately the tetanus morbidity and mortality rates. This is because the number of notifications is clearly dependent, not only on the accuracy of diagnosis, but also on the consistency with which notification is made. Difficulties of this sort mean that the vital statistics of tetanus often show a lower morbidity than mortality. The overall global mortality rates of tetanus in various countries are summarized in *Table 7.1*. In his excellent review of tetanus as a world problem, Bytchenko (1966) concluded that the disease causes more than 50 000 deaths each year.

Table 7.1 Global mortality rates of tetanus in various countries

High-incidence area	*Mortality per 100 000 population*	*Low-incidence area*	*Mortality per 100 000 population*
Haiti	46.2	Italy	0.90
Malaya	31.4	France	0.78
Panama	17.7	Japan	0.63
Nigeria	14.6	New Zealand	0.45
Philippines	9.0	Switzerland	0.35
Mexico	6.7	Australia	0.33
Venezuela	6.7	West Germany	0.30
Fiji	4.6	USA	0.13
Kenya	4.6	Finland	0.09
Thailand	4.3	England	0.05
Ceylon	4.0	Canada	0.03

(Reproduced from Rubbo (1966), *Lancet*, **2**, 449, by courtesy of the author and Editor)

In compiling the figures for deaths from accidental violence, it has been the practice in England and Wales to ignore tetanus should it supervene and cause death of injured persons, and to assign such deaths to the original trauma. Moreover, until October 1968, tetanus was not a statutorily notifiable disease. It is clear from these considerations that published figures relating to tetanus are frequently conservative, and that the incidence of the disease is greater than is generally supposed.

In England and Wales, the Report of the Department of Health and Social Security for 1969 shows that for the period 1960–69 there was an average of 17 fatal cases per annum for those deaths assigned exclusively to tetanus (Department of Health and Social Security, 1970). For those deaths complicated by tetanus but assigned to injury or other conditions, the average number of fatal cases per annum was 10 (*Table 7.2*). If all these deaths are attributed to tetanus, the average annual mortality rate was only 27, which represents a total case incidence of only 81 per annum, if the mortality rate is assumed to be 30%. In the first 4 year period during which tetanus has been a notifiable disease (1969–72) the average reported incidence was 30 cases per year, and the average annual mortality rate was 14 (Department of Health and Social Security, 1973). During the preceding decade (1950–59) the average annual fatality rate was as high as 45 (Ministry of

Table 7.2 Deaths from tetanus in England and Wales, 1960–69

	1960	*1961*	*1962*	*1963*	*1964*	*1965*	*1966*	*1967*	*1968*	*1969*
Deaths assigned to tetanus	18	24	19	13	21	21	18	18	13	10
Deaths complicated by tetanus assigned to injury or other conditions	14	17	10	8	8	12	9	7	9	9
Total	32	41	29	21	29	33	27	25	22	19

Reproduced from the Annual Report of the Chief Medical Officer of the Department of Health and Social Security for the Year 1969, by courtesy of the Department and HMSO

Health, 1961). This notable reduction in the mortality rate during the past decade is doubtless partly attributable to improved methods in treatment. It seems likely, however, that the incidence of the disease has also declined. This is probably due to greater protection of the general population through wider use of active immunization and to more thoughtful surgical prophylaxis in patients with tetanus-prone wounds.

MORTALITY RATE

The mortality rate of tetanus depends on the nature of the wound, the site of infection and the treatment instituted. Severe wounds carry a higher mortality, while wounds of the upper extremities are more dangerous than those of the lower. In the pre-serum days, tetanus carried a mortality rate of about 85%. Today, with full modern treatment, this has fallen to about 30%. In general, the shorter the incubation period the worse the prognosis. The danger of tetanus developing is greatly reduced if full prophylactic treatment is instituted early after wounding. Severe tetanus, though terrible and often fatal, leaves no sequelae; those who recover do so completely.

Action of tetanus neurotoxin

Tetanus neurotoxin acts on the central nervous system, where it appears to diminish or abolish synaptic inhibition (Brooks, Curtis and Eccles, 1955, 1957; Wilson, Diecke and Talbot, 1960; *see also* van Heyningen, 1968). The route by which the toxin reaches the susceptible cells has been the subject of dispute for many years. In *ascending tetanus* (the form of the experimental disease commonly seen in laboratory animals), it now seems clear that the toxin passes from the lesion to the central nervous system by way of the regional motor nerve trunks, and then up the spinal cord. In *descending tetanus* (the form of the natural disease in man and animals), although there is general agreement that the toxin operates at some site within the central nervous system, there is no clear indication as to how the circulating toxin reaches the vulnerable nervous structures. The pathogenesis of tetanus and pharmacological effects of tetanus neurotoxin have been well reviewed by Wright (1955, 1959) and Lamanna and Carr (1966).

Diagnosis

The diagnosis of tetanus is always made on clinical grounds before it is confirmed bacteriologically. Not infrequently bacteriological confirmation is impossible, as for example in idiopathic cases, in which the presumed lesion has been so slight as to be undetectable by the time clinical tetanus develops.

Microscopic examination of pathological material

Microscopic examination is made for the typical 'drum-stick' bacilli. Their presence in material from a wound, however, is not in itself indicative of *Cl. tetani* infection, since pathogenic strains of the organism may be present in contaminated wounds from which patients recover uneventfully without prophylactic treatment. Thus, the Committee upon Anaerobic Bacteria and Infections (1919) isolated *Cl. tetani* from 19 of 100 wounds, the patients showing no clinical evidence of tetanus. On the other hand, in cases of clinical tetanus, material from a wound which is the obvious focus of infection may contain such small numbers of tetanus bacilli that they escape detection in direct films. I have examined the wounds of a number of cases of clinical tetanus, but in very few of them were organisms resembling *Cl. tetani* seen in direct smears, though in most cases the organism was subsequently isolated. The fluorescent labelled antibody technique of staining *Cl. tetani* (Batty and Walker, 1964, 1967) may find a useful application here.

Culture of material from the wound

Culture of wound material is much more likely to succeed; for this purpose, the best specimen is the excised wound. In the case of small wounds, the whole excised area should be submitted for bacteriological examination. The material is cut up into small pieces with sterile instruments, and these are inoculated into air-free cooked meat broth, and on to suitable agar plate media such as fresh blood agar and heated blood agar. A small segment only of each plate is inoculated, so that the spreading edge of a tetanus culture is more easily recognized should it develop. These cultures are incubated in the anaerobic jar. Similar plates are inoculated in the usual way for aerobic incubation. All the plate cultures are examined after 18–24 h incubation, and are reincubated if necessary. The enrichment broth culture is subcultured every 24 h for 4 days.

These procedures not only provide the greatest chance of isolating *Cl. tetani*, but also enable other organisms to be isolated and identified. The value of concentrated agar media and the use of tetanus antitoxic serum for the suppression of swarming growth of *Cl. tetani* have been noted elsewhere (p. 115). Methods of purification and the criteria of identification of *Cl. tetani* are discussed on p. 118–119.

Animal inoculation

Animal inoculation of a saline suspension of some ground-up wound tissue may induce tetanus in guinea pigs or mice, but this is not recommended. It is better to use as much material as possible in the enrichment and other cultures, thereby increasing the chance of isolating the organism. Animal inoculation is best performed with pure cultures.

Prophylaxis

During the past decade there has been increasing concern about anti-tetanus immunization, and in particular about the use of prophylactic heterologous tetanus antitoxin. Not unreasonably, perhaps, heterologous antitoxin has been viewed with a good deal of suspicion, due chiefly to the risks of severe serum reactions (Hunter, 1959; Catzel, 1961; Tisdall, 1961; and many others). Indeed, to many, the very existence of heterologous antitoxin has posed an inescapable dilemma, for it has been felt that sooner or later one may be called upon under legal circumstances to justify one's actions in the management of tetanus prophylaxis in a patient at risk from this disease; that the complications of serum therapy may lead to the courts as surely as the withholding of antitoxin from a patient who subsequently develops tetanus (*see* Bennett, 1939; Morley, 1953).

This dilemma was the direct result of two surgical misconceptions that grew out of the undeniable efficacy of equine antitoxin in tetanus prophylaxis: (1) that any accidental break in the skin surface placed the patient at risk from tetanus, and (2) that such lesions provide an absolute indication for the administration of tetanus antitoxin (*see* Ackland, 1959). Both of these dicta are now known to be untrue, and, at least in the United Kingdom, it was the attack by Cox, Knowelden and Sharrard (1963) on the indiscriminate use of antitoxin that led to a reappraisal of the use of heterologous tetanus antitoxin in particular, and of tetanus prophylaxis in general. The main recommendation that emerged from the work of Cox and her colleagues was that 'adequate wound treatment and antibiotics be substituted for tetanus antitoxin as the modern prophylaxis against tetanus, together with active immunization by toxoid against future injury'. In retrospect, it seems likely that this recommendation, which is now widely practised in Britain, caused the pendulum to swing too far the other way, for there remains a small group of patients at risk from tetanus for whom tetanus antitoxin is an important part of the prophylactic regimen.

Since these important re-evaluations of the indications for heterologous serum prophylaxis were made, human antitetanus immunoglobulin has become available, to which there are no serious disadvantages. The mere availability of this virtually side-effect-free antitoxin, however, should never be allowed to lead to its indiscriminate use; and the important tenet emphasized by Cox and her colleagues, of adequate and careful wound treatment must remain the corner stone of modern tetanus prophylaxis.

Interesting and practical, though sometimes conflicting, accounts of the modern approach to tetanus prophylaxis are given by Cox, Knowelden and Sharrard (1963), Rubbo (1965, 1966), Laurence, Evans and Smith (1966). *Annals of internal Medicine* (1966). Eckmann (1967). Furste, Skudder and Hampton (1967). Adams, Laurence and Smith (1969), Batten (1969), Furste (1969), Robles and Walske (1969), Willis (1969, 1972–73), Hellberg (1970), Giliberty (1972), Furste and Wheeler (1972), Furste (1974), Smith (1975) and Fraser (1976).

Although tetanus is not common in modern communities, the possibility of its development in traumatized patients dictates that prophylactic measures must always be considered and commonly taken. The prophylactic measures available to a wounded patient are:

1. surgical treatment,
2. active immunization,
3. passive immunization,
4. antibiotic therapy.

Surgery

It is not the purpose here to consider surgical treatment except to note that it is essential for all wounds, no matter how trivial they may be. It may range from the major procedures of radical excision or amputation to simple scrubbing or cleansing by washing. Surgery aims at removal of dirt, foreign bodies, necrotic tissue and blood clot, and at restoration of the blood supply to the part. As part of the surgical toilet, antibiotics may be applied to the wound.

Active immunization

Because its elective nature enables establishment of basal immunity before exposure to the risk of tetanus, active immunization is universally accepted as the prophylactic measure of choice.

Bensted (1941) reported upon the incidence of tetanus among members of the British Expeditionary Force in France in 1940: of the troops who had been actively immunized, many of whom did not receive tetanus antiserum at the time of wounding, none developed tetanus. On the other hand, eight cases of tetanus occurred amongst a group of wounded men who had not received prophylactic toxoid. Similar findings for the United States Armed Forces during the Second World War have been recorded by Long and Sartwell (1947), who reported that amonst 2 734 819 hospital admissions for wounds and injuries, only 12 cases of tetanus occurred – 6 of these had received no basic immunization. Newell *et al.* (1967) studied the effect of toxoid immunization of Colombian Negro women on the incidence of neonatal tetanus among their offspring. Among 347 infants of women who received no toxoid there were 27 cases of tetanus; there were no cases of tetanus, however, among 341 infants of women who received courses of adsorbed toxoid either during pregnancy or up to 5 years before.

Active immunization against tetanus may be instituted at any age by the intramuscular administration of two doses of adsorbed toxoid at an interval of 4–6 weeks, followed by a third dose 6–12 months later. Reinforcing doses may then be given every 5–10 years, and when the individual presents with a tetanus-prone wound. The dose of toxoid depends on the potency of the particular preparation. Adsorbed toxoid is commonly dispensed in 0.5 ml doses. White *et al.* (1973) advocated the use of a tetanus vaccine containg 10 Lf (limit of flocculation) of toxoid rather than 20 Lf in order to minimize the risk of untoward reactions (*see below*).

Table 7.3 Geometric mean of tetanus antitoxin titres of sera from 11 persons receiving a third dose of tetanus toxoid (adapted from Evans, 1943c)

Antitoxin titre (units/ml)							
Before first inoculation	*9 weeks after first inoculation*	*After second inoculation*		*After third inoculation*			
		4 months	*10 months*	*5 days*	*12 days*	*30 days*	*18 months*
<0·01	0·03	0·21	0·08	1·1	10·1	9·0	0·37

Evans (1943c) studied the antitoxin response in man following active immunization with tetanus toxoid. He examined two groups of patients, one of 16 persons who received 2 doses of fluid toxoid (1 ml at an interval of 9 weeks), and the other of 11 persons who received 3 doses of toxoid, the first 2 at an interval of 9 weeks and the third dose 10 months after the second. *Table 7.3* and *Figure 7.1* summarize Evans'

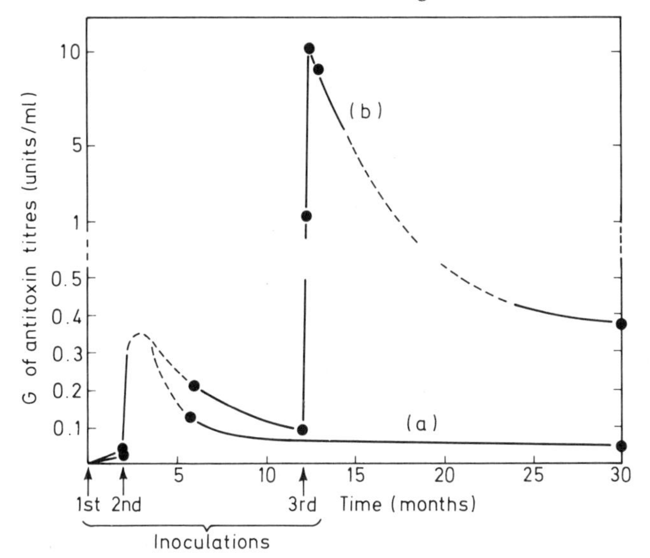

Figure 7.1. Antitoxin titres in the blood of persons receiving (a) two doses, and (b) a third dose of tetanus toxoid. (Reproduced by courtesy of D. G. Evans and the Editor of the *Lancet*)

findings. Further similar work was reported by Sachs (1952), Barr and Sachs (1955), Takayama, Sakurai and Nakayama, (1971) and Solomonova and Vizev (1973).

ALLERGIC REACTIONS

Although allergic reactions to tetanus toxoid do occasionally occur, the risk of sensitization is small (Brindle and Twyman, 1962; Deliyannakis, 1971; Facktor, Bernstein and Fireman, 1973; White *et al.* 1973). Reactions to tetanus toxoid rarely occur in children; their incidence in adults increases with age. In a careful study of over 65 000 industrial workers, White and his co-workers (1973) reported an incidence of local reactions of 2.6%, and of general reactions of 0.3%; they encountered local reaction rates of 0.9%, 2.7%, 7.4% and 1.6% after the first, second, third and booster doses of toxoid respectively. Local reactions, which may develop immediately following injection or up to several days later, consist of excessive pain, erythema and swelling about the injection site; these symptoms may persist for 3–4 days. The much less common general reactions consist of headache, lethargy and aching limbs, the symptoms developing a few hours after the injection. Delayed illness, such as serum sickness and peripheral neuropathy, is rare (Blumstein and Kreithen, 1966; Daschbach, 1972). Reactions are not common after the first dose of toxoid, but their incidence increases according to the number of previously administered doses (Edsall *et al.* 1967; Relihan, 1969). In general it may be said that 'reactions to tetanus toxoid do not endanger life, do not leave any sequelae, and do not occur in more than about 1% of adults, mainly the overimmunized' (*British medical Journal*, 1974). As is the case with the injection of any foreign protein, the prophylactic use of oral antihistaminic is rational, and adrenaline 1 in 1000 should be immediately available in the unlikely event of a severe immediate general reaction.

The case for universal active immunization of the population with tetanus toxoid is overwhelming (Ministry of Health, 1964; Robles, Walske and Tella, 1968; Furste, 1969; *Lancet*, 1969). It not only provides a sound and desirable basal immunity that can be boosted as necessary, but it also precludes the necessity of giving prophylactic heterologous antitoxin, thus avoiding the risks of serum therapy.

Experience in Poland, Hungary, Bulgaria and Haiti underlines the efficacy of this form of prophylaxis (Petrilla, 1960; Galazka, 1968; Solomonova and Kebedjiev, 1972; Berggren, 1974). Compulsory immunization against tetanus, which was instituted in Hungary in 1953, resulted in a reduction in the tetanus morbidity rate of 80%; and in Bulgaria, mass immunization led to more than a six-fold reduction in the incidence of the disease. It is, of course, an oversimplification to relate this reduced incidence only to the use of toxoid. Improved social and economic conditions and improved medical care are also important influencing factors.

The free availability of immunizing agents, and modern attitudes to the prevention of infectious disease, ensure that the great majority of children in the United Kingdom receive protection against tetanus as part of the mass immunization programme (*see* Emond, 1972–73). Unfortunately, less attention is paid to the tetanus immune status of the adult population, especially of the older age groups whose childhood years were before the introduction of mass immunization (*British medical Journal*, 1971; *Health Notes*, 1972). Moreover, there is at present no functional system of immunization follow-up once adulthood is reached. In the United Kingdom efforts to rectify this deficiency have been made (*Department of Health and Social Security Circular 10/70*, 1970; *Community Medicine*, 1972), but they do not seem to have been sustained. In a study of 132 agricultural accidents in Cambridge, Cooper (1969) reported that 56% of patients had never received a course of prophylactic adsorbed toxoid. In surveys of tetanus immunity among populations in Western Australia and Victoria (Chapman and Davey, 1973; Trinca and Fraser, 1968) it was shown that although the state of active immunization among infants and children was excellent, among the adult populations it was unsatisfactory. In Victoria, only one in every 2 males, and one in every 4 females over the age of 20 years had significant active immunity. In the USA and Canada, too, there is a large reservoir of as yet unprotected persons, especially among those past the age of 60 years (Ashley and Bell, 1969; Christensen, 1972).

'The tragedy of tetanus is not its dreadful symptoms nor high mortality, but the fact that it could have been eradicated within the past 25 years by a vigorous programme of immunization with tetanus toxoid' (Creech, Glover and Ochsner, 1957).

The logistics of widespread administration of tetanus adsorbed toxoid are beyond the scope of this book; Fulford (1960) has indicated how this might be accomplished (*see also* Cotter and Wilson, 1975). It is pertinent here, however, to summarize the categories of persons who should receive full active immunization in any event.

1. All infants born into the community should be immunized according to an accepted schedule, such as that suggested by Emond (1972–73).
2. All people with a history of allergy.
3. People whose occupation places them at greater risk from tetanus than others. Such occupations would include agricultural workers, sewer workers, garage mechanics, sportsmen, builders, verterinary surgeons and laboratory workers. Drug addicts are also included in this category.
4. All people requiring elective surgery, those with chronic ulcers and those required to wear a plaster case.
5. All pregnant women. Apart from the active immunity developed by the expectant mother, active immunization during pregnancy provides neonates with a substantial protection against the disease. Women whose newborn children are likely to be exposed to the risk of neonatal tetanus are virtually non-existent in Great Britain, but are commonly encountered in underdeveloped communities such as occur in New Guinea, Nigeria, Thailand and the South Americas. Following a suggestion of the World Health Organization (1950) that experiments be made to

determine whether active immunization of pregnant women would prevent neonatal tetanus, data have been collected by Katitch (1960), Schofield, Tucker and Westbrook (1961), MacLennan *et al.* (1965), Newell *et al.* (1966, 1967) and Garnier *et al.* (1975) which show that such a procedure provides neonates with a substantial protection against the risk of tetanus. Rubbo (1958) considered that all pregnant women should be actively immunized against tetanus.

6. All patients who develop clinical tetanus. Active immunization of these people is essential for two reasons. Patients who suffer from clinical tetanus acquire no significant immunity to the disease, and are therefore as vulnerable to a second episode as they were to the first. Moreover, the use of heterologous tetanus antitoxin in the treatment of tetanus induces hypersensitivity to the horse serum, so that subsequent exposure to serum may be followed by catastrophic reactions, and by accelerated elimination of the serum (*see below*). Active immunization eliminates the necessity of exposure to horse serum should the patient be in danger of tetanus again. Active immunization in tetanus patients is conveniently initiated when the patient first presents with the disease by giving a first dose of adsorbed toxoid intramuscularly. The simultaneous use of therapeutic antitoxin (heterologous or human) does not interfere with the immunological response to the toxoid provided the preparations are given by different syringes into different sites (Smith *et al.*, 1963).
7. All patients who have received heterologous antitoxin as a prophylactic measure against tetanus. This is important for the reasons of hypersensitivity discussed in 6 above. In this connection it should be remembered that sensitized persons eliminate foreign serum rapidly, so that serum administered for prophylactic purposes is largely ineffective. It is probably this accelerated elimination of the serum from sensitized individuals which explains cases of tetanus which have developed in patients at risk after they have been passively protected (*see* Littlewood, Mant and Wright, 1954; Woodward, 1960). 'It is of paramount importance to think of heterologous tetanus antitoxin as a preparation to be administered only once in a lifetime. If it is used, subsequent active immunization must be mandatory, and should be the responsibility of the physician who orders the antitoxin to be given' (Filler and Ellerbeck, 1960).
8. All patients who have received diphtheria antitoxin (*see* 6 and 7 above).

When the above categories are summated it is evident that the recommendation is that virtually the whole population should be actively immunized. 'Tetanus can and should become a disease of historical significance only' (Furste, 1969).

CONTRAINDICATIONS TO ACTIVE IMMUNIZATION

There are few contraindications to active immunization in persons not at risk from tetanus (*see* Cvjetanovic, Bytchenko and Edsall, 1976). In those uncommon instances when patients develop hypersensitivity to toxoid during the course of basal immunization, unless hypersensitivity is severe, immunization may be completed safely by injecting a smaller dose of adsorbed toxoid intramuscularly (Trinca, 1963).

Active immunization does not provide protection in non-immune patients who require immediate prophylaxis (*see* Thompson and Harper, 1974), but in such cases active immunization is conveniently initiated with adsorbed toxoid by combining active with passive immunization.

BOOSTER DOSES OF TOXOID

In recent years there has been a good deal of discussion about the frequency with which booster doses of tetanus toxoid should be administered. A number of studies have shown that a single dose of toxoid given as many as 25 years after previous active immunization acts as an adequate recall stimulus (Moss, Waters and Brown, 1955; Eckman, 1964; Scheibel *et al*, 1966; Gottlieb *et al*, 1967; Trinca, 1967, 1974). These observations suggest that many people have been grossly 'overimmunized' in the past, with consequent waste of time and money, and with the attendant slight risk of unpleasant hypersensitivity reactions.

Current recommendations concerning the frequency of booster injections of adsorbed toxoid vary from between one every 10 years to their not being required at all (Rubbo, 1966; White *et al.*, 1969; Peebles *et al.*, 1969).

The following is a reasonable compromise. Actively immunized persons should receive booster doses of adsorbed toxoid at 5–10 year intervals. In immune patients who present with a tetanus-prone wound a booster dose of toxoid is recommended only for those whose final basal course injection or last booster dose of toxoid was given more than 12 months previously. This means that full advantage is taken of their basal immunity, and the necessity of exposure to serum is precluded. Indeed, in actively immunized persons, prophylactic antitoxin is contraindicated, except, perhaps, in severely shocked patients who have seriously infected untreated wounds of some days' standing. In these cases adsorbed toxoid should be given at the same time as antitoxin (*see* p. 300).

A patient who is receiving a course of active immunization may sustain an injury which exposes him to the risk of tetanus. The problem then arises as to which type of prophylaxis should be given, antitoxin or toxoid. If only one dose of toxoid has been given, the patient is regarded as non-immune and should be given antitoxin and adsorbed toxoid (*see* Ullberg-Olsson and Eriksson, 1975). If two doses of toxoid have been given, the second only recently, the patient cannot be regarded as fully immune. In such cases one must balance the risks by considering the nature, extent and age of the injury, and the likely effectiveness of surgical toilet. If, on the other hand, the second dose of toxoid was given some months previously, adequate protection is provided by the third dose of toxoid, given at the time of wounding.

Passive immunization

Generally speaking, administration of antitoxin to the non-immunized is indicated for persons with soiled wounds, including women who have submitted to a criminal abortion. *Tetanus proneness should be assessed after the wound has been treated surgically.*

REFINED HORSE SERUM (HETEROLOGOUS) ANTITOXIN

A dose of 1500 units (1950) is injected subcutaneously as early as possible. For severe or grossly contaminated or infected wounds, the dose of antitoxin may be increased to 3000 units. Simultaneously, adsorbed toxoid is administered by a different syringe and at a different site, to initiate an active immunity. In view of the reactions which may occur following administration of heterologous serum, careful appraisal of each individual case is essential before the decision to give antitoxin is taken. Antitoxin is not indicated in actively immunized persons, in those with clean superficial injuries and in those whose wounds receive early and adequate surgical attention.

The main disadvantages of heterologous antitetanus serum are the allergic reactions which are likely to follow its administration, and the accelerated elimination of antiserum from sensitized patients. The latter is of some importance, since in these cases quite substantial doses of antitoxin provide inadequate protection.

An unusual case of accidentally acquired hypersensitivity to heterologous antitoxin was reported by Sneddon (1960), in which a casualty officer developed severe symptoms of sensitivity merely from handling the serum. This patient, who had not previously been exposed to horse serum, was probably sensitized by the tetanus antitoxin entering the conjunctival sacs; during administration of antitoxin to others it was this patient's practice to hold the syringe at eye level and express antitoxin through the needle to obtain the correct dose.

It is appropriate to consider here some of the reactions which might follow administration of heterologous antitoxin.

Anaphylactic shock. The onset occurs in a few minutes, or may be delayed up to 2 h. It is characterized by pallor, dyspnoea and collapse, and sometimes also by the appearance of oedema and a rash. The condition is treated by immediate intramuscular adrenaline (0.5–1.0 ml of 1 in 1000). The patient is treated for shock, and is given full doses of intramuscular antihistamine. The adrenaline is repeated in 15 min if recovery is delayed. Usually, however, recovery is quite rapid.

Serum sickness. The common 'delayed' type of reaction occurs 7–12 days after heterologous serum therapy. The 'accelerated' type, which occurs typically in persons who have had serum before, develops in 3–4 days. It is characterized by oedema, pyrexia, joint pains and urticarial rashes. The condition is treated by oral antihistamine, which may be given by injection in serious cases.

Local reactions. These occur after 7–10 days and consist of local erythema and urticaria. They are satisfactorily treated by oral antihistamine.

Neurological complications. These have been described by Allen (1931), Bennett (1939) and others. Within a few hours after the onset of serum sickness, severe neuritic pains develop, involving the neck and pectoral girdle. Flaccid paralysis of some shoulder muscles may follow after a few days, and the muscles gradually atrophy. There may be motor paralysis, fibrillations or sensory loss. After recovery from the serum sickness stage, treatment is largely symptomatic. Complete recovery may be delayed up to six months.

Some of these complications of serum therapy are fortunately rare. Thus, Laurent and Parish (1952), basing their calculations on reports in the literature, estimated an incidence of one death in every 50 000–200 000 injections of heterologous antitoxin given. Local reactions, and serum sickness are, however, all too common. Scheibel (1955) estimated that patients receiving prophylactic heterologous serum run a 20–30% risk of delayed serum reactions, but Moynihan (1955), in Great Britain, reported an incidence of only 5.3% reactions among 7580 injections of antitoxin. Toogood (1960), who estimated a similar incidence of serum reactions in Canada suggested that allergic reactions to prophylactic antitoxin caused more morbidity and mortality than the disease which the serum was intended to prevent. From data accumulated over one year in Ontario, he calculated that the risk of disability due to severe serum reactions was 265 times greater than that from tetanus itself.

With these risks of heterologous serum therapy in mind, it is easy to understand why there has been such an exaggerated swing away from the use of antiserum prophylaxis (*see* Caro and Shaw, 1974), despite the fact that there remain a few absolute indications for its use. In the last few years too great a reliance has been placed on antibiotics as substitutes for antitoxin in tetanus prophylaxis, although in appropriate cases antibiotics do find an important place in the prophylactic regimen. In countries where human antitetanus immunoglobulin (*see* p. 299) is commercially available, there is never any indication for the use of heterologous antitoxin. In those parts of the world, however, where human antitoxin cannot be obtained, heterologous serum remains an important agent in tetanus prophylaxis. The use of serum prophylaxis (homologous or heterologous) should always be combined with adsorbed toxoid active immunization – a matter of particular importance when heterologous serum is used.

In an analysis of 3455 patients who had received tetanus prophylaxis over a period of nine months, Binns (1961), in Great Britain, made the startling and alarming observation that little or no effort had been made to provide active immunity for patients who had received serum, had asthma or infantile eczema or had shown a previous reaction to serum.

Before administration of heterologous antiserum, patients should be tested for sensitivity to the horse serum. There can be no doubt that intradermal testing is unreliable, both 'positive' and 'negative' results providing little guide to the sensitivity of the patient (Moynihan, 1956; Laurent and Parish, 1962). Some clinicians prefer to omit sensitivity testing altogether, but keep the patient under observation for half to one hour after administration of the full dose of antiserum, and should an immediate reaction develop, this is controlled with parenteral adrenaline. A better procedure is the 'trial dose' method recommended by Laurent and Parish (1952, 1962) and Moynihan (1956). Here, a trial dose of serum (0.1 ml) is injected subcutaneously, and the patient's general condition is observed before a larger dose is given. If the patient is hypersensitive further heterologous serum must not be administered. In view of the sensitization which is likely to result from treatment with tetanus antiserum, and from the prophylactic or therapeutic administration of gas gangrene and botulinum antitoxin, it is important to repeat here that passively immunized patients must be actively immunized with adsorbed toxoid.

HOMOLOGOUS ANTITOXIN

The use of convalescent serum or human gamma globulin in the prophylaxis of such conditions as measles, hepatitis and vaccinia is well known, and although the use of homologous antitoxin in passive immunization against tetanus had been suggested some years earlier, the possibility of its use was first seriously considered by Suri and Rubbo (1961) and Rubbo and Suri (1962). In a series of animal experiments, these workers showed that homologous serum was at least one hundred times more effective in protecting guinea pigs against tetanus than heterologous antitoxin, and that in man it was superior to heterologous immunization in every respect. Using gamma globulin prepared from actively immunized Europeans, they found that high antitoxin levels in man could be obtained with relatively low doses, and that the rate of disappearance of the antitoxin was slow. Rubbo and Suri recommended that a protective dose of human immune globulin for an average adult should be 400 units, and for children 200 units. They found that these doses gave serum levels of antitoxin of not less that 0.05 units per ml for at least 14 days, levels which are considered to be protective for passively induced homologous antitoxin.

The advantages of homologous serum protection in man are enormous, for no untoward serum reactions occur, the rate of elimination of the antibody is much slower than with heterologous serum, and accelerated elimination due to sensitization is not induced.

Tetanus gamma globulin (human) was made available in America and in the United Kingdom some years ago, but its use was restricted to prophylaxis in patients who were allergic to heterologous antitoxin (*Journal of the American medical Association*, 1959; Department of Health and Social Security, 1971). It is now generally agreed that human antitetanus immunoglobulin should replace heterologous serum in the modern prophylaxis of tetanus (Furste, 1969; *British medical Journal*, 1967; *Lancet* 1974; Smith, 1974; Smith, Laurence and Evans, 1975; Willis, 1975; *Drug and Therapeutics Bulletin*, 1975), and it is now generally available commercially in Great Britain as Humotet (Wellcome). For unimmunized persons at risk from tetanus, who would also receive a first dose of adsorbed toxoid, an intramuscular dose of 250–500 units of human antitetanus immunoglobulin is recommended. The commercial material is expensive – a dose of 250 units costs £2.56, compared with about 7p for an equivalent dose of equine antitoxin, but this is regarded as a small price to pay for sound, immediate and safe protection.

A wealth of valuable information about antisera and toxoids has been published by Parish and Cannon (1961). A brief, but interesting review of the development of knowledge of antitoxins has been written by Pope (1963), and Smith (1969, 1974) has provided informative accounts of tetanus toxoid and human tetanus immunoglobulin.

Antibiotic prophylaxis

In view of the problems of serum therapy, it was recommended by Cox, Knowelden and Sharrard (1963) that antibiotic therapy, combined with surgical treatment of wounds, should replace heterologous antitoxin as the modern prophylaxis against tetanus.

The efficacy of antibiotic therapy is dependent not only on the sensitivity of contaminating strains to the drug, but also on the concentration of it attainable at the site of contamination. It seems likely that most strains of *Cl. tetani* are highly sensitive to penicillin, tetracycline and erythromycin. Experiments with animals indicate that provided large enough doses are given, therapeutic concentrations of these drugs may be obtained in some necrotic foci that are the seat of *Cl. tetani* infection. Work on the action of penicillin in mice exposed to *Cl. tetani* has shown that a blood serum level of free penicillin of 0.1 unit per ml is protective (Smith, 1964). In order to be effective, penicillin prophylaxis must be commenced soon after injury, within six hours, and continued for at least five days.

It is difficult to compare the relative efficacy of antitoxin and penicillin as prophylactic agents against tetanus in man, so that the effectiveness and limitations of antibiotic prophylaxis must be judged by laboratory experiments. Although the available evidence supports the view that antibiotics and surgical treatment may be as efficacious as prophylactic antitoxin and surgical treatment, the effectiveness of antibiotics for prophylaxis against tetanus in man remains unproven

(Lucas and Willis, 1965; Eckmann, 1967). In discussing the success of the regimen of tetanus prophylaxis used at Sheffield, where wound surgery and antibiotics are used instead of wound surgery and antitoxin, Sharrard (1964) concluded that 'since the regimen of treatment now differs only in that we have stopped giving serum, the inference must be that the antibiotic, in wounds possibly containing *Cl. tetani*, has been the effective agent . . .'. This might be so. It seems more likely, however, that surgery alone more often than not provides the essential protection. After a particular wound has received surgical toilet, a judgment is then made on the tetanus-proneness of the treated wound. A judgment of tetanus-proneness is based on two conclusions; (1) that the wound is still likely to be contaminated with tetanus bacilli, and (2) that anaerobic conditions will prevail in the wound. Under these circumstances it is difficult to see how antibiotics can act as reliable prophylactics. In order to be effective the antibiotics must kill the growing organisms, or at least inhibit their growth; that is, the drugs must penetrate and reach a therapeutic level within a region of tissue that has no blood supply – it will be recognized that antibiotics are without effect on tetanus toxin, but can only prevent an intoxication by preventing its formation by the organisms that produce it. Thus, despite the evidence of animal experimental work, the value of antibiotics as *total* replacement agents for antitoxin in the management of patients at risk from tetanus is doubtful. There seems little doubt,

Table 7.4 Alternative antibiotic schedules for tetanus prophylaxis (after Adams *et al.*, 1969)

Penicillin	1 Phenoxymethyl penicillin, 250 mg 6-hourly for 5 days
	OR
	2 Fortified procaine penicillin (BP), 2.0 ml intramuscularly daily for 5 days
	OR
	3 Cloxacillin, 500 mg 6-hourly (before meals) for 5 days
	OR
Tetracycline	250 mg 6-hourly for 5 days
	OR
Erythromycin	500 mg 6-hourly for 5 days

however, that chemoprophylaxis, used with care and discretion, can play a role in modern tetanus prophylaxis. A number of alternative antibiotic schedules for tetanus prophylaxis are outlined in *Table 7.4* (Adams, Laurence and Smith, 1969).

Summary of tetanus prophylaxis

1. Active immunization provides the best and safest protection against tetanus.
2. Wounded patients who are known to be actively immune do not require passive protection, but may be given a booster dose of adsorbed toxoid at the time of wounding.
3. All wounded patients are treated surgically. Then an assessment is made of the tetanus-proneness of the treated wound.
4. Tetanus-proneness of a wound is determined after surgical toilet. Consideration is given to such factors as the success of surgery, the extent and degree of remaining contamination, the age of the injury, the presence of wound infection due to other organisms, the possibility of a retained foreign body and the immune status of the patient.
5. For non-immune patients with a tetanus-prone wound, passive protection is required and is best provided by human antitetanus immunoglobulin. If human antitoxin is not available, heterologous antitoxin should be used.
6. When heterologous antitoxin is used, its administration is preceded by a test dose. Should a hypersensitivity reaction develop, further horse or other foreign serum should not be given.
7. When passive protection is given, it is accompanied by a first dose of adsorbed toxoid, administered quite separately from the antitoxin.
8. Tetanus toxoid, injected for the first time into a non-immune individual at risk from tetanus provides no protection against the risk at this time.
9. If antibiotics are used as prophylactic agents, they must be given in adequate doses for an adequate time. It is probably wise to give parenteral antibiotics prior to surgical toilet.

Treatment of tetanus

The treatment of established tetanus falls under five headings:

1. Prevention of the absorption of further toxin by the administration of antitoxin, and by surgery of the wound.

2. Control of reflex spasms.
3. Prevention of intercurrent pulmonary infection.
4. Prevention and control of other intercurrent infections.
5. Control of fluid and electrolyte balance, and maintenance of nutrition.

Therapeutic antitoxin

The therapeutic administration of antitoxin to patients who have already developed clinical tetanus is of disputed value. Judging by the poor results obtained, it seems likely that by the time tetanus is clinically evident, the neurotoxin is 'fixed' to the nervous tissue. However, there is little doubt that toxin present in the blood and other tissues is readily neutralized by antitoxin. Serum therapy must, therefore, be considered an important part of treatment and should be directed at producing an effective concentration of antitoxin in the blood. For equine tetanus antitoxin, all available evidence suggests that a dose of 10 000 units, given by a single subcutaneous or intramuscular injection, is sufficient for tetanus of any severity. Human antitetanus immunoglobulin, e.g. Humotet (Wellcome), is the antitoxin of choice, however, the recommended dosage being 30–300 units per kg of body weight, given intramuscularly (*see Lancet*, 1976).

The production of toxin by the infecting organism may be a continuous process, which can be interrupted only by removal or destruction of the tetanus bacilli (*see* Smith and MacIver, 1969; 1974). This is best accomplished by thorough surgical toilet of the wound, but in idiopathic cases this is clearly not possible.

General management

General measures for the management of tetanus may include nasogastric feeding, and tracheostomy which enables artificial ventilation of the lungs to be properly controlled. Adequate surgical toilet of the wound is important and should be delayed only until antitoxin has been administered. Muscular relaxants and sedatives are given in appropriate cases, and this is continued until all signs of tetanic spasms have passed.

Treatment of human tetanus with hyperbaric oxygen drenching in a compression chamber has been reported upon by a number of workers (Rhea *et al*., 1967; Milledge, 1968; Lockwood and Langston, 1970; Hill and Osterhout, 1973). Its value is doubtful.

Bacteriological control is important during the management of the acute phase of the illness. Infection of the tracheostomy wound and respiratory passages may develop, and must be controlled by antibiotic therapy. Intercurrent chest infection may cause the death of a patient who might have recovered from tetanus.

Following recovery from a clinical attack of tetanus, the patient should receive a full course of active immunization with adsorbed tetanus toxoid. This is necessary because the natural immunity conferred by the disease may be feeble and of short duration. Active immunization is conveniently initiated early in treatment of the disease, the second and third doses of toxoid being given subsequently.

Tetanus treatment units

Modern experience dictates that the only place for the effective management of tetanus is the fully equipped and experienced tetanus unit or intensive therapy unit. A report in the early 1960s on the treatment of 36 consecutive cases of tetanus without a single death at the Tetanus Unit in Leeds, attributed the success chiefly to increased skill in management of the disease (Ellis, 1963). An American report similarly concluded that 'the common denominator of success in the reported larger series of tetanus with highest survival rates is emphasis on comprehensive nurse and physician care, under supervision of interested and experienced staff' (Box, 1964). This policy finds more recent support in a *Lancet* annotation in which it is emphasized that 'treatment demands close collaboration between physicians, anaesthetists, and surgeons, careful laboratory supervision, and, perhaps most important of all, the highest standard of nursing. The trend in Britain has been for treatment to be concentrated in centres, such as that in Leeds, where intensive care units have been established for patients whose treatment places a great strain on the resources of traditional wards . . . . Every effort should be made to deliver patients with tetanus to these units, if necessary by helicopter.' (*Lancet,* 1967).

Useful accounts of the management of tetanus have been published by Ablett (1967), Adams, Laurence and Smith (1969), Cole and Youngman (1969), Furste (1969), Furste and Wheeler (1972), Christensen (1969), Macrae (1973), Smythe, Bowie and Voss (1974), Sanders, Strong and Peacock (1969), and Busuttil, Pace and Muscat (1974).

GAS GANGRENE (CLOSTRIDIAL MYONECROSIS)

The organisms associated with gas gangrene are discussed in Chapter 4; the three most important are *Cl. perfringens* type A, *Cl. septicum* and

Table 7.5 Clostridia isolated from gas-gangrenous wounds by different workers

Organism	*Percentage of cases*				
	MacLennan (1943c) 146 cases	*Cooke and his colleagues (1945) 72 cases*	*Smith and George (1946) 110 cases*	*Stock (1947) 30 cases*	*Dhayagude and Purandare (1949) 25 cases*
Cl. perfringens	56	48	39	80	52
Cl. novyi	37	0	32	50	0
Cl. septicum	19	0	0	3	36
Cl. histolyticum	6	0	0	0	16
Cl. sporogenes	37	11	54	63	0
Cl. bifermentans	4	2	54	23	0
Cl. tetani	13	0	4	10	0
Cl. tertium	30	1	3	13	0
Cl. butyricum	13	2	3	7	0
Cl. cadaveris	5	3	3	0	0
Cl. fallax	1	6	3	3	0
Cl. cochlearium	9	0	2	3	0
Cl. lentoputrescens	19	0	2	0	0
Cl. sphenoides	3	0	2	0	0
Cl. tetanomorphum	2	0	0	0	0
Cl. hastiforme	3	0	0	0	0
Cl. regulare	0	0	2	0	0
Cl. multifermentans	0	0	5	0	0
Unidentified clostridia	16	3	53	13	0

Cl. novyi type A, while *Cl. histolyticum, Cl. sporogenes, Cl. sordellii, Cl. bifermentans* and *Cl. fallax* are of less importance. To this list must be added the anaerobic streptococci (peptostreptococci) which are also sometimes present in gangrenous infections (MacLennan, 1943a, b, c, and d) (*see below*). We may note here the average incubation periods of the three main types of gas gangrene: infection with *Cl. perfringens,* 10–48 h; with *Cl. septicum*, 2–3 days; and with *Cl. novyi*, 5–6 days.

In gas gangrenous infections of wounds, usually more than one anaerobic species is involved. Monomicrobic infections occur, however; and in both types of infection facultative organisms, such as *Proteus* species, are frequently present also. Surveys of the anaerobic flora of gas gangrenous wounds have been made by MacLennan (1943c), Cooke *et al.*, (1945), Smith and George (1946), Stock (1947) and Dhayagude and Purandare (1949). Their findings, summarized in *Table 7.5*, are in general agreement, excepting those of Dhayagude and Purandare, who recorded a much higher incidence of *Cl. septicum* and *Cl. histolyticum* and failed altogether to recover *Cl. novyi* from their patients.

The genesis of gas gangrene

The mode of development of anaerobic wound infections has already been briefly discussed. Gas gangrene is considered to have commenced 'not when a wound has become infected with the pathogenic anaerobes, but from the moment when a group of these bacteria have been enabled to surround themselves with a toxin sufficiently concentrated to abolish the local defences of the tissues' (Committee upon Anaerobic Bacteria, 1919). Further multiplication and toxin production leads to rapid invasion of healthy tissue and the development of clinical gas gangrene.

Lecithinases, such as *Cl. perfringens* α-toxin and *Cl. novyi* γ- toxin, may damage or destroy cell membranes. The effect of such enzymes on capillaries may render them freely permeable to fluid and protein, leading to increased tension in affected muscles, with a resultant increased anoxia. Lysis of erythrocytes by the α-toxin is probably the cause of the haemolytic anaemia and haemoglobinuria associated with *Cl. perfringens* gas gangrene. There is adequate evidence that in *Cl. perfringens* type A infections the lecithinase is the most important lethal factor, and that α-antitoxin is the only significant antibody in the control of the disease (Evans, 1943a, b, 1945, 1947; Kass, Lichstein and Waisbren, 1945).

The collagenases of *Cl. perfringens* (κ-toxin) and *Cl. histolyticum* (β-toxin) are able to destroy collagen barriers in the tissues, which might otherwise tend to localize the infection, while destruction of reticulin around capillaries leads to haemorrhage and thrombosis, with resulting increased anoxia.

The hyaluronidase of *Cl. perfringens* (μ-toxin) greatly facilitates the spread of the organism through the tissues, but whether this is mainly due to the removal of a physical barrier, or to the liberation of fermentable carbohydrates following breakdown of hyaluronic acid, is uncertain.

While there is ample evidence to show that the exotoxins produced by the clostridia are largely responsible for the nature of the lesions and the systemic manifestations, the modes of action of the toxins, and the individual importance of each, are far more complex and uncertain than the outline given above would suggest. The literature on this subject has been reviewed by Oakley (1954), Macfarlane (1955) and Willis (1969), and the pathology of gas gangrenous infections has been reported upon by Robb-Smith (1945), Govan (1946) and Aikat and Dible (1956, 1960).

Diagnosis

As with tetanus, the diagnosis of gas gangrene is made on clinical grounds. The presence of pathogenic clostridia, especially *Cl. perfringens*, in a wound is, in itself, of no diagnostic significance (*see* Lowbury and Lilly, 1955). The term 'gas gangrene' is, to some extent, misleading, because gas may not be an obvious feature of the disease. Moreover, the presence of gas in a wound or other infected tissue may be due to a variety of causes other than clostridial infection (Filler, Griscom and Pappas, 1968; Chopra and Mukherjee, 1970; Brightmore, 1971; Brightmore and Greenwood, 1974; VanBeek *et al.* 1974; Bessman and Wagner, 1975). In this connection it seems clear that most of the 12 cases of gas gangrene of the scrotum and perineum described by Himal and his colleagues (1974) were not true clostridial infections.

Anaerobic myonecrosis must be distinguished from clostridial cellulitis, and from anaerobic streptococcal myositis (Anderson, Marr and Jaffe, 1972) (*see* p. 234).

Microscopic examination of pathological material

Tentative confirmation of the clinical diagnosis may be aided considerably by microscopic examination of pathological material. Thus, the presence of large numbers of regularly shaped Gram-positive bacilli without spores is strongly suggestive of a *Cl. perfringens* infection, and Butler (1945) drew attention to the correlation between the severity of *Cl. perfringens* infections on the one hand, and the degree of capsulation of the organisms and the extent of leucocyte damage on the other (*see* Stratford, 1973). In anaerobic streptococcal myositis, Gram-stained films may be diagnostic.

Culture from pathological material

The most important step in identifying the infecting organisms is by culture, and this should include cultures for both aerobes and anaerobes. The object is not merely to identify the bacterial species present in the wound, but to assess their relative numbers and significance (Lindsey, 1959). For this reason, primary direct plating is of the utmost importance. Preliminary enrichment without recourse to direct plating may give a completely false impression of the relative importance of the anaerobes ultimately isolated. The following examination, carried out by the author on an excised wound from a patient with clinical tetanus illustrates this important factor.

No organisms were seen in Gram-stained films of the material. The material was plated on fresh blood agar, heated blood agar and on an egg yolk agar medium with and without neomycin. The remainder (and bulk) of the specimen was inoculated into a number of tubes of cooked meat broth, some of which were heated to 100 °C for varying times. After 24 h incubation, all the anaerobic plate cultures showed abundant growth of *Cl. tetani*; on the egg yolk and fresh blood agar plates, 12, 11 and 7 colonies of *Cl. perfringens* were present respectively, and one colony of *Cl. sporogenes* was present on the neomycin egg yolk plate. On the heated blood agar plate, *Cl. tetani* appeared to be present in pure culture. Aerobic plate cultures were sterile. Plating from the enrichment cultures after 16 h incubation produced the following results: three unheated samples gave apparently pure growths of *Cl. perfringens*, two different strains being obtained; one of the heated samples gave an apparently pure growth of *Cl. sporogenes*, while the others were sterile. After further incubation of the enrichment cultures, an aerobic spore-forming bacillus was obtained from them all, and *Cl. tetani* was isolated from the unheated samples.

Method of examination of wounds

The following method of examination of wounds for clostridia is recommended (*Table 7.6*). It is based on that of Hayward (1945). This

Table 7.6 Scheme for examination of wounds for clostridia

Pathological material				
Films				*Cultures*
Aerobic fresh blood and heated blood agar, and lactose egg yolk milk agar	Anaerobic fresh blood and heated blood agar	Anaerobic half-antitoxin lactose egg yolk milk agar with and without neomycin	Cooked meat broth	Cooked meat broth
			Heat to 80 °C for	Plate out at
(24 h)	(24 and 48 h)	(24 and 48 h)	5 10 15 20 min	1 2 4 7 days
Aerobic organisms	Anaerobes (use concentrated agar as indicated)	Egg yolk positive clostridia and *Cl. histolyticum*	Plate out for *Cl. novyi* and *Cl. septicum* using half-antitoxin plates	Early subcultures for rapidly growing clostridia. Later ones for slow growing ones. Use differential heating to reduce aerobic contaminants and fast growing anaerobes. Use concentrated agar as indicated

procedure aims at obtaining all the relevant organisms in pure culture. Its full execution, however, is not always necessary, and must not delay the issue of a preliminary report, which usually can be made with confidence (and by telephone) within 24 h.

1. Direct films.
2. Inoculate exudate or tissue onto: (a) aerobic and anaerobic horse blood agar; (b) aerobic and anaerobic heated blood agar; (c) anaerobic half-antitoxin egg yolk agar, with and without neomycin; (d) aerobic egg yolk agar. Lactose egg yolk milk agar cultures give more definitive information than those on plain egg yolk plates, but require more experience for their interpretation. The complex egg medium is included in *Table 7.6*.
3. Inoculate material into four tubes of cooked meat broth for anaerobic incubation, and heat to 80 °C for 5, 10, 15 and 20 min respectively. The primary object here is to isolate *Cl. novyi* and *Cl. septicum*, the tubes being subcultured after 24 and 48 h incubation to aerobic and anaerobic fresh blood agar and to aerobic and anaerobic half-antitoxin egg yolk agar plates, neomycin being present in the latter medium for anaerobic incubation.
4. Inoculate material into cooked meat broth for anaerobic incubation. This enrichment broth is subcultured to plates as in 2 above, at 1, 2, 4 and 7 days. It is useful to make these subcultures in duplicate, the inoculum in one case being heated to 80 °C for 10 min before transfer.

It is often advisable to use concentrated agar plates for anaerobic culture in order to prevent swarming growth. These plates may be inoculated in addition to, or instead of, those mentioned above. The procedures mentioned in 2c and 3 are carried out to isolate particular species. The half-antitoxin egg yolk agar plates, using a mixture of *Cl. perfringens* type A and *Cl. novyi* type A antitoxic sera, indicate the presence of *Cl. perfringens, Cl. bifermentans, Cl. sordellii, Cl. novyi* and *Cl. sporogenes*. An incubation period of 48 h is required for full development of the cultural characters of the last two clostridia. A separate anaerobic jar should be used for examining cultures at 24 h, since a brief exposure to air at this time may prevent further development of tiny, and as yet unrecognizable, colonies of *Cl. novyi*. If the amount of specimen submitted for examination is limited, step 3 may be omitted from the above scheme.

Blood cultures

Blood cultures are often positive, especially in cases of *Cl. perfringens* and *Cl. septicum* gas gangrene, though *Cl. perfringens* bacteraemia can occur in the absence of overt infection (Butler, 1937; Rathbun, 1968; Ellner and O'Donnell, 1969). It is important, therefore, that blood cultures should be made, suitable media being cooked meat broth, glucose broth and Thiol broth (Difco). These are incubated anaerobically and are dealt with in the usual way.

Prophylaxis and treatment

Without doubt, surgery is by far the most important prophylactic measure. Early radical excision is imperative; all damaged muscle is removed, leaving only well-vascularized tissue and eliminating most of the contaminating organisms. The efficacy of early surgery is substantiated by experience during recent military operations. During the Korean War, Howard and Inui (1954) showed that severe clostridial infections in battle casualties were rare when early and adequate surgical treatment was available; Latta (1951) reported only 3 cases among 1850 wounded. During the military operations in Borneo (1963–65) there was no reported gas gangrene among 119 wounded (Wheatley, 1967), and in Vietnam, Moffat (1967) recorded only 2 cases among 60 wounded personnel. Surgical debridement is combined with local and parenteral antibiotic therapy. Any effective drug may be used, (*see* Dornbusch, Nord and Dahlback 1975; Dornbusch, Nord and Olson 1975), but mixtures of drugs, such as benzylpenicillin and sulphonamides, appear to give better results. Chemoprophylaxis alone is of no value, but its use in conjunction with surgical toilet may be expected to reduce the risk of an anaerobic infection developing, especially if treatment is initiated early after wounding. An appropriate prophylactic regime utilizing soluble penicillin would be at least 2 million units daily given by intramuscular injection in divided doses (Garrod, 1958; Garrod *et al.*, 1973). Parenteral antibiotics should be commenced before surgical toilet is undertaken. These days, there appear to be few indications for the use of prophylactic gas gangrene antitoxin. In exceptional cases, such as those with extensive soiled wounds of long standing, in which adequate surgical toilet cannot be effected, intramuscular administration of polyvalent gas gangrene antitoxin might be considered. In clinical experience, there is no clear evidence to suggest that antitoxin combined with surgical and antibiotic prophylaxis is more effective than surgical and antibiotic prophylaxis alone.

Owen-Smith and Matheson (1968) and Boyd and his colleagues (1972) respectively, studied antibiotic and antitoxin prophylaxis of gas gangrene in sheep subjected to high velocity missile wounds contaminated with gas gangrene organisms. Penicillin alone or antitoxin alone were found to be completely effective in preventing gas gangrene if given within 9 h of wounding and challenge. Although there is no acceptable way of determining the efficacy of non-surgical prophylaxis in man, these observations add considerable weight to the importance of combining antibiotic therapy with surgery.

It is pertinent to reiterate here that patients submitted to elective lower limb surgery are at risk from endogenously derived gas gangrene, especially if there is obliterative arterial disease. Prophylaxis in these patients is provided by thorough preoperative skin preparation (Ayliffe and Lowbury, 1969), and by the use of prophylactic benzylpenicillin.

The treatment of established gas gangrene has been revolutionized in recent years by the introduction of hyperbaric oxygen therapy. The early pioneer reports of its value in the management of gas gangrene by Brummelkamp and his colleagues (1961, 1963), and Smith *et al.* (1962) have been fully substantiated by subsequent experience (Irvin and Smith, 1968; Chew, Hanson and Slack, 1969; Slack, Hanson and Chew, 1969; Johnson *et al.*, 1969; Rodin, Groeneveld and Boerema, 1972; *British medical Journal,* 1972; Schweigel and Shim, 1973; Jackson and Waddell, 1973; Hitchcock, Demello and Haglin, 1975; Holland *et al.*, 1975). The effects of hyperbaric oxygen on micro-organisms, and the factors affecting the response of clostridial infection to hyperbaric oxygen therapy have been reviewed by Gottlieb (1971). The modern management of gas gangrene is along the following lines.

When the patient is admitted to the unit, any sutures are removed from the wound which is widely opened; no surgical excision is performed at this time. After collection of wound swabs for bacteriological examination, and of blood for culture and biochemical analyses, antibiotic therapy is commenced. Slack, Hanson and Chew (1969) advocated a combination of ampicillin 500 mg intramuscularly 6-hourly and benzylpenicillin 1 mega unit intramuscularly 6-hourly for at least 5 days. Treatment sessions with hyperbaric oxygen are then commenced. Since hyperbaric oxygen drenching does not rejuvenate necrotic tissue, surgical toilet of the wound is performed as soon as the patient's general condition has improved. There is no place for the use of gas gangrene antitoxin in this regime.

Since untreated gas gangrene is often fulminating and leads to a rapidly fatal intoxication, it is imperative that afflicted patients be transferred early and speedily to a hyperbaric oxygen unit. Clinical over-reaction is not misplaced if gas gangrene is suspected; it is better to send a misdiagnosed case for hyperbaric oxygen therapy than to procrastinate with a genuine one.

Active immunization

Penfold and Tolhurst (1937, 1938) and Penfold, Tolhurst and Wilson (1941) showed that formol–toxoid gave protective antitoxic immunity to guinea pigs and rabbits against experimental infections with *Cl. perfringens*, and that in man also there was a good antigenic response to this toxoid. Subsequent work in this field was reviewed and summarized by Oakley (1943, 1945).

Recent studies by Boyd and his colleagues (1972a, b) showed that active immunization of sheep with a mixed *perfringens–septicum–novyi* toxoid provided sound protection against infection by *Cl. novyi* and *Cl. perfringens*. When immunized animals were wounded and challenged in the thigh muscle by a high velocity bullet together with spores of the organisms, protection was almost completely effective in preventing the onset of *Cl. novyi* infection, even one year after immunization; the survival rate in animals challenged with *Cl. perfringens* was between 83 and 94%. In these experiments, in which no post-traumatic treatment was given, it seems likely that protection was due solely to circulating antitoxin present as a result of immunization, and not to an anamnestic response to toxin produced in the wounded animals. These results of prophylactic immunization are most promising, but there are at present no data by which its efficiency can be estimated in man. It seems unlikely that immunized patients would benefit from a booster dose of toxoid at the time of wounding, since the disease commonly develops much more rapidly than does the secondary antibody response. In guinea pigs, the response to a booster dose develops after 3–10 days (Robertson and Keppie, 1943); the incubation period of gas gangrene in man is commonly less than two days, and may be a matter of hours. Nevertheless, the remarkably high degree of protection reported by Boyd and his colleagues in sheep, led them to suggest that gas gangrene toxoids may find a place, both in the protection of military personnel engaged in combat, and in the protection of sections of the civilian population, such as those submitted to lower limb surgery for obliterative arterial disease. There is clearly great potential here, and it is to be hoped that the matter will not now be passed over, as were the early observations of Penfold and his colleagues 40 years ago.

UTERINE GAS GANGRENOUS INFECTIONS

Anaerobic infections of the uterus may occur in the puerperium or following abortion. These are frequently due to non-clostridial anaerobes, notably *Bacteroides* and anaerobic cocci (*see* p. 217). The rare cases of uterine gas gangrene are usually due to *Cl. perfringens*, and most often develop as postabortal infections (Hill, 1964). Of special interest from the bacteriological point of view are the post-abortal infections studied by Butler (1941, 1942).

In a study of 20 patients with severe clinical *Cl. perfringens* infections of the uterus, cervical smears showed the presence of heavily capsulated *Cl. perfringens* and damaged leucocytes, and absence of phagocytosis. On the other hand, of 64 control patients, in 20 of whom *Cl. perfringens* infection was considered possible on clinical grounds, but in which a severe infection did not develop, capsulation of the organisms was not seen, leucocytes were intact and in some cases active phagocytosis was evident. As a result of her investigations, Butler concluded that the presence of capsulated *Cl. perfringens* in pathological material from obstetrical cases was almost diagnostic of *Cl. perfringens* infection. The detection of non-capsulated strains in the uterine contents or in the vagina was of no significance.

Cl. perfringens bacteraemia is not uncommonly associated with postabortal and puerperal infections of the uterus, and this may lead rapidly to generalized intravascular haemolysis with haemoglobinaemia, haemoglobinuria, and rapidly deepening jaundice and cyanosis (Strum *et al.*, 1968; Pritchard and Whalley, 1971; Smith, MacLean and Maughan, 1971). Under these conditions the organism is not only recoverable from the blood by culture, but can sometimes be seen in Gram-stained films of the blood. Since *Cl. perfringens* infections of the uterus are likely to be very fulminating and rapidly fatal, early diagnosis and treatment are essential. Disconcerting instances of delay and gross mismanagement have been published by Jewett (1972, 1973). As with other forms of gas gangrene, the diagnosis of uterine infection is essentially clinical; examination of Gram-stained films of vaginal discharge, which may expedite a clinical diagnosis, and culture of the discharge and blood are the essential bacteriological procedures. In patients with marked intravascular haemolysis, Gram-stained films of the blood should also be examined.

The treatment of uterine gas gangrene may involve the application of a number of measures – administration of antitoxin, antimicrobial therapy (penicillin is the drug of choice), treatment of shock, exchange transfusion, hyperbaric oxygen therapy, uterine curettage and/or hysterectomy, and management of renal failure. (*see* Hanson *et al.*, 1966; O'Neill, Niall and O'Sullivan, 1972).

BOTULISM

Botulism is an intoxication that is almost always produced by the ingestion of preformed botulinum neurotoxin. Characteristically, the causative organism, *Cl. botulinum*, multiplies in food before it is eaten, and there produces its powerful exotoxin. On ingestion, the toxin is absorbed through the gastric and upper intestinal mucosa, after which its presence can be demonstrated in the blood. It is then presumably absorbed by the peripheral nervous system, on which it exerts its toxic effects. It acts widely on all those parts of the peripheral nervous system that are cholinergic, irrespective of whether they form components of the autonomic nervous system or are somatic nerves supplying skeletal muscle. The toxin acts at the tips of the motor nerve endings at the neuromuscular junction, where it interferes with the release of acetylcholine. As with tetanospasmin, botulinum neurotoxin appears to be fixed to the susceptible tissues, so that antitoxin administered after fixation produces no beneficial effects. The toxin produces no permanent damage to the nervous structures it attacks, and does not appear to exert any directly injurious effects on the brain or spinal cord (Wright, 1955; van Heyningen and Arseculeratne, 1964; Lamanna and Carr, 1966).

Cl. botulinum neurotoxin is the most powerful poison known, its toxicity being comparable to that of tetanospasmin. A purified sample of type D toxin, prepared by Wentzel, Sterne and Polson (1950), had the remarkably high toxicity of 4×10^{12} mouse MLD/mg N, while a purified preparation of type A toxin had a toxicity of 2.4×10^8 LD_{50}/mg N in mice (Lamanna, McElroy and Eklund, 1946). The neurotoxin of type F strains has a low lethality and is relatively unstable (Dolman and Murakami, 1961). Theoretically, and put into popular terms, it can be said that, weight for weight, the toxicity of botulinum neurotoxin is about 10^3 that of *Cl. perfringens* α-toxin, 10^5 that of crotactin (rattlesnake venom) and 10^6 that of strychnine (van Heyningen, 1968, 1970). A fatal dose of toxin for man has been estimated from accidental cases of botulism to be between 0.1 and 1.0 μg (Schantz and Sugiyama, 1974).

The incubation period of the disease is usually less than 24 h. Vomiting, thirst, pharyngeal and ocular pareses occur. Gradually, all the voluntary muscles weaken, and death, which may occur within 24 h of the onset, generally results from paralysis of the muscles of respiration. The symptoms and signs of intoxication have been described by Dickson (1918), Sutherland (1960), Koenig *et al* (1964), Petty (1965), Donadio, Gangarosa and Faich (1971) and many others. In those patients who survive, convalescence is slow, and many months may pass before control of ocular movements is regained. Although the botulinum neurotoxins are good antigens, natural immunity rarely, if ever, occurs, because the dose of toxin to cause death is less than that to elicit an antibody response. The pharmacological effects of botulinum toxin have been reviewed by Wright (1955).

Classic botulism is thus one of the few examples of a bacterial disease due, not to infection, but to absorption of a bacterial metabolic product from the alimentary tract. The toxin can be absorbed also through the respiratory mucous membranes – a fact to be borne in mind by laboratory workers handling dried toxin preparations.

Cl. botulinum sometimes occurs as a wound contaminant, and 15 cases of botulism following wound infection had been reported up to the end of 1974. All 15 cases occurred in the USA, and in most of them *Cl. botulinum* type A was isolated from the patients' wounds, and/or type A toxin was detected in the patients' blood. Nine of these cases have been usefully reviewed by Merson and Dowell (1973) (*see also* Weekly Report, 1974; 1975; Wapen and Gutmann, 1974; Cherington and Ginsburg, 1975). The rarity of botulism following a true wound infection with the organism is presumably due to the fact that its spores fail to germinate readily in the tissues (Keppie, 1951). The development of the disease is similar to tetanus in that there is an incubation period of 4–14 days between wounding and the onset of neurological symptoms.

Man is susceptible to the toxins of type A, B, E and F; types A, B, C and D cause intoxications in animals. Generally speaking, human botulism due to type A strains is commoner than that due to type B, while mixed intoxications, and those due to type F are rare. This incidence is reflected in the last column of *Table 7.7*, which records the number of cases of botulism and their toxin types which occurred in the USA from 1900–1967 (Botulism in the United States, 1968; Gangarosa *et al.*, 1971). Aggregate figures of this sort are, of course, misleading for a variety of reasons, of which an obvious one is the large proportion of cases (1019 of 1669, 61%) in which a toxin type was not identified. Apart from the decline in the incidence of botulism since 1940 (doubtless attributable to improved canning methods), one of the most striking features of this analysis is the increase, decade by decade, in the incidence of type E intoxication. In the period 1960–67, type E emerged as the commonest cause of botulism in the United States, accounting for no less than 45% of all cases of known toxin

Table 7.7 Cases of botulism in the USA due to different types of *Cl. botulinum*, 1900–1967 (Adapted from Botulism in the United States, 1968)

Type	*Number of cases during years*							
	1900–09	*1910–19*	*1920–29*	*1930–39*	*1940–49*	*1950–59*	*1960–69*	*Totals*
A	0	44	156	94	110	39	20	463
B	0	10	33	33	22	4	19	121
E	0	0	0	6	3	14	34	57
F	0	0	0	0	0	0	3	3
Mixed A/B	0	0	1	5	0	0	0	6
Unknown	10	189	138	245	181	176	79	1018
Totals	10	243	328	383	316	233	155	1668

type. This state of affairs was paralleled in other parts of the world, such as Japan, where episodes of type E intoxication appeared in 1951 (Nakamura *et al.* 1956; Iida *et al.*, 1958), and in Canada. The Canadian figures for the period 1919–1973 are summarized in *Table 7.8* (Dolman, 1974). It seems, therefore, that on a world-wide basis, the commonest cause of botulism today is type E (*see* Dolman, 1960; Hauschild and Gauvreau, 1976).

Table 7.8 Outbreaks of botulism in Canada due to different toxin types – 1919–1973 (Compiled from Dolman, 1974)

Period	*Number of incidents due to type*				*Total*
	A	*B*	*E*	*Unknown*	
1919–1960	5	1	7	11	24
1961–1973	2	4	23	9	38
Total	7	5	30	20	62

There are three reports in the literature of type C botulism in man– from the USA (Meyer *et al.*, 1953), from France (Prévot *et al.*, 1955) and from Russia (Matveev *et al.*, 1967), – and one report of a human type D intoxication (Demarchi *et al.*, 1958). Although none of these incidents is fully documented, the possibility of food poisoning due to these types must be considered. Two recent reports (Sebald, Jouglard and Gilles, 1974; Kauf *et al.*, 1974) described a most unusual 'epidemic' of type B botulism, in which French soft cheese (Brie) appeared to be the vehicle of intoxication. There were two simultaneous outbreaks in Marseilles and in Switzerland, involving 32 and 45 cases respectively. Severe epizootics of type C botulism in broiler chickens have occurred recently in the United Kingdom (Blandford and Roberts, 1970; Roberts, Thomas and Gilbert, 1973), and *Cl. botulinum* type C was detected in carcase swabs from 4 of 1249 broiler chickens prepared for national distribution (*Veterinary Record*, 1971).

No cases of human or animal botulism have been attributed to type G.

Botulism is a rare disease. In Great Britain only six episodes have been recorded. The first of these was the famous Loch Maree tragedy, in which eight people died following consumption of type A-contaminated duck paste (Leighton, 1923; Dolman, 1964). The other British episodes were those reported by Lane and Jones-Davies (1935) and Lance (1935) – one fatal case due to type A-contaminated nut brawn; by Templeton (1935) – two, or possibly, three fatal cases, following consumption of nut brawn, but the toxin type not determined; by Kitcat (1935) – one fatal case following consumption of nut meat brawn, but the toxin type not determined; by Aitken, Barling and Miles (1936) – one fatal case due to type B-contaminated meat and potato pie; and by Mackay-Scollay (1958) – two non-fatal cases due to type A-contaminated pickled fish imported from Mauritius.

Since botulism is due literally to poisoning of the food, it occurs as limited sporadic outbreaks, and secondary cases do not occur. The morbidity rate is nearly 100%, and the mortality rate 50–70% (Meyer, 1956). In general, the longer the incubation period the better the prognosis. Botulism is associated almost exclusively with preserved food, such as sausages, meats, brawn, and canned fruit and vegetables. During the period 1958–67 in America, home canned foods accounted for 43 of 52 outbreaks, 34 of which were associated with vegetable produce; the remaining 9 outbreaks were due to commercially prepared or canned foods, chiefly sea foods. More recent reports of botulism in American commercially canned foods implicated mushrooms (Wood, 1973) and vichyssoise soup (*Lancet*, 1971). Of 16 outbreaks of botulism of known origin that occurred in the United States during 1974, 15 of them were due to home canned or home preserved foods (Weekly Report, 1975) – *see also* Lynt, Kautter and Read (1975). No cases of botulism have been reported following the consumption of fresh food, cooked or uncooked.

The wide distribution of *Cl. botulinum* in soil accounts for its common association with vegetable produce. In order that a preserved food may become the seat of botulinum toxin production, the spores of the organisms present on the fresh food must survive the preserving process, and must then germinate and multiply. Contamination of preserved food with botulinum toxin is thus the outcome of an unsatisfactory preserving technique.

In canned foods, several factors predispose to botulinum toxin production. The presence of spores in suitable numbers is important, for, as Bigelow and Esty (1920) showed, the smaller the number of spores in food, the shorter is the heating time necessary to kill them. Insufficient heating leads to survival of the spores; the high thermal resistance of *Cl. botulinum* spores is well known. The presence of anaerobic conditions following the sterilization process greatly favours germination, multiplication and toxin formation. The danger from home canned or bottled food is much greater than from commercially prepared food, and it is for this reason that cookery books often advise the housewife not to attempt preservation, except by freezing, of vegetables, meat, fish and poultry (*see British medical Journal*, 1968).

Table 7.9 Some features of the different types of *Cl. botulinum* (After Dolman and Murakami, 1961; Dolman, 1964)

Type	*Investigator*	*Species mainly affected*	*Common vehicles*	*Geographic incidence*
A	Leuchs (1910)	Man; chickens	Home canned vegetables and fruits; meat and fish	Western United States; Soviet Ukraine
B	Leuchs (1910)	Man; horses; cattle	Prepared meats, notably pork products	France; Norway; Eastern United States
C	Bengston (1922)	Aquatic wild birds	Fly larvae (*Lucilia caesar*); rotting vegetation of alkaline ponds	Western United States and Canada; South America; South Africa; Australia
	Seddon (1922)	Cattle; horses; mink	Toxic forage; carrion	Australia; South Africa; Europe; North America
D	Theiler and Robinson (1927); Meyer and Gunnison (1929)	Cattle	Carrion	South Africa; Australia
E	Gunnison *et al*. (1935)	Man	Uncooked products of fish and marine mammals	North Japan; Labrador; British Columbia; Alaska; Great Lakes, Sweden; Denmark; USSR
F	Moeller and Scheibel (1960) Dolman and Murakami (1961)	Man	Home-made liver paste	Denmark
G	Gimenez and Ciccarelli (1967)	None recorded	–	Organism isolated in Argentina (only report)

Some of the epidemiological features of the different types of *Cl. botulinum* are summarized in *Table 7.9*. Spencer (1969) published an excellent review of the factors affecting the survival and growth of *Cl. botulinum* in cured foods (*see also* Gilbert, 1974).

Botulinum toxin is destroyed in a few minutes by boiling, the thermal destruction being dependent on the pH and other characteristics of the medium in which the toxin is present (Schoenholz and Meyer, 1924). Though adequate cooking before consumption thus renders contaminated food harmless, it is clearly unwise to consume preserved foods that are noticeably spoiled. While *Cl. botulinum* contamination is usually associated with spoiling, the food may sometimes appear well preserved; this is especially the case with type E strains which are non-proteolytic. Interesting accounts of the neurotoxins of *Cl. botulinum* are given by Lamanna (1959) and Schantz and Sugiyama (1974).

Diagnosis

As with tetanus and gas gangrene, botulism is diagnosed clinically, and the disease may be confirmed bacteriologically. The demonstration of botulinum toxin in suspected food, and in the contents of the alimentary tract, and the patient's blood is of first importance. This is carried out as described on p. 126. Culture of suspected food, and of the patient's faeces and vomit, may give positive results; but the isolation of the organism is, alone, of no significance. For full clinical accounts of patients with the classic symptoms of botulism *see* Sutherland (1960), and Koenig *et al.* (1964).

Prophylaxis and treatment

The prevention of botulism

Adequate methods of food preservation are important in the prevention of botulism. These must be such that any spores of *Cl. botulinum* present are killed, or, if not destroyed, are prevented from germinating and multiplying. This has been largely accomplished by modern methods of commercial processing, but, as has been noted, home processing is unreliable.

PROPHYLACTIC ADMINSTRATION OF BOTULINUM ANTITOXIN

The consumption of suspected food is an indication for the prophylactic administration of botulinum antitoxin. Parish and Cannon (1961) recommended a prophylactic dose of at least 10 000 units of the polyvalent antitoxin (A, B, and E) intramuscularly, to be followed by larger intravenous doses should symptoms appear. Since botulinum antitoxin is a horse serum preparation, its administration carries the same risks as have been outlined for heterologous tetanus antiserum therapy (*see* p. 296); the main dose of botulinum antitoxin must always be preceded by a trial dose, and one must be prepared to deal with an immediate serious serum reaction should one occur.

Therapeutic administration of antitoxin

Because toxin has already been absorbed and 'fixed' when treatment is started, the therapeutic administration of antitoxin will not reverse the intoxication. There is no doubt, however, about the efficacy of antitoxin in preventing the fixation of further toxin (Dolman and Iida, 1963). The sooner antitoxin is given, the better the prognosis. Parish and Cannon (1961) suggest that at least 50 000 units of polyvalent antitoxin should be given intravenously, and that this dose should be repeated every 6–12 h until recovery of the patient (*see also* Werner and Chin, 1973; Dolman, 1974).

Other prophylactic and therapeutic measures include high enemas of soap and olive oil, since soap neutralizes the toxin and olive oil prevents its absorption from the gut. Iodine, potassium permanganate and sodium bicarbonate also destroy the toxin *in vivo* and may be given by mouth or used in gastric lavage. It is doubtful whether any of these measures are of therapeutic value, since by the time symptoms have developed, much of the toxin in the intestine will have been absorbed. Their value as prophylactic measures is probably much greater. The general management of botulism is similar to that already outlined for tetanus, and has been discussed by Dolman (1974) and Werner and Chin (1973).

In recent years, guanidine hydrochloride has been used successfully in the treatment of human and animal botulism (Cherington and Ryan, 1968, 1970; Scaer, Tooker and Cherington, 1969; Cherington, 1974; Oh, Halsey and Briggs, 1975). This drug appears to act at a presynaptic site to antagonize the block caused by the neurotoxin (Otsuka and Endo, 1960). The drug is associated with a variety of unpleasant side effects, and its use is still regarded as experimental (Faich, Graebner and Sato, 1971).

Owing to the rarity of botulism, general active immunization of the community is not indicated. It seems desirable, however, that laboratory workers in contact with the organism or its toxin should receive a prophylactic course of toxoid.

Botulinum toxin and bacteriological warfare

The high lethality of the botulinum toxins has suggested their use in biological warfare (Rosebury and Kabat, 1947). The danger from this type of attack would not be serious, however, since adequate defence measures against it could be easily and cheaply instituted (Wright, 1955), and no secondary cases would occur.

CL. PERFRINGENS FOOD POISONING

Some strains of *Cl. perfringens* type A cause a characteristic form of food poisoning. The condition is mild, and of short duration, characterized by acute abdominal pain and diarrhoea developing 8–24 h after consumption of food. Nausea is common, but vomiting is rare. Its main features were summarized by Cruickshank (1955). The causal relation of *Cl. perfringens* to epidemic outbreaks of diarrhoea has long been suspected (Klein, 1895; Larner, 1922; Knox and MacDonald, 1943; McClung, 1945). The work of Hobbs *et al.* (1953) and Dische and Elek (1957) finally established *Cl. perfringens* food poisoning as a definite clinical entity.

The causal organism studied by Hobbs and her colleagues, and the one subsequently encountered commonly in food poisoning episodes in England is a type A variant, which differs from typical type A strains in the greater heat resistance of its spores, its feeble toxigenicity and its slight or non-haemolytic effect on horse blood agar; these organisms have been referred to as type A2 by Sterne and Warrack (1964), in order to distinguish them from classic type A strains (*Table 4.5* p. 134). It soon became clear, however, that these heat-resistant, non-haemolytic variants were not the only, and indeed not even the commonest, strains of *Cl. perfringens* that cause food poisoning in man, for both non-haemolytic non-heat resistant strains, and haemolytic non-heat resistant (classic) strains were incriminated in outbreaks of food poisoning both in the United Kingdom and in America (McKillop, 1959; Hall *et al.*, 1963; Taylor and Coetzee, 1966; Hauschild and Thatcher, 1967, 1968; Sutton and Hobbs, 1968).

Infection is usually due to meat dishes which, after cooking, have been allowed to cool slowly and are eaten the following day either cold or reheated. Contamination of the meat dishes (roasts, stews, pies, poultry, soups and so on) may occur before or after cooking. During subsequent slow cooling, the prevailing anaerobic conditions in the food allow rapid multiplication of organisms so that the prepared food becomes virtually a pure culture of *Cl. perfringens*. Consumption of the food at this stage, or after inadequate reheating, causes the development of the typical mild gastroenteritis of *Cl. perfringens* food poisoning. The number of bacterial cells per g of food required to induce food poisoning is of the order of $10^8 - 10^9$ (Dische and Elek, 1957; Hauschild and Thatcher, 1968).

The genesis of *Cl. perfringens* food poisoning has been painstakingly unravelled by the careful work of Hauschild and his colleagues, conducted over a number of years in Ottawa, Canada. Those interested in the details of this fascinating work should refer to the reviews of Hauschild (1971, 1973, 1974). The genesis of the condition may be briefly summarized as follows.

Upon entry into the gastrointestinal tract of food containing $10^6 - 10^8$ vegetative cells of *Cl. perfringens* per g, there occurs further multiplication, and then sporulation of the organism in the small intestine. During spore formation an enterotoxic endotoxin is produced in the sporulating cells, which is then released into the intestinal lumen as the vegetative cell remnants lyse. Although the mode of action of this enterotoxin is not fully established, it is clear that it causes excess fluid movement into the intestinal lumen which results in diarrhoea.

Cl. perfringens food poisoning is thus properly regarded as a toxic infection; it is initiated by live bacterial cells, and the causative enterotoxin is subsequently produced *in vivo*. In contrast to botulism (and staphylococcal food poisoning), *Cl. perfringens* food poisoning is not due to the ingestion of a pre-formed toxin, so that its bacteriological diagnosis is accomplished by isolation of the organism from food and faeces rather than by detection of the enterotoxin.

Cl. perfringens food poisoning is not uncommon. In an analysis of food poisoning and *Salmonella* infections in England and Wales for 1967, Vernon (1969) noted that more than one quarter of the cases of food poisoning in which a causal agent was established were due to *Cl. perfringens*. All of 47 confirmed outbreaks were associated with meat products, mostly reheated meat, precooked meat served cold, and meat pies, and in 6 of them more than 100 persons were ill. There are very many more recent reports of *Cl. perfringens* food poisoning episodes involving larger or smaller numbers of persons (*see for example British medical Journal*, 1971c, 1972b, 1973; Sanders *et al.*, 1974; Anderson, Pedersen and Beare, 1973; Basarsky and Schnee, 1974; *Communicable Diseases, Scotland*, 1974a, b).

Bacteriological diagnosis

The criteria generally accepted for implicating *Cl. perfringens* in food poisoning episodes are the clinical history and epidemiology of the outbreak, together with the demonstration of large numbers of *Cl. perfringens* in the suspected food and in the patients' faeces (Sutton, 1966; Sutton and Hobbs, 1968; Hobbs and Sutton, 1968). Ideally, the isolates from the food and from faecal samples in a single outbreak should be of the same somatic serotype.

Since a diversity of strains of *Cl. perfringens* type A may cause food poisoning, the demonstration of large numbers of any type A strain present in food or faeces is likely to be significant. It is now no longer sufficient to look for heat resistant strains, although this has been a common practice in the United Kingdom. The following technique, based on that suggested by Sutton and Hobbs (1968), is well suited to the facilities of the clinical laboratory, and ensures isolation of all appropriate strains of the organism.

Examination of faeces

1. A thick emulsion of faeces (about 1/10) is made in quarter-strength Ringer's solution.
2. Using 10-fold dilutions of this emulsion, semi-quantitative counts are performed on neomycin horse blood agar, using a '50-dropper' pipette (total count).
3. The counts outlined in 2 are repeated using a sample of the faecal emulsion that has been heated at 80 °C for 10 min (spore count).
4. One ml of the emulsion is inoculated into a tube of cooked meat broth and heated at 100 °C for 30–60 min (heat resistant strains only).
5. Plate count cultures and the enrichment culture are incubated anaerobically for 24 h, and counts obtained. The enrichment culture is subcultured to neomycin horse blood agar; this detects the presence or absence of heat resistant spores of *Cl. perfringens*.

This isolation procedure ensures that neither heat sensitive nor heat resistant strains of *Cl. perfringens* are excluded, that the various haemolytic variants of the organism are recognized, and that an assessment is obtained of whether *Cl. perfringens* is present in relatively small or large numbers. For samples of faeces collected and examined soon after the illness, counts below 10^5 organisms per g are regarded as low. The faeces of patients afflicted with *Cl. perfringens* food poisoning usually give counts of the order of $10^5 - 10^7$ cells per g.

In a recent report Dowell *et al.* (1975) described a haemagglutination technique for the detection of *Cl. perfringens* enterotoxin in human faeces.

Examination of food

Food reaching the laboratory within 24 h of an outbreak is likely to contain large numbers of vegetative cells of *Cl. perfringens*; spores are rarely present.

The food, homogenized if necessary, is prepared in 10-fold dilutions as described in 2 for faeces, and semi-quantitative counts are performed on neomycin horse blood agar, using a '50-dropper' pipette. Duplicate plate count cultures are prepared on horse blood agar plates without neomycin for aerobic incubation.

Isolates from food and faeces are checked for identity by showing that they are lactose fermenters, and produce a lecithinase C reaction on egg yolk agar that is inhibited by *Cl. perfringens* type A antitoxin. This confirmation is conveniently effected by subculturing isolates to half-antitoxin lactose egg yolk agar.

Instead of using plain horse blood agar (with and without neomycin as required), it is often useful to add 0.5% $CaCl_2$ to the medium. This enhances the activity of *Cl. perfringens* α-toxin, and thus makes α-haemolysis more conspicuous (*see* p. 131).

Serological (somatic) typing of isolates which is performed by slide agglutination, is not within the scope of the routine laboratory, since appropriate antisera are not commercially available. In any one outbreak, the same serotype should be isolated from the majority of faecal samples and from the suspected food.

In North America, the cultural method most commonly used for the investigation of *Cl. perfringens* food poisoning outbreaks employs the highly selective sulphite polymyxin sulphadiazine agar medium of Angelotti *et al*. (1962). This method is fully described by Thatcher and Clark (1968) (*see also* Hauschild, 1970). Recently, Hauschild and Hilscheimer (1974) recommended a modification of the tryptose sulphite cycloserine agar medium of Harman, Kautter and Peeler (1971), which allows quantitative recovery of *Cl. perfringens*; and Handford (1974) described an oleandamycin polymyxin sulphadiazine sulphite agar medium for the specific quantitative isolation of *Cl. perfringens* from foods.

Useful general papers on *Cl. perfringens* food poisoning have been published by Bryan (1969), Hobbs (1969), Duncan (1970), Nakamura and Schulze (1970), Bryan and Kilpatrick (1971), Smith (1973), Hobbs (1974), Genigeorgis (1975) and Walker (1975).

ENTERITIS NECROTICANS (NECROTIZING JEJUNITIS)

Necrotizing jejunitis is a severe, and often fatal, disease due to strains of *Cl. perfringens* type C. The first recorded cases of this condition were reported by Zeissler and Rassfeld-Sternberg (1949) from North-west Germany. Some hours after eating rabbit, tinned meat or fish paste, the patients studied by these workers developed severe lower abdominal pain and diarrhoea. Some cases showed blood and mucosal sloughs in their stools; and a few deaths occurred due to peripheral circulatory failure, or to intestinal obstruction following massive oedema of the intestinal mucosa. The area affected was principally the jejunum. These organisms, the spores of which are highly heat-resistant, are now distinguished toxicologically as type C_4 (Sterne and Warrack, 1964).

Although sporadic cases of this condition have occurred, the second and only other recorded outbreak was reported from New Guinea by Murrell and his co-workers (*see* Murrell *et al.*, 1966; Murrell, 1967). In this episode, strains of *Cl. perfringens* type C were again implicated, but since they showed minor toxicological differences to the German isolates, they have been designated as type C_5 (Egerton and Walker, 1964). Skjelkvale and Duncan (1975) have shown that type C_5 strains elaborate a major enterotoxin that is serologically and pharmacologically identical to that produced by type A food poisoning strains.

Necrotizing jejunitis in New Guinea has been referred to as the 'pig-bel' syndrome (New Guinea pidgin English – *pig-bel*, abdominal discomfort following a large pork meal), because of its aetiological relationship to the widespread practice of pork feasting in the locality. The disease, which appears to be endemic, is characterized by severe upper abdominal pain, bloody diarrhoea with melena, and nausea with occasional vomiting. The signs are those of dehydration from severe enteritis, shock, and toxaemia, or those of acute small bowel obstruction. If untreated, the disease progresses to complete segmental gangrene of parts of the small intestine with ileus, shock, and severe toxaemia.

NECROTIZING COLITIS

Killinback and Lloyd Williams (1961–62) described six patients with extensive gangrene of the colon. There was no evidence of vascular obstruction in these examples of necrotizing colitis. Although bacteriological proof was lacking, the authors considered that the condition was probably due to *Cl. perfringens* infection, since in each case the bowel wall showed extensive invasion by Gram-positive bacilli resembling *Cl. perfringens*. Subsequently Tate, Thomson and Willis (1965) described a similar case in which a heat-resistant strain of *Cl. perfringens* type A was incriminated. In this patient, laparotomy showed that the whole of the colon, from the caecum to the upper sigmoid, was gangrenous, and there was no evidence of mesenteric vascular occlusion. A heat-resistant *Cl. perfringens* type A was isolated from material obtained from the lumen of the large bowel during operation, and again from a piece of necrotic large bowel post mortem.

OTHER CLOSTRIDIAL INFECTIONS

Rarely, accidental penetrating wounds in various parts of the body may result in clostridial infections, such as brain abscess, acute purulent meningitis, panophthalmitis, endocarditis and intrapleural infections; the organism most commonly responsible is *Cl. perfringens*. These infections have been reviewed by Willis (1969).

Some recent reports of these uncommon infections are as follows: Empyema (Bentley and Lepper, 1969; Bayer *et al.*, 1975), panophthalmitis (Bhargava and Chopdar, 1971; Bristow, Kasser and Sevel, 1971), septic arthritis (Torg and Lammot, 1968; Korn *et al.*, 1975), and meningitis (Mackay *et al.*, 1971). Some virtually unique cases of clostridial infection include *Cl. perfringens* infection of the urinary tract (Nielsen and Laursen, 1972), *Cl. perfringens* endocarditis (More, 1943; Felner and Dowell, 1970; Case, Goforth and Silva, 1972), and *Cl. perfringens* myocardial abscess (Tennant and Parkes, 1959; Guneratne, 1975). Spontaneous *Cl. septicum* infections, which are rare, usually present as a bacteraemia in debilitated patients with underlying malignant disease (Alpern and Dowell, 1969). Such seems to be the case in the unusual *Cl. septicum* infection of the thyroid gland described by Warren and Mason (1970).

Acute appendicitis

Although the bacteriology of acute appendicitis is very confused, it seems likely that some clostridia, probably in association with other anaerobic species, such as *Bacteroides fragilis* and anaerobic cocci, are responsible for the gangrene and general toxaemia of some severe cases. Weinberg and his colleagues (1926, 1928a, b) isolated a number of clostridia from cases of acute appendicitis, including *Cl. perfringens, Cl. fallax* and *Cl. histolyticum*. As might be expected, clostridia are also met with in some cases of acute peritonitis, especially when there is a perforation of the large intestine.

More recently, the non-clostridial anaerobes have come into prominence as important causes of infections that are associated with the gastrointestinal tract. Thus, although it seems reasonable to conclude that *Cl. perfringens* may participate in the pathogenesis of some severe cases of appendicitis, the bacteroides must also be regarded as significant pathogens in this condition (*see* p. 221).

Non-specific urethritis

Recently, Hafiz *et al.* (1975) drew attention to *Cl. difficile* as a possible cause of urethritis in males. Following the observation that *Cl. difficile* was commonly present in vaginal specimens from patients attending a venereal disease clinic, they found that the organism was also present in the urethral discharge of urine of all of 42 male patients with non-specific urethritis. It is not yet clear whether the organism is an opportunist, causing infection of an already damaged urethra, or whether it is a primary cause of disease.

REFERENCES

Abeyatunge, J. L. (1969). 'Tetanus following jejuno-jejuno-gastric intussusception with gangrenous jejunum.' *British Journal of clinical Practice,* **23,** 468

Ablett, J. J. L. (1967). 'An analysis and main experiences in 82 patients treated in the Leeds Tetanus Unit.' In *Symposium on Tetanus in Great Britain, Leeds, 1967,* p. 1. Ed. By M. Ellis. Leeds: General Infirmary

Ackland, T. H. (1959). 'Tetanus prophylaxis.' *Medical Journal of Australia,* **1,** 185

Adadevoh, B. K. and Akinla, O. (1970). 'Postabortal and postpartum tetanus.' *Journal of Obstetrics and Gynaecology of the British Commonwealth,* **77,** 1019

Adams, E. B. (1968). 'The prognosis and prevention of tetanus.' *South African medical Journal,* **42,** 739

Adams, E. B., Laurence, D. R. and Smith, J. W. G. (1969). *Tetanus.* Oxford; Blackwell

Adams, J. Q. and Morton, R. F. (1955). 'Puerperal tetanus.' *American Journal of Obstetrics and Gynecology,* **69,** 169

Afonja, A. O., Jaiyeola, B. O. and Tunwashe, O. L. (1973). 'Tetanus in Lagos: A review of 228 adult Nigerian patients.' *Journal of Tropical Medicine and Hygiene,* **76,** 171

Aguileiro Moreira, J. M., Braneiro, R. L. and Ghigliazza, H. M. (1960). 'A case of a second attack of clinical tetanus.' *Semana Medica,* **117,** 969 (Spanish)

Aikat, B. K. and Dible, J. H. (1956). 'The pathology of *Clostridium welchii* infection.' *Journal of Pathology and Bacteriology,* **71,** 461

Aikat, B. K. and Dible, J. H. (1960). 'The local and general effects of cultures and culture-filtrates of *Clostridium oedematiens, Cl. septicum, Cl. sporogenes* and *Cl. histolyticum.' Journal of Pathology and Bacteriology,* **79,** 227

Aitken, R. S., Barling, B. and Miles, A. A. (1936). 'A case of botulism.' *Lancet,* **2,** 780

Alhady, S. M. A. (1961). 'Recurrent tetanus: Report of a case.' *Medical Journal of Australia,* **2,** 219

Allen, J. M. (1931). 'The neurological complications of serum treatment.' *Lancet,* **2,** 1128

Alpern, R. J. and Dowell, V. R. (1969). '*Clostridium septicum* infections and malignancy.' *Journal of the American medical Association,* **209,** 385

Anderson, C. B., Marr, J. and Jaffe, B. N. (1972). 'Anaerobic streptococcal infections simulating gas gangrene.' *Archives of Surgery,* **104,** 186

Anderson, H. W., Pedersen, A. H. B. and Beare, J. A. (1973). '*Clostridium perfringens* gastroenteritis – Washington.' *Morbidity and Mortality Weekly Report,* **22,** 3 and 8

Angelotti, R., Hall, H. E., Foter, M. J. and Lewis, K. H. (1962). 'Quantitation of *Clostridium perfringens* in foods.' *Applied Microbiology,* **10,** 193

Annals of internal Medicine, (1966). 'Prevention of tetanus.' **65,** 1079

Anusz, Z. (1967). 'Tetanus in Poland in the years 1961–1965.' *Epidemiological Review,* **21,** 179

Ashley, M. J. and Bell, J. S. (1969). 'Tetanus in Ontario: A review of the epidemiological and clinical features of 102 cases occurring in the 10-year period 1958–1967.' *Canadian medical Association Journal,* **100,** 798

Ayliffe, G. A. J. and Lowbury, E. J. L. (1969). 'Sources of gas gangrene in hospital.' *British medical Journal,* **2,** 333

Bamford, J. A. C. (1967). 'Tetanus prevention: active immunization in Western Australia.' *Medical Journal of Australia,* **2,** 601

Barr, M. and Sachs, A. (1955). *Army Pathology Advisory Committee Report on the Investigation into the Prevention of Tetanus in the British Army*

Basarsky, N. and Schnee, P. (1974). '*Clostridium welchii* food poisoning at a service club dance.' *Epidemiological Bulletin, Ottawa,* **18,** 47

Batten, R. L. (1969). 'Tetanus prophylaxis: clinical aspects.' *Anglo-German medical Review,* **5,** 83

Batty, I. and Walker, P. D. (1964). 'The identification of *Clostridium novyi* (*oedematiens*) and *Clostridium tetani* by the use of fluorescent labelled antibodies.' *Journal of Pathology and Bacteriology,* **88,** 327

Batty, I. and Walker, P. D. (1967). 'The use of fluorescent labelled antibody technique for the detection and differentiation of bacterial species.' In *International Symposium of Immunological Methods of Biological Standardization, Royaumont 1965*, Vol. 4, p. 73. Basel; Karger

Bayer, A. S., Nelson, S. C., Galpin, J. E., Chow, A. W. and Guze, L. B. (1975). 'Necrotizing pneumonia and empyema due to *Clostridium perfringens.*' *American Journal of Medicine,* **59,** 851

Beare, F. (1953). 'Some observations on tetanus.' *Medical Journal of Australia,* **2,** 949

Bennett, A. E. (1939). 'Horse serum neuritis with report of five cases.' *Journal of the American medical Association,* **112,** 590

Bensted, H. J. (1941). 'Modern practice in war-time – immunization of soldiers and civilians; immunization against bacterial toxins.' *Journal of the Royal Institute of Public Health,* **4,** 17

Bentley, D. W. and Lepper, M. H. (1969). 'Empyema caused by *Clostridium perfringens.*' *American Review of respiratory Disease,* **100,** 706

Berggren, W. L. (1974). 'Control of neonatal tetanus in rural Haiti through the utilization of medical auxiliaries.' *Bulletin of the Pan American Health Organization,* **8,** 24

Bessman, A. N. and Wagner, W. (1975). 'Nonclostridial gas gangrene.' *Journal of the American medical Association,* **233,** 958

Bhargava, S. K. and Chopdar, A. (1971). 'Gas gangrene panophthalmitis.' *British Journal of Ophthalmology,* **55,** 136

Bigelow, W. D. and Esty, J. R. (1920). 'The thermal death point in relation to time of typical thermophilic organisms.' *Journal of infectious Diseases,* **27,** 602

Binns, P. M. (1961). 'An analysis of tetanus prophylaxis in 3455 cases.' *British Journal of preventive and Social Medicine,* **15,** 180

Bishop, R. F. and Marshall, V. (1960). 'The enhancement of *Clostridium welchii* infection by adrenaline-in-oil.' *Medical Journal of Australia,* **2,** 656

Blake, P. A. and Feldman, R. A. (1975). 'Tetanus in the United States, 1970–1971.' *Journal of infectious Diseases,* **131,** 745

Blandford, T. B. and Roberts, T. A. (1970). 'An outbreak of botulism in broiler chickens.' *Veterinary Record,* **87,** 258

Blumstein, G. I. and Kreithen, H. (1966). 'Peripheral neuropathy following tetanus toxoid administration.' *Journal of the American medical Association,* **198,** 1030

Botulism in The United States (1968). *Review of cases, 1899-1967 and handbook for epidemiologists, clinicians, and laboratory workers.* Washington; US Department of Health

Box, Q. T. (1964). 'The treatment of tetanus.' *Pediatrics,* **33,** 872

Boyd, J. S. K. (1946). 'Tetanus in the African and European theatres of war, 1939–45.' *Lancet,* **1,** 113

Boyd, N. A., Thomson, R. O. and Walker, P. D. (1972a). 'The prevention of experimental *Clostridium novyi* and *Cl. perfringens* gas gangrene in high-velocity missile wounds by active immunization.' *Journal of medical Microbiology,* **5,** 467

Boyd, N. A., Walker, P. D. and Thomson, R. O. (1972b). 'The prevention of experimental *Clostridium novyi* gas gangrene in high-velocity missile wounds by passive immunization.' *Journal of medical Microbiology,* **5,** 459

Brightmore, T. (1971). 'Non-clostridial gas infection.' *Proceedings of the Royal Society of Medicine,* **64,** 1084

Brightmore, T. and Greenwood, T. W. (1974). 'The significance of tissue gas and clostridial organisms in the differential diagnosis of gas gangrene.' *British Journal of clinical Practice,* **28,** 43

Brindle, M. J. and Twyman, D. G. (1962). 'Allergic reactions to tetanus toxoid. A report of four cases.' *British medical Journal,* **1,** 1116

Bristow, J. H., Kasser, B. and Sevel, D. (1971). 'Gas gangrene panophthalmitis treated with hyperbaric oxygen.' *British Journal of Ophthalmology,* **55,** 139

British medical Journal (1959). 'Occurrence of tetanus spores in materials used for dressing wounds; Report of the Public Health Laboratory Service (Medical Research Council, London).' **1,** 1150

British medical Journal (1961). 'Gas gangrene from adrenaline.' **1,** 730

British medical Journal (1964). 'Clostridial infection.' **2,** 1150

British medical Journal (1967). 'Tetanus prophylaxis.' **4,** 635

British medical Journal (1968). 'Home preservation of food.' **4,** 6

British medical Journal (1969a). 'Postoperative gas gangrene.' **2,** 328

British medical Journal (1969b). 'Postoperative gas gangrene.' **3,** 665

British medical Journal (1971a). 'Gas gangrene.' **4,** 819

British medical Journal (1971b). 'Tetanus.' **4,** 372

British medical Journal (1971c). 'Food poisoning due to *Clostridium welchii.*' **4,** 758
British medical Journal (1972a). 'Gas gangrene and hyperbaric oxygen.' **3,** 715
British medical Journal (1972b). 'Food-poisoning from school meals.' **4,** 307
British medical Journal (1973) 'An outbreak of food poisoning.' **1,** 559
British medical Journal (1974). 'Reactions to tetanus toxoid.' **1,** 48
Brooks, V. B., Curtis, D. R. and Eccles, J. C. (1955). 'Mode of action of tetanus toxin.' *Nature, London,* **175,** 120
Brooks, V. B., Curtis, D. R. and Eccles, J. C. (1957). 'The action of tetanus toxin on the inhibition of motor neurones.' *Journal of Physiology,* **135,** 655
Brummelkamp, W. H., Boerema, I. and Hoogendyk, L. (1963). 'Treatment of clostridial infections with hyperbaric oxygen drenching. A report on 26 cases.' *Lancet,* **1,** 235
Brummelkamp, W. H., Hogendijk, J. and Boerema, I. (1961). 'Treatment of anaerobic infections (clostridial myositis) by drenching the tissues with oxygen under high atmospheric pressure.' *Surgery,* **49,** 299
Bryan, F. L. (1969). 'What the sanitarian should know about *Clostridium perfringens* foodborne illness.' *Journal of Milk and Food Technology,* **32,** 381
Bryan, F. L. and Kilpatrick, E. G. (1971). '*Clostridium perfringens* related to roast beef cooking, storage, and contamination in a fast food service restaurant.' *American Journal of public Health,* **61,** 1869
Bullock, W. E. and Cramer, W. (1919). 'On a new factor in the mechanism of bacterial infection.' *Proceedings of the Royal Society,* B, **90,** 513
Busuttil, A., Pace, J. B. and Muscat, J. A. (1974). 'Traditional treatment of 194 cases of tetanus.' *British Journal of Surgery,* **61,** 731
Butler, H. M. (1937). *Blood Cultures and their Significance.* London; Churchill
Butler, H. M. (1941). 'The bacteriological diagnosis of severe *Clostridium welchii* infection following abortion.' *Medical Journal of Australia,* **1,** 33
Butler, H. M. (1942). 'The examination of cervical smears as a means of rapid diagnosis in severe *Clostridium welchii* infections following abortion.' *Journal of Pathology and Bacteriology,* **54,** 39
Butler, H. M. (1945). 'Bacteriological studies of *Clostridium welchii* infections in man.' *Surgery, Gynecology and Obstetrics,* **81,** 475
Bytchenko, B. (1966). 'Geographical distribution of tetanus in the World, 1951–60.' *Bulletin of the World Health Organization,* **34,** 71
Caro, D. and Shaw, E. (1974). 'Current practice in tetanus prophylaxis.' *British medical Journal,* **1,** 389
Case, D. B., Goforth, J. M. and Silva, J. (1972). 'A case of *Clostridium perfringens* endocarditis.' *Johns Hopkins medical Journal,* **130,** 54
Catzel, P. (1961). 'Antitetanus immunization.' *British medical Journal,* **2,** 1434
Chapman, W. G. and Davey, M. G. (1973). 'Tetanus immunity in Busselton, Western Australia, 1969.' *Medical Journal of Australia,* **2,** 316
Cherington, M. (1974). 'Botulism. Ten-year experience.' *Archives of Neurology,* **30,** 432
Cherington, M. and Ginsburg, S. (1975). 'Wound botulism.' *Archives of Surgery,* **110,** 436
Cherington, M. and Ryan, D. W. (1968). 'Botulism and guanidine.' *New England Journal of Medicine,* **278,** 931
Cherington, M. and Ryan, D. W. (1970). 'Treatment of botulism with guanidine.' *New England Journal of Medicine,* **282,** 195
Cherubin, C. E. (1967). 'Urban tetanus. The epidemiologic aspects of tetanus in narcotic addicts in New York City.' *Archives of environmental Health,* **14,** 802

Cherubin, C. E. (1971). 'Infectious disease problems of narcotic addicts.' *Archives of internal Medicine,* **128,** 309

Chew, H. E. R., Hanson, G. C. and Slack, W. K. (1969). 'Hyperbaric oxygenation.' *British Journal of Diseases of the Chest,* **63,** 113

Chopra, I. B. and Mukherjee, P. (1970). 'Nonclostridial crepitant cellulitis.' *Journal of the Indian medical Association,* **54,** 177

Christensen, N. A. (1969). 'Treatment of the patient with severe tetanus.' *Surgical Clinics of North America,* **49,** 1183

Christensen, N. A. (1972). 'Potential for tetanus unchanged; Are your patients all properly immunized?' *Journal of the American medical Association,* **222,** 578

Cole, L. and Youngman, H. (1969). 'Treatment of tetanus.' *Lancet,* **1,** 1017

Cole, L. B. (1951). 'Tetanus immunization.' *Practitioner,* **167,** 247

Committee upon Anaerobic Bacteria and Infections (1919). *Special Report Series of the Medical Research Council, London,* No. 39. London; HMSO

Communicable Diseases, Scotland (1974a). '*Clostridium welchii* food poisoning.' **74/3,** 3

Communicable Diseases, Scotland. (1974b). '*Clostridium welchii* food poisoning.' **74/15,** 3

Community Medicine (1972). 'Scottish move to prevent tetanus.' **128,** 357

Cooke, W. T., Frazer, A. C., Govan, A. D. T., Peeney, A. L. P., Barling, S. G., Thomas, G., Leather, J. B., Elkes, J. J. and Scott Mason, R. P. (1945). 'Clostridial infections in war wounds.' *Lancet,* **1,** 487

Cooper, D. K. C. (1969). 'Agricultural accidents: A study of 132 patients seen at Addenbrooke's Hospital, Cambridge, in 12 months.' *British medical Journal,* **4,** 193

Cormie, J. (1962). 'Unusual presentation of tetanus.' *British medical Journal,* **1,** 31

Cotter, E. H. J. and Wilson, K. V. (1975). 'Preventing tetanus in adults in general practice.' *Journal of the Royal College of General Practitioners,* **25,** 812

Cox, C. A., Knowelden, J. and Sharrard, W. J. W. (1963). 'Tetanus prophylaxis.' *British medical Journal,* **2,** 1360

Creech, O., Glover, A. and Ochsner, A. (1957). 'Tetanus; Evaluation of treatment at Charity Hospital, New Orleans, Louisiana.' *Annals of Surgery,* **146,** 369

Critchely, A. M. (1958). 'Tetanus neonatorum.' *Public Health,* **71,** 459

Cruickshank, J. C. (1955). 'The bacteriology of food poisoning.' *Practitioner,* **174,** 664

Cvjetanovic, B., Bytchenko, B. and Edsall, G. (1976). 'Guidelines for the prevention of tetanus.' *WHO Chronicle,* **30,** 201

Daramola, T. (1968). 'Tetanus in Lagos.' *West African medical Journal,* **17,** 136

Daschbach, R. J. (1972). 'Serum sickness and tetanus immunization.' *Journal of the American medical Association,* **220,** 1619

Dean, H. R. (1917). 'A report of twenty-five cases of tetanus.' *Lancet,* **1,** 673

Deliyannakis, E. (1971). 'Peripheral nerve and root disturbances following active immunization against smallpox and tetanus.' *Military Medicine,* **136,** 458

Demarchi, J., Mourgues, C., Orio, J. and Prévot, A. R. (1958). 'The occurrence of human type D botulism.' *Bulletin de L'Academie Nationale de Medicine,* **142,** 580 (French)

Department of Health and Social Security (1970). 'Active Immunization against tetanus.' Circular 10/70

Department of Health and Social Security (1970). *The Annual Report of the Chief Medical Officer of the Department of Health and Social Security for the Year 1969.* London; HMSO

Department of Health and Social Security (1971). *Supply of certain prophylactic and therapeutic agents. HM (71) 64*. London; HMSO

Department of Health and Social Security (1973). *The Annual Report of the Chief Medical Officer of the Department of Health and Social Security for the year 1972*. London; HMSO

Deveridge, R. J. and Unsworth, I. P. (1973). 'Gas Gangrene.' *Medical Journal of Australia,* **1,** 1106

Dhayagude, R. G. and Purandare, N. M. (1949). 'Studies on anaerobic wound infections.' *Indian Journal of medical Research,* **37,** 283

Dickinson, K. M. and Edgar, W. M. (1963). 'Anaerobic cellulitis of the abdominal wall after prostatectomy and orchidectomy.' *Lancet,* **1,** 1139

Dickson, E. C. (1918). *Botulism*. New York; Rockefeller Institute for Medical Research. Monograph No. 8

Dische, F. E. and Elek, S. D. (1957). 'Experimental food-poisoning by *Clostridium welchii.' Lancet,* **2,** 71

Doane, J. C. (1924). 'Tetanus as a complication in drug inebriety.' *Journal of the American medical Association,* **82,** 1105

Dolman, C. E. (1960). 'Type E botulism: A hazard of the North.' *Arctic,* **13,** 230

Dolman, C. E. (1964). 'Botulism as a world health problem.' In *Botulism, Proceedings of a Symposium*. p. 5. Ed. by Lewis, K. H. and Cassel, K. Cincinnati; US Department of Health

Dolman, C. E. (1974). 'Human botulism in Canada (1919–1973). *'Canadian medical Association Journal,* **110,** 191

Dolman, C. E. and Iida, H. (1963). 'Type E botulism: its epidemiology, prevention, and specific treatment.' *Canadian Journal of Public Health*, **54,** 293

Dolman, C. E. and Murakami, L. (1961). *'Clostridium botulinum* type F with recent observations on other types.' *Journal of infectious Diseases,* **109,** 107

Donadio, J. A., Gangarosa, E. J. and Faich, G. A. (1971). 'Diagnosis and treatment of botulism.' *Journal of infectious Diseases,* **124,** 108

Dornbusch, K., Nord, C-E. and Dahlback, A. (1975). 'Antibiotic susceptibility of *Clostridium* species isolated from human infections.' *Scandinavian Journal of infectious Diseases,* **7,** 127

Dornbusch, K., Nord, C-E. and Olsson, B. (1975). 'Regression line analysis of five antibiotics with strains of *Clostridium* species.' *Scandinavian Journal of infectious Diseases,* **7,** 135

Dowell, V. R., Torres-Anjel, M. J., Riemann, H. P., Merson, M., Whaley, D. and Darland, G. (1975). 'A new criterion for implicating *Clostridium perfringens* as the cause of food poisoning.' *Revista Latinoamericana de Microbiologia,* **17,** 137

Drewett, S. E., Payne, D. J. H., Tuke, W. and Verdon, P. E. (1972). 'Skin distribution of *Clostridium welchii*: Use of iodophor as sporicidal agent.' *Lancet,* **1,** 1172

Drug and Therapeutics Bulletin (1975). 'The use of human antitetanus immunoglobulin.' **13,** 77

Duncan, C. L. (1970). '*Clostridium perfringens* food poisoning.' *Journal of Milk and Food Technology,* **33,** 35

Eckmann, L. (1964). 'Active and passive tetanus immunization.' *New England Journal of Medicine,* **271,** 1087

Eckman, L. (1967). *Principles on Tetanus. Proceedings of the International Conference on Tetanus, Bern, July 15–19, 1966.* Bern; Huber

Edsall, G., Elliot, M. W., Peebles, T. C. Levine, J. and Eldred, M. C. (1967). *Journal of the American medical Association,* **202,** 17

Egerton, J. R. and Walker, P. D. (1964). 'The isolation of *Clostridium perfringens* type C from necrotic enteritis of man in Papua – New Guinea.' *Journal of Pathology and Bacteriology,* **88,** 275

Elliot-Smith, A. and Ellis, H. (1957). '*Clostridium welchii* infection following cholecystectomy.' *Lancet,* **2,** 723

Ellis, M. (1963). 'Human antitetanus serum in the treatment of tetanus.' *British medical Journal,* **1,** 1123

Ellner, P. D. and O'Donnell, E. D. (1969). 'Non-fatal *Clostridium perfringens* bacteraemia.' *Journal of the American Geriatrics Society,* **17,** 644

Emond, R. (1972–73). 'Immunization.' In *Infectious Diseases 1, Medicine* **5,** p. 348. Ed. by H. Smith. London; Medical Education (International)

Engeset, J., MacIntyre, J., Smith, G. and Welch, G. (1973). '*Clostridium welchii* infection: An unusual case.' *British medical Journal,* **2,** 91

Evans, D. G. (1943a). 'The protective properties of the alpha antitoxin and theta antihaemolysin occurring in *Cl. welchii* type A antiserum.' *British Journal of experimental Pathology,* **24,** 81

Evans, D. G. (1943b). 'The protective properties of the alpha antitoxin and antihyaluronidase occurring in *Cl. welchii* type A antiserum.' *Journal of Pathology and Bacteriology,* **5,** 427

Evans, D. G. (1943c). 'Persistence of tetanus antitoxin in man following active immunization.' *Lancet,* **2,** 316

Evans, D. G. (1945). 'The *in vitro* production of α-toxin, θ-haemolysin and hyaluronidase by strains of *Cl. welchii* type A, and the relationship of *in vitro* properties to virulence for guinea pigs.' *Journal of Pathology and Bacteriology,* **57,** 75

Evans, D. G. (1947). 'Anticollagenase in immunity to *Cl. welchii* type A infection.' *British Journal of experimental Pathology,* **28,** 24

Facktor, M. A. Bernstein, R. A. and Fireman, P. (1973). 'Hypersensitivity to tetanus toxoid.' *Journal of Allergy and clinical Immunology,* **52,** 1

Faich, G. A., Graebner, R. W. and Sato, S. (1971). 'Failure of guanidine therapy in botulism A.' *New England Journal of Medicine,* **285,** 773

Felner, J. M. and Dowell, V. R. (1970). 'Anaerobic bacterial endocarditis.' *New England Journal of Medicine,* **283,** 1188

Ferguson, T. (1958). 'Infantile tetanus in some Western Isles in the second half of the nineteenth century.' *Scottish medical Journal,* **3,** 140

Fildes, P. (1929). 'Tetanus. VIII. The positive limit of oxidation-reduction potential required for the germination of spores of *B. tetani in vitro.*' *British Journal of experimental Pathology,* **10,** 151

Filler, R. M. and Ellerbeck, W. (1960). 'Tetanus prophylaxis.' *Journal of the American medical Association,* **174,** 1

Filler, R. M., Griscom, N. T. and Pappas, A. (1968). 'Post-traumatic crepitation falsely suggesting gas gangrene.' *New England Journal of Medicine,* **278,** 758

Fraser, D. W. (1976). 'Preventing tetanus in patients with wounds.' *Annals of internal Medicine,* **84,** 95

Fraser-Moodie, A. (1973). 'An unusual presentation of gas gangrene.' *British Journal of Surgery,* **60,** 621

Fulford, G. E. (1960). 'The prevention of tetanus.' *Lancet,* **1,** 1121

Furste, W. (1969). 'Current status of tetanus prophylaxis.' *Industrial Medicine,* **38,** 251

Furste, W. (1974). 'Four keys to 100 per cent success in tetanus prophylaxis.' *American Journal of Surgery,* **128,** 616

Furste, W., Skudder, P. A. and Hampton, O. P. (1967). 'Prophylaxis against tetanus in wound management.' *American College of Surgeons Committee on Trauma.* Chicago

Furste, W. and Wheeler, W. L. (1972). 'Tetanus: A team disease.' *Current Problems in Surgery,* October 3rd

Futter, B. V. and Richardson, G. (1971). 'Anaerobic jars in the quantitative recovery of clostridia.' In *Isolation of Anaerobes.* Society for Applied Bacteriology Technical Series No. 5. p. 81. Ed. by D. A. Shapton and R. G. Board. London: Academic Press

Galazka, A. (1968). 'Specific prevention of tetanus.' *Epidemiological Review,* **22,** 87

Gangarosa, E. J., Donadio, J. A., Armstrong, R. W., Meyer, K. F., Brachman. P.S. and Dowell, V. R. (1971). 'Botulism in the United States, 1899–1969.' *American Journal of Epidemiology,* **93,** 93

Garnier, M. J., Marshall, F. N., Davison, K. J. and Lepreau, F. J. (1975). 'Tetanus in Haiti.' *Lancet,* **1,** 383

Garrod, L. P. (1958). 'The chemoprophylaxis of gas gangrene.' *Journal of the Royal Army Medical Corps,* **104,** 209

Garrod, L. P., Lambert, H. P., O'Grady, F. and Waterworth, P. M. (1973). *Antibiotic and Chemotherapy.* 4th edn. Edinburgh; Churchill Livingstone

Genigeorgis, C. (1975). 'Public Health importance of *Clostridium perfringens.' Journal of the American Veterinary medical Association,* **167,** 821

Gilbert, R. J. (1974). 'Staphylococcal food poisoning and botulism.' *Postgraduate medical Journal,* **50,** 603

Giliberty, R. P. (1972). 'Tetanus toxoid boosters and human immune globulin for tetanus prophylaxis.' *New York State Journal of Medicine,* **72,** 1131

Glenn, F. (1946). 'Tetanus – a preventable disease; including an experience with civilian casualties in the battle for Manila, 1945.' *Annals of Surgery,* **124,** 1030

Gorbach, S. L. and Thadepalli, H. (1975). 'Isolation of *Clostridium* in human infection. Evaluation of 114 cases.' *Journal of infectious Diseases,* **131,** S81

Gottlieb, S. F. (1971). 'Effect of hyperbaric oxygen on micro-organisms.' *Annual Review of Microbiology,* **25,** 111

Gottlieb, S., Martin, M., McLaughlin, F. X., Panaro, R. J., Levine, L. and Edsall, G. (1967). 'Long-term immunity to diphtheria and tetanus: A mathematical model.' *American Journal of Epidemiology,* **85,** 207

Govan, A. D. T. (1946). 'An account of the pathology of some cases of *Cl. welchii* infection.' *Journal of Pathology and Bacteriology,* **58,** 423

Guneratne, F. (1975). 'Gas gangrene (Abscess) of heart.' *New York State Journal of Medicine,* **75,** 1766

Gye, R., Rountree, P. M. and Loewenthal, J. (1961). 'Infection of surgical wounds with *Clostridium welchii.' Medical Journal of Australia,* **1,** 761

Hafiz, S., McEntegart M. G., Morton, R. S. and Waitkins, S. A. (1975). *Clostridium difficile* in the urogenital tract of males and females.' *Lancet,* **1,** 420

Hall, H. E., Angelotti, R., Lewis, K. H. and Foter, M. J. (1963). 'Characteristics of *Clostridium perfringens* strains associated with food and food-borne disease.' *Journal of Bacteriology,* **85,** 1094

Handford, P. M. (1974). 'A new medium for the detection and enumeration of *Clostridium perfringens* in foods.' *Journal of applied Bacteriology,* **37,** 559

Hanke, M. E. and Bailey, J. H. (1945). 'Oxidation-reduction requirements of *Cl. welchii* and other clostridia.' *Proceedings of the Society of Experimental Biology, New York,* **59,** 163

Hanke, M. E. and Tuta, J. (1928). 'Studies on the oxidation-reduction potential of blood.' *Journal of biological Chemistry,* **78,** Proceedings xxxvi

Hanson, G. C., Slack, W. K., Chew, H. E. R. and Thomas, D. A. (1966). 'Clostridial infection of the uterus – a review of treatment with hyperbaric oxygen.' *Postgraduate medical Journal,* **42,** 499

Harman, S. M., Kautter, D. A. and Peeler, J. T. (1971). 'Improved medium for enumeration of *Clostridium perfringens.' Applied Microbiology,* **22,** 688

Harvey, P. W. and Purnell, G. V. (1967). 'Fatal case of gas gangrene associated with intramuscular injections.' *British medical Journal,* **1,** 744

Hauschild, A. H. W. (1970). 'Enumeration of *Clostridium perfringens* in foods– A review.' *Canadian Institute of Food Technology Journal,* **3,** 82

Hauschild, A. H. W. (1971). '*Clostridium perfringens* enterotoxin.' *Journal of Milk and Food Technology,* **34,** 596

Hauschild, A. H. W. (1973). 'Food poisoning by *Clostridium perfringens.' Canadian Institute of Food Science and Technology,* **6,** 106

Hauschild, A. H. W. (1974). 'Enterotoxin of *Clostridium perfringens.'* In *Anaerobic Bacteria: Role in Disease.* p. 149. Ed. by A. Balows, R. M. DeHaan, V. R. Dowell and L. B. Guze. Springfield; Thomas

Hauschild, A. and Gauvreau, L. (1976). 'Botulism in Canada – Summary for 1975.' *Health and Welfare Canada,* **2,** 49

Hauschild, A. H. W. and Hilsheimer, R. (1974). 'Evaluation and modifications of media for enumeration of *Clostridium perfringens.' Applied microbiology,* **27,** 78

Hauschild, A. H. W. and Thatcher, F. S. (1967). 'Experimental food poisoning with heat-susceptible *Clostridium perfringens* type A.' *Journal of Food Science,* **32,** 407

Hauschild, A. H. W. and Thatcher, F. S. (1968). 'Experimental gas gangrene with food-poisoning *Clostridium perfringens* type A. '*Canadian Journal of Microbiology,* **14,** 705

Hayward, N. J. (1945). 'The examination of wounds for clostridia.' *Proceedings of the Association of clinical Pathology,* **1,** 5

Health Notes (1972). 'Tetanus.' *Health Notes, Wakefield,* **4,** 17

Hellberg, B. W. (1970). 'Tetanus prophylaxis – antibiotics versus antitetanus serum.' *South African medical Journal,* **44,** 496

Heyningen, W. E. van (1968). 'Tetanus.' *Scientific American,* **218,** 69

Heyningen, W. E. van (1970). 'The pathogenic actions of exotoxins.' *Zentralblatt für Bacteriologie, Parasitenkunde, Infektionskrankheiten und Hygiene, Abteilung I. Originale,* **212,** 191

Heyningen, W. E. van and Arseculeratne, S. N. (1964). 'Exotoxins.' *Annual Review of Microbiology,* **18,** 195

Hill, A. M. (1964). 'Why be morbid? Paths of progress in the control of obstetric infection, 1931 to 1960.' *Medical Journal of Australia,* **1,** 101

Hill, G. B. and Osterhout, S. (1973). 'Exposure to hyperbaric oxygen not beneficial for murine tetanus.' *Journal of infectious Diseases,* **128,** 238

Himal, H. S., McLean, A. P. H. and Duff, J. H. (1974). 'Gas gangrene of the scrotum and perineum.' *Surgery, Gynecology and Obstetrics,* **139,** 176

Hitchcock, C. R. Demello, F. J. and Haglin, J. J. (1975). 'Gas gangrene. New approaches to an old disease.' *Surgical Clinics of North America,* **55,** 1403

Hobbs, B. C. (1969). 'Staphylococcal and *Clostridium welchii* food poisoning.' In *Bacterial Food Poisoning.* Ed. by J. Taylor. London; Royal Society of Health

Hobbs, B. C. (1974). '*Clostridium welchii* and *Bacillus cereus* infection and intoxication.' *Postgraduate medical Journal,* **50,** 597

Hobbs, B. C., Smith, M. E., Oakley, C. L. Warrack, G. H. and Cruickshank, J. C. (1953). '*Clostridium welchii* food poisoning.' *Journal of Hygiene, Cambridge,* **51,** 75

Hobbs, B. C. and Sutton, R. G. A. (1968). '*Clostridium perfringens* food poisoning.' *Annales de L'Institut Pasteur de Lille,* **19,** 29

Holland, J. A., Hill, G. B., Wolfe, W. G., Osterhout, S., Saltzman, H. A. and Brown, I. W. (1975). 'Experimental and clinical experience with hyperbaric oxygen in the treatment of clostridial myonecrosis.' *Surgery, St. Louis,* **77,** 75

Holmdahl, M. H. S. and Thoren, L. (1962). 'Tetanus in pregnancy. Report of a case of severe tetanus with survival of mother and child following tracheotomy and artificial respiration.' *American Journal of Obstetrics and Gynecology,* **84,** 339

Howard, J. M. and Inui, K. K. (1954). 'Clostridial myositis – gas gangrene. Observations of battle casualties in Korea.' *Surgery,* **36,** 1115

Hunter, W. F. (1959). 'Tetanus prophylaxis.' *Medical Journal of Australia,* **2,** 98

Iida, H., Nakamura, Y., Nakagawa, I. and Karashimada, T. (1958). 'Additional type E botulism outbreaks in Hokkaido, Japan.' *Japan Journal of medical Sciences and Biology,* **11,** 215

Irvin, T. T. and Smith, G. (1968). 'Treatment of bacterial infections with hyperbaric oxygen.' *Surgery,* **63,** 363

Jackson, R. W. and Waddell, J. P. (1973). 'Hyperbaric oxygen in the management of clostridial myonecrosis (gas gangrene).' In *Clinical Orthopaedics and Related Research.* No. 96, p. 271. Ed. by M. R. Urist. Philadelphia; Lippincott

Januszkiewicz, J., Galazka, A., Adamczyk, J. and Sporzynska, Z. (1973). 'Severe tetanus in late pregnancy.' *Scandinavian Journal of infectious Diseases,* **5,** 233

Jelliffe, D. B. (1950). 'Tetanus neonatorum.' *Archives of Diseases of Childhood,* **25,** 190

Jewett, J. F. (1972). 'Clostridial endometritis.' *New England Journal of Medicine,* **286,** 264

Jewett, J. F. (1973). 'Septic induced abortion.' *New England Journal of Medicine,* **289,** 748

Johnson, D. W. (1956). 'Epidemiology of tetanus in Queensland.' *Medical Journal of Australia,* **2,** 710

Johnson, J. T., Gillespie, T. E., Cole, J. R. and Markowitz, H. A. (1969). 'Hyperbaric oxygen therapy for gas gangrene in war wounds.' *American Journal of Surgery,* **118,** 839

Johnstone, D. D. (1958). 'Tetanus in Nigeria. Review of 100 cases treated in Ibadan.' *British medical Journal,* **1,** 12

Journal of the American medical Association, (1959). 'Tetanus immune globulin.' **171,** 2156

Journal of the American medical Association, (1968). 'Tetanus in addicts.' **205,** 584

Journal of Obstetrics and Gynaecology of the British Empire (1941). 'Report of the tetanus committee of the Royal College of Obstetricians and Gynaecologists.' **48,** 394

Kapusta, M. A., Mendelson, J. and Niloff, P. (1972). 'Non-traumatic gas gangrene: Report of a case with long term survival.' *Canadian medical Association Journal,* **106,** 863 and 866

Kass, E. H., Lichstein, H. C. and Waisbren, B. A. (1945). 'Occurrence of hyaluronidase and lecithinase in relation to virulence in *Clostridium welchii.*' *Proceedings of the Society of experimental Biology, New York,* **58,** 172

Katitch, R. V. (1960). 'The strength of immunity to tetanus following the prophylactic injection of antitetanus vaccine during pregnancy.' *Revue d'Immunologie et de Thérapie Antimicrobienne,* **24,** 521 (French)

Kauf, C., Lorent, J. P., Mosimann, J., Schlatter, I., Somaini, B. and Velvart, J. (1974). 'An epidemic of type B botulism.' *Schweizerische Medizinische Wochenschrift,* **104,** 677 (German)

Keppie, J. (1951). 'The pathogenicity of the spores of *Clostridium botulinum.' Journal of Hygiene, Cambridge,* **49,** 36

Killinback, M. J. and Lloyd Williams, K. (1961–62). 'Necrotizing colitis.' *British Journal of Surgery,* **49,** 175

Kitcat, C. de W. (1935). 'A fatal case of botulism.' *British medical Journal,* **2,** 580

Klein, E. (1895). 'On a pathogenic anaerobic intestinal bacillus, *Bacillus enteritidis sporogenes.' Zentralblatt für Bakteriologie, Parasitenkunde, Infektionskrankheiten und Hygiene (Abteilung I)* **18,** 737 (German)

Knight, B. C. J. G. and Fildes, P. (1930). 'Oxidation-reduction studies in relation to bacterial growth. III. The positive limit of oxidation-reduction potential required for germination of *B. tetani* spores *in vitro.' Biochemical Journal,* **24,** 1496

Knott, F. A. and Cole, L. B. (1950). 'Tetanus.' In *British Encyclopaedia of medical Practice,* **12,** 40. Ed. by Lord Horder. London; Butterworths

Knox, R. and MacDonald, E. K. (1943). 'Outbreaks of food poisoning in certain Leicester institutions.' *Medical Officer,* **69,** 21

Koenig, M. G., Spickard, A., Cardella, M. A. and Rogers, D. E. (1964). 'Clinical and laboratory observations on type E botulism in man.' *Medicine,* **43,** 517

Korn, J. A., Gilbert, M. S., Siffert, R. S. and Jacobson, J. H. (1975). '*Clostridium welchii* arthritis.' *Journal of Bone and Joint Surgery,* **57A,** 555

Lamanna, C. (1959). 'The most poisonous poison. What do we know about the toxin of botulism? What are the problems to be solved?' *Science,* **130,** 763

Lamanna, C. and Carr, C. J. (1966). 'The botulinal, tetanal, and entero-staphylococcal toxins: a review.' *Clinical Pharmacology and Therapeutics,* **8,** 286

Lamanna, C., McElroy, O. E. and Eklund, H. W. (1946). 'Purification and crystallization of *Clostridium botulinum* type A toxin.' *Science,* **103,** 613

Lancet (1967). 'Tetanus in Britain.' **1,** 886

Lancet (1969). 'Tetanus.' **1,** 763

Lancet (1971). 'Botulism from canned soup.' **2,** 536

Lancet (1974). 'Human antitoxin for tetanus prophylaxis.' **1,** 51

Lancet (1976). 'Antitoxin in treatment of tetanus.' **1,** 944

Lane, C. R. (1935). 'Botulism.' *Lancet,* **2,** 1377

Lane, C. R. and Jones-Davies, T. E. (1935). 'A case of botulism with a note on the bacteriological examination of the suspected food.' *Lancet,* **2,** 717

Larkin, J. M. and Moylan, J. A. (1975). 'Tetanus following a minor burn.' *Journal of Trauma,* **15,** 546

Larner, H. B. (1922). '*Bacillus welchii* in a public water supply as a possible cause of intestinal disease.' *Journal of the American medical Association,* **78,** 276

Latta, R. M. (1951). 'Management of battle casualties from Korea.' *Lancet,* **1,** 228

Laurence, D. R., Evans, D. G. and Smith, J. W. G. (1966). 'Prevention of tetanus in the wounded.' *British medical Journal,* **1,** 33

Laurent, L. J. M. and Parish, H. J. (1952). 'Serum reactions and serum sensitivity tests.' *British medical Journal,* **1,** 1294

Laurent, L. J. M. and Parish, H. J. (1962). 'Unreliability of local reactions to serum as tests for general sensitivity.' *British Journal of preventive and Social Medicine,* **16,** 111

Leighton, G. (1923). *Botulism and Food Preservation (The Loch Maree Tragedy).* London; Collins

Levinson, A., Marske, R. L. and Shein, M. K. (1955). 'Tetanus in heroin addicts.' *Journal of the American medical Association,* **157,** 658

Lindsey, D. (1959). 'Gas gangrene. Clinical inferences from experimental data.' *American Journal of Surgery,* **97,** 582

Littlewood, A. H. M., Mant, A. K. and Wright, G. P. (1954). 'Fatal tetanus in a boy after prophylactic tetanus antitoxin.' *British medical Journal,* **2,** 444

Lockwood, W. R. and Langston, L. L. (1970). 'Hyperbaric treatment of experimental tetanus in Swiss mice.' *Current therapeutic Research,* **12,** 311

Long, A. P. and Sartwell, P. E. (1947). 'Tetanus in United States Army in World War II.' *Bulletin of the US Army medical Department,* **7,** 371

Lowbury, E. J. L. and Lilly, H. A. (1955). 'A selective plate medium for *Cl. welchii.*' *Journal of Pathology and Bacteriology,* **70,** 105

Lowbury, E. J. L. and Lilly, H. A. (1958a). 'The sources of hospital infection of wounds with *Clostridium welchii.*' *Journal of Hygiene, London,* **56,** 169

Lowbury, E. J. L. and Lilly, H. A. (1958b). 'Contamination of operating-theatre air with *Cl. tetani.*' *British medical Journal,* **2,** 1334

Lowbury, E. J. L., Lilly, H. A. and Bull, J. P. (1964). 'Methods for disinfection of hands and operation sites.' *British medical Journal,* **2,** 531

Lowry, K. F. and Curtis, G. M. (1947). 'Diagnosis of clostridial myositis.' *American Journal of Surgery,* **74,** 752

Lucas, A. O. and Willis, A. J. P. (1965). 'Prevention of tetanus.' *British medical Journal,* **2,** 1333

Lynt, R. K., Kautter, D. A. and Read, R. B. (1975). 'Botulism in commercially canned foods.' *Journal of Milk and Food Technology,* **38,** 546

McClung, L. S. (1945). 'Human food poisoning due to growth of *Clostridium perfringens* (*Cl. welchii*) in freshly cooked chicken: Preliminary note.' *Journal of Bacteriology,* **50,** 229

Macfarlane, M. G. (1955). 'On the biochemical mechanism of action of gas gangrene toxins.' *Fifth Symposium of the Society for General Microbiology.* Cambridge; Cambridge University Press

Mackay, N. N. S., Gruneberg, R. N., Harries, B. J. and Thomas, P. K. (1971). 'Primary *Clostridium welchii* meningitis.' *British medical Journal,* **1,** 591

Mackay-Scollay, E. M. (1958). 'Two cases of botulism.' *Journal of Pathology and Bacteriology,* **75,** 482

McKillop, E. J. (1959). 'Bacterial contamination of hospital food with special reference to *Clostridium welchii* food poisoning.' *Journal of Hygiene, Cambridge,* **57,** 31

McKinnon, D. and McDonald, P. (1973). 'Gas gangrene – A ten-year survey from the Royal Adelaide Hospital.' *Medical Journal of Australia,* **1,** 1087

MacLennan, J. D. (1943a). 'Streptococcal infection of muscle.' *Lancet,* **1,** 582

MacLennan, J. D. (1943b). 'Anaerobic infections of war wounds in the Middle East.' *Lancet,* **2,** 63

MacLennan, J. D. (1943c). 'Anaerobic infections of war wounds in the Middle East.' *Lancet,* **2,** 94

MacLennan, J. D. (1943d). 'Anaerobic infections of war wounds in the Middle East.' *Lancet,* **2,** 123

MacLennan, R., Schofield, F. D., Pittman, M., Hardegree, M. C. and Barile, M. F. (1965). 'Immunization against neonatal tetanus in New Guinea: antitoxin response of pregnant women to adjuvant and plain toxoids.' *Bulletin of the World Health Organization,* **32,** 683

McNally, M. J. and Crile, G. (1964). 'Diagnosis and treatment of gas gangrene of the abdominal wall.' *Surgery, Gynecology and Obstetrics,* **118,** 1046

Macrae, J. (1973). 'Tetanus.' *British medical Journal,* **1,** 730

Marshall, J. H., Bromberg, B. E., Adrizzo, J. R., Heurich, A. E. and Samet, C. M. (1972). 'Fatal tetanus complicating a small partial-thickness burn.' *Journal of Trauma,* **12,** 91

Marshall, V. and Simms, P. (1960). 'Gas gangrene after the injection of adrenaline-in-oil, with a report of three cases.' *Medical Journal of Australia,* **2,** 653

Martin, H. L. and McDowell, F. (1954). 'Recurrent tetanus: Report of a case.' *Annals of internal Medicine,* **41,** 159

Marty, A. T. and Filler, R. M. (1969). 'Recovery from non-traumatic localized gas gangrene and clostridial septicaemia.' *Lancet,* **2,** 79

Matheson, J. M. (1968). 'Infection in missile wounds.' *Annals of the Royal College of Surgeons of England,* **42,** 347

Matveev, K. I., Nefedjeva, N. P., Bulatova, T. I. and Sokolov, I. S. (1967). 'Epidemiology of botulism in the USSR.' In *Botulism 1966.* p. 1. Ed. by M. Ingram and T. A. Roberts. London; Chapman and Hall

Merson, M. H. and Dowell, V. R. (1973). 'Epidemiologic, clinical and laboratory aspects of wound botulism.' *New England Journal of Medicine,* **289,** 1005

Meyer, K. F. (1956). 'The status of botulism as a world health problem.' *Bulletin of the World Health Organization,* **15,** 281

Meyer, K. F., Eddie, B., York, G. K., Collier, C. P. and Townsend, C. T. (1953). 'Clostridium botulinum type C and human botulism.' In *Proceedings of the 6th International Congress of Microbiology, Rome,* September, 1953. p. 18

Milledge, J. S. (1968). 'Hyperbaric oxygen therapy in tetanus.' *Journal of the American medical Association,* **203,** 875

Mills, K. L. G. (1973). 'Wound contamination with skin organisms.' *Journal of the Royal College of Surgeons of Edinburgh,* **18,** 321

Ministry of Health (1961). 'General Epidemiology.' Report of the Ministry of Health, London, for 1960, Part 2, p. 31 London; HMSO

Ministry of Health (1964). 'Protection against tetanus.' *British medical Journal,* **2,** 243

Moffat, W. C. (1967). 'Recent experience in the management of battle casualties.' *Journal of the Royal Army medical Corps,* **113,** 25

More, R. H. (1943). 'Bacterial endocarditis due to *Clostridium welchii.*' *American Journal of Pathology,* **19,** 413

Morgan, A., Morain, W. and Eraklis, A. (1971). 'Gas gangrene of the abdominal wall: Management after extensive debridement.' *Annals of Surgery,* **173,** 617

Morley, A. H. (1953). 'Prophylaxis of tetanus.' *British medical Journal,* **1,** 730

Moss, G. W. O., Waters, G. G. and Brown, M. H. (1955). 'The efficacy of tetanus toxoid.' *Canadian Journal of public Health,* **46,** 142

Moynihan, N. H. (1955). 'Serum-sickness and local reactions in tetanus prophylaxis. A study of 400 cases.' *Lancet,* **2,** 264

Moynihan, N. H. (1956). 'Tetanus prophylaxis and serum sensitivity tests.' *British medical Journal,* **1,** 260

Murrell, T. G. C. (1967). 'Pig-bel – epidemic and sporadic necrotizing enteritis in the highlands of New Guinea.' *Australian Annals of Medicine,* **16,** 4

Murrell, T. G. C., Roth, L., Egerton, J. Samels, J. and Walker, P. D. (1966). 'Pig-bel: enteritis necroticans. A study in diagnosis and management.' *Lancet,* **1,** 217

Mzabi, R., Himal, H. S. and MacLean, L. D. (1975). 'Gas gangrene of the extremity: the presenting clinical picture in perforating carcinoma of the caecum.' *British Journal of Surgery,* **62,** 373

Nakamura, M. and Schulze, J. A. (1970). '*Clostridium perfringens* food poisoning'. *Annual Review of Microbiology,* **24,** 359

Nakamura, Y., Iida, H., Saeki, K., Kanzawa, K. and Karashimada, Γ. (1956). 'Type E botulism in Hokkaido, Japan.' *Japan Journal of medical Science and Biology,* **9,** 45

Newell, K. W., Lehmann, A. D., LeBlanc, D. R. and Osorio, N. G. (1966). 'The use of toxoid for the prevention of tetanus neonatorum. Final report of a double-blind controlled field trial.' *Bulletin of the World Heath Organization,* **35,** 863

Newell, K. W., Lehmann, A. D., LeBlanc, D. R. and Osorio, N. G. (1967). 'Tetanus neonatorum: Epidemiology and prevention.' In *Principles on Tetanus.* Ed by L. Eckmann, p. 261. *Proceedings of the International Conference on Tetanus,* Bern, July 15–19, 1966. Bern; Huber

Nielsen, M. L. and Laursen, H. (1972). 'Clostridial infection in the urinary tract.' *Scandinavian Journal of Urology and Nephrology,* **6,** 120

North, J. P. (1947). 'Clostridial wound infections and gas gangrene.' *Surgery,* **21,** 364

Oakley, C. L. (1943). 'Toxins of *Clostridium welchii*; critical review.' *Bulletin of Hygiene, London,* **18,** 781

Oakley, C. L. (1954). 'Gas gangrene.' *British medical Bulletin,* **10,** 52

Oh, S. J., Halsey, J. H. and Briggs, D. D. (1975). 'Guanidine in Type B botulism.' *Archives of internal Medicine,* **135,** 726

O'Neill, J. P., Niall, J. F. and O'Sullivan, E. F. (1972). 'Severe postabortal *Clostridium welchii* infection: Trends in management.' *Australian and New Zealand Journal of Obstetrics and Gynaecology,* **12,** 157

Orr, K. B. and Coffey, R. (1969). 'A case of puerperal tetanus with recover.' *Medical Journal of Australia,* **2,** 557

Osbourne, C. A. P. (1892). 'Tetanus due to puncture of a hypodermic needle.' *British medical Journal,* **2,** 75

Otsuka, M. and Endo, M. (1960). 'The effect of guanidine on neuro-muscular transmission.' *Journal of Pharmacology and experimental Therapeutics,* **128,** 273

Owen-Smith, M. S. and Matheson, J. M. (1968). 'Successful prophylaxis of gas gangrene of the high-velocity missile wound in sheep.' *British Journal of Surgery,* **55,** 36

Pace, J. B. and Busuttil, A. (1971). 'Recurrent tetanus.' *St. Luke's Hospital Gazette,* **6,** 48

Parish, H. J. and Cannon, D. A. (1961). *Antisera, Toxoids, Vaccines and Tuberculins in Prophylaxis and Treatment.* 5th edn. Edinburgh; Livingstone

Parker, M. T. (1969). 'Postoperative clostridial infections in Britain.' *British medical Journal,* **3,** 671

Patel, J. C., Dhirawani, M. K., Mehta, B. C. and Agarwal, K. K. (1960b). 'Tetanus following intramuscular injection.' *Journal of the Indian medical Association,* **35,** 505

Patel, J. C., Dhirawani, M. K., Mehta, B. C. and Verdhachari, N. S., (1960a) 'Tetanus following vaccination against small-pox.' *Indian Journal of Pediatrics,* **27,** 251

Patel, J. C., Mehta, B. C., Dhirawani, M. K. and Mehta, V. R. (1961). 'Relapse and occurrence of tetanus.' *Journal of the Association of Physicians, India,* **1,** 1

Peebles, T. C., Levine, L., Eldred, M. C. and Edsall, G. (1969). 'Tetanus-toxoid emergency boosters. A reappraisal.' *New England Journal of Medicine,* **280,** 575

Penfold, W. J. and Tolhurst, J. C. (1937). 'Formol-toxoids in the prophylaxis of gas gangrene.' *Medical Journal of Australia,* **1,** 982

Penfold, W. J. and Tolhurst, J. C. (1938). 'The prophylaxis of gas gangrene in man.' *Medical Journal of Australia,* **1,** 604

Penfold, W. J., Tolhurst, J. C. and Wilson, D. (1941). 'Active immunization against gas gangrene and tetanus.' *Journal of Pathology and Bacteriology,* **52,** 187

Petrilla, A. (1960). 'Results of active immunization of civilian population against tetanus.' *Acta Microbiologica,* **7,** 65

Petty, C. S. (1965). 'Botulism: The disease and the toxin.' *American Journal of medical Science,* **249,** 345

Phadke, M. V., Godbole, R. E. and Pande, S. S. (1958). 'Tetanus in paediatric practice.' *Indian Journal of Child Health,* **7,** 675

Phillips, J., Heimbach, D. M. and Jones, R. C. (1974). 'Clostridial myonecrosis of the abdominal wall.' *American Journal of Surgery,* **128,** 436

Pitt, P. C. C. (1971). 'Tetanus.' *Journal of the Royal Army medical Corps,* **117,** 130

Pope, C. G. (1963). 'Development of knowledge of antitoxins.' *British medical Bulletin,* **19,** 230

Prévot, A. R., Terrasse, J., Daumail, J., Cavaroc, M., Riol, J. and Sillioc, R. (1955). 'Occurrence of human type C botulism in France.' *Bulletin de L'Academie Nationale de Medecine,* **139,** 355 (French)

Pritchard, J. A. and Whalley, P. G. (1971). 'Abortion complicated by *Clostridium perfringens* infection.' *American Journal of Obstetrics and Gynecology,* **111,** 484

Pulvertaft, R. J. V. (1937). 'Post-hysterectomy and puerperal tetanus. A study of the bacteriology of obstetrical and gynaecological dressings.' *British medical Journal,* **1,** 441

Pyrtek, L. J. and Bartus, S. H. (1962). '*Clostridium welchii* infection complicating biliary-tract surgery.' *New England Journal of Medicine,* **266,** 689

Ramsay, A. M., France, E. M. and Dempsey, B. M. (1956). 'Puerperal tetanus. Treatment with muscular relaxants and by assisted respiration.' *Lancet,* **2,** 548

Rathbun, H. K. (1968). 'Clostridial bacteremia without haemolysis.' *Archives of internal Medicine,* **122,** 496

Relihan, M. (1969). 'Reactions to tetanus toxoid.' *Journal of the Irish medical Association,* **62,** 430

Rey, M., Armengaud, M. and Mar, I. D. (1967). 'Tetanus in Dakar – epidemiological considerations.' In *Proceedings of the International Conference on Tetanus, Bern 1966 – Principles on Tetanus*, p. 49. Ed. by L. Eckman. Bern; Huber

Rhea, J. W., Graham, A. W., Akhnoukh, F. and Parthew, C. T. (1967). 'Effect of hyperbaric oxygenation on neonatal tetanus.' *Journal of Pediatrics,* **71,** 33

Riis, P. (1958). 'Tetanus and chronic ulcers of the legs.' *Nordisk Medicin,* **24,** 1048 (Danish)
Robb-Smith, A. H. T. (1945). 'Tissue changes induced by *Cl. welchii* type A filtrates.' *Lancet,* **2,** 362
Roberts, T. A., Thomas, A. I. and Gilbert, R. J. (1973). 'A third outbreak of type C botulism in broiler chickens.' *Veterinary Record,* **92,** 107
Robertson, M. and Keppie, J. (1943). 'Gas gangrene. Active immunization by means of concentrated toxoids.' *Lancet,* **2,** 311
Robles, N. L. and Walske, B. R. (1969). 'Current concepts of tetanus prophylaxis.' *American Journal of Surgery,* **118,** 835
Robles, N. L., Walske, B. R. and Tella, A. R. (1968). 'Tetanus prophylaxis and therapy.' *Surgical Clinics of North America,* **48,** 799
Rodin, B., Groeneveld, P. H. A. and Boerema, I. (1972). 'Ten years experience in the treatment of gas gangrene with hyperbaric oxygen.' *Surgery, Gynecology and Obstetrics,* **134,** 579
Rosebury, T. and Kabat, E. A. (1947). 'Bacterial warfare. A critical analysis of the available agents, their possible military applications, and the means for protection against them' *Journal of Immunology,* **56,** 7
Rubbo, S. D. (1958). 'Prevention of postoperative tetanus.' *Lancet,* **2,** 268
Rubbo, S. D. (1965). 'A re-evaluation of tetanus prophylaxis in civilian practice.' *Medical Journal of Australia,* **2,** 105
Rubbo, S. D. (1966). 'New approaches to tetanus prophylaxis.' *Lancet,* **2,** 449
Rubbo, S. D. and Suri, J. C. (1962). 'Passive immunization against tetanus with human immune globulin.' *British medical Journal,* **2,** 79
Sachs, A. (1952). 'Modern views on the prevention of tetanus in the wounded.' *Proceedings of the Royal Society of Medicine,* **45,** 641
Sanders, R. K. M., Strong, T. N. and Peacock, M. L. (1969). 'The treatment of tetanus with special reference to betamethasone.' *Transactions of the Royal Society of Tropical Medicine and Hygiene,* **63,** 746
Sanders, R. S., Fields, A., Catignani, J. and Hutcheson, R. H. (1974). '*Clostridium perfringens* food poisoning – Tennessee.' *Morbidity and Mortality Weekly Report,* **23,** 19
Scaer, R. C., Tooker, J. and Cherington, M. (1969). 'Effect of guanidine on the neuromuscular block of botulism.' *Neurology (Minneapolis),* **19,** 1107
Schantz, E. J. and Sugiyama, H. (1974). 'Toxic proteins produced by *Clostridium botulinum.*' *Agricultural and Food Chemistry,* **22,** 26
Scheibel, I. (1955). 'The uses and results of active tetanus immunization.' *Bulletin of the World Health Organization,* **13,** 381
Scheibel, I., Bentzon, M. W., Christensen, P. E. and Biering, A. (1966). 'Duration of immunity after tetanus immunization.' In *Principles on Tetanus.* p. 245. Ed. by L. Eckman. Bern; Huber
Schoenholz, P. and Meyer, K. F. (1924). 'Effect of direct sunlight, diffuse daylight and heat on potency of botulinus toxin in culture medium and vegetable products. XXIV.' *Journal of infectious Diseases,* **35,** 361
Schofield, F. D., Tucker, V. M. and Westbrook, G. R. (1961). 'Neonatal tetanus in New Guinea. Effect of active immunization in pregnancy.' *British medical Journal,* **2,** 785
Schweigel, J. F. and Shim, S. S. (1973). 'A comparison of the treatment of gas gangrene with and without hyperbaric oxygen.' *Surgery, Gynecology and Obstetrics,* **136,** 969
Sebald, D. M., Jouglard, J. and Gilles, G. (1974). 'Type B botulinum in man due to cheese.' *Annales de Microbiologie,* 125A, 349 (French)

Sevitt, S. (1949). 'Source of two hospital-infected cases of tetanus.' *Lancet,* **2,** 1075

Sevitt, S. (1953). 'Gas gangrene infection in an operating theatre.' *Lancet,* **2,** 1121

Shah, N. J. (1955). 'Study of otogenic tetanus.' *Indian Journal of medical Science,* **9,** 52

Sharrard, W. J. W. (1964). 'Tetanus prophylaxis.' *British medical Journal,* **1,** 55

Shaw, J., Vellar, I. D. and Vellar, D. (1973). 'Clostridial (gas gangrene) infection in a general hospital.' *Medical Journal of Australia,* **1,** 1080

Sherman, R. T. (1970). 'The prevention and treatment of tetanus in the burn patient.' *Surgical Clinics of North America,* **50,** 1277

Simpson, W. I. (1907). 'The evidence and conclusions relating to the Mulkowal tetanus case.' *Practitioner,* **78,** 796

Singleton, A. R. and Witt, R. W. (1956). 'Tetanus complicating pregnancy. Report of a case treated with chlorpromazine with survival of both mother and infant.' *Obstetrics and Gynecology,* **7,** 540

Skjelkvale, R. and Duncan, C. L. (1975). 'Enterotoxin formation by different toxigenic types of *Clostridium perfringens*.' *Infection and Immunity,* **11,** 563

Slack, W. K., Hanson, G. C. and Chew, H. E. R. (1969). 'Hyperbaric oxygen in the treatment of gas gangrene and clostridial infection.' *British Journal of Surgery,* **56,** 505

Smith, G. Sillar, W., Norman, J. N., Ledingham, I. McA., Bates, E. H. and Scott, A. C. (1962). 'Inhalation of oxygen at 2 atmospheres for *Clostridium welchii* infections.' *Lancet,* **2,** 756

Smith, J. W. G. (1964). 'Penicillin in prevention of tetanus.' *British medical Journal,* **2,** 1293

Smith, J. W. G. (1969). 'Diphtheria and tetanus toxoids.' *British medical Bulletin,* **25,** 177

Smith, J. W. G. (1974). 'Human tetanus immunoglobulin.' *Bulletin of the Hoffkine Institute,* **2,** 94

Smith, J. W. G. (1975). 'A new look at tetanus.' *General Practitioner,* February 14, p. 14

Smith, J. W. G., Jones, D. A., Gear, M. W. L., Cunliffe, A. C. and Barr, M. (1963). 'Simultaneous active and passive immunization against tetanus.' *British medical Journal,* **1,** 238

Smith, J. W. G., Laurence, D. R. and Evans, D. G. (1975). 'Prevention of tetanus in the wounded.' *British medical Journal,* **3,** 453

Smith, J. W. G. and MacIver, A. G. (1969). 'Studies in experimental tetanus infection.' *Journal of medical Microbiology,* **2,** 385

Smith, J. W. G. and MacIver, A. G. (1974). 'Growth and toxin production of tetanus bacilli *in vivo,* ' *Journal of medical Microbiology,* **7,** 497

Smith, L. DS. (1955). *Introduction to the Pathogenic Anaerobes*. Chicago; University of Chicago Press

Smith, L. DS. (1973). '*Clostridium perfringens* food poisoning.' *Memorias Institute Oswaldo Cruz,* **71,** 183

Smith, L. DS. and George, R. L. (1946). 'The anaerobic bacterial flora of clostridial myositis.' *Journal of Bacteriology,* **51,** 271

Smith, L. P., McLean, A. P. H. and Maughan, G. B. (1971). '*Clostridium welchii* septicotoxaemia.' *American Journal of Obstetrics and Gynecology,* **110,** 135

Smythe, P. M., Bowie, M. D. and Voss, T. J. V. (1974). 'Treatment of tetanus neonatorum with muscle relaxants and intermittent positive-pressure ventilation.' *British medical Journal,* **1,** 223

Sneddon, I. B. (1960). 'Accidental acquired hypersensitivity to tetanus antitoxin.' *British medical Journal,* **1,** 1468

Solomonova, K. and Kebedjiev, G. (1972). 'Tetanus in Bulgaria.' *Zeitschrift für Immunitätsforschung,* **144,** 242

Solomonova, K. and Vizev, St. (1973). 'Immunological reactivity of senescent and old people actively immunized with tetanus toxoid.' *Zeitschrift für Immunitätsforschung,* **146,** 81

Spann, J. L. and McGill, R. A. (1957). 'Clostridial myositis following resection of bowel. Case report.' *Annals of Surgery,* **146,** 98

Spencer, R. (1969). 'Food poisoning due to clostridia. The factors affecting the survival and growth of food poisoning clostridia in cured foods.' *British Food Manufacturing Industries Research Association, Scientific and Technical Surveys, Number 58.*

Stahlie, T. D. (1960). 'The role of tetanus neonatorum in infant mortality in Thailand.' *Journal of tropical Pediatrics,* **6,** 15

Sterne, M. and Warrack, G. H. (1964). 'The types of *Clostridium perfringens.*' *Journal of Pathology and Bacteriology,* **88,** 279

Stock, A. H. (1947). 'Clostridia in gas gangrene and local anaerobic infections during the Italian campaign.' *Journal of Bacteriology,* **54,** 169

Stratford, B. C. (1973). 'Gas gangrene.' *Medical Journal of Australia,* **2,** 47

Strum, W. B., Cade, J. R., Shires, D. L. and Quesada, A, de (1968). 'Postabortal septicaemia due to *Clostridum welchii.*' *Archives of internal Medicine,* **122,** 73

Suri, J. C. and Rubbo, S. D. (1961). 'Immunization against tetanus.' *Journal of Hygiene, Cambridge,* **59,** 29

Sutherland, H. P. (1960). 'Report of a case of botulinus poisoning.' *Journal of the American medical Association,* **172,** 1266

Sutton, R. G. A. (1966). 'Enumeration of *Clostridium welchii* in the faeces of varying sections of the human population.' *Journal of Hygiene, Cambridge,* **64,** 367

Sutton, R. G. A. and Hobbs, B. C. (1968). 'Food poisoning caused by heat-sensitive *Clostridium welchii.* A report of five recent outbreaks.' *Journal of Hygiene, Cambridge,* **66,** 135

Takayama, N., Sakurai, N. and Nakayama, M. (1971). 'Secondary response to booster injection of alum-precipitated tetanus toxoid.' *Japanese Journal of Microbiology,* **15,** 273

Tate, G. T., Thomson, H. and Willis, A. T. (1965). '*Clostridium welchii* colitis.' *British Journal of Surgery,* **52,** 194

Taylor, C. E. D., and Coetzee, E. F. C. (1966). 'Range of heat resistance of *Clostridium welchii* associated with suspected food poisoning.' *Monthly Bulletin of the Ministry of Health and the Public Health Laboratory Service,* **25,** 142

Taylor, G. W. (1960). 'Preventive use of antibiotics in surgery.' *British medical Bulletin,* **16,** 51

Templeton, W. L. (1935). 'Two fatal cases of botulism.' *British medical Journal,* **2,** 500

Tennant, R. and Parkes, H. W. (1959). 'Myocardial abscess. A study of pathogenesis with report of a case.' *Archives of Pathology, Chicago,* **68,** 456

Tetanus Surveillance (1974). *Center for Disease Control, Report No. 4.* Atlanta; US Department of Health, Education and Welfare Publication No. CDC 74.8274

Thatcher, F. S. and Clark, D. S. (1968). *Microorganisms in Foods. Their Significance and Methods of Enumeration*. Toronto; University of Toronto Press

Thompson, R. G. and Harper, I. A. (1974). 'Lawn mower injuries.' *British medical Journal,* **3,** 687

Tisdall, M. W. (1961). 'Antitetanus immunization.' *British medical Journal,* **2,** 1647

Tompkins, A. B. (1958). 'Neonatal tetanus in Nigeria.' *British medical Journal,* **1,** 1382

Tonge, J. I. (1957). 'Gas gangrene following the injection of adrenaline in oil.' *Medical Journal of Australia,* **2,** 936

Toogood, J. H. (1960). 'Allergic reactions to antitetanus serum.' *Canadian medical Association Journal,* **82,** 907

Torg, J. S. and Lammot, T. R. (1968). 'Septic arthritis of the knee due to *Clostridium welchii.*' *Journal of Bone and Joint Surgery,* **50A,** 1233

Trinca, J. C. (1963). 'Active tetanus immunization: effect of reduced reinforcing dose of adsorbed toxoid on the partly immunized reactive patient.' *Medical Journal of Australia,* **2,** 389

Trinca, J. C. (1967). 'Problems in tetanus prophylaxis. The immune patient.' *Medical Journal of Australia,* **2,** 153

Trinca, J. C. (1974). 'Antibody response to successive booster doses of tetanus toxoid in adults.' *Infection and Immunity,* **10,** 1

Trinca, J. C. and Fraser, A. N. (1968). 'Tetanus immunity in Victoria, 1967.' *Medical Journal of Australia,* **2,** 300

Ullberg-Olsson, K. and Eriksson, E. (1975). 'Active immunization against tetanus in man. I. Duration of anamnestic reaction after one dose of vaccine.' *European surgical Research,* 7, 249

Vaillard, L. and Rouget, J. (1892). 'Contribution to the study of tetanus.' II. Aetiology.' *Annales de L'Institut Pasteur,* **6,** 385 (French)

VanBeek, A., Zook, E., Yaw, P., Gardner, R., Smith, R. and Glover, J. L. (1974). 'Nonclostridial gas-forming infections.' *Archives of Surgery,* **108,** 552

Vernon, E. (1969). 'Food poisoning and Salmonella infections in England and Wales, 1967.' *Public Health, London,* **83,** 205

Walker, H. W. (1975). 'Food borne illness from *Clostridium perfringens.*' *Critical Reviews in Food Science and Nutrition,* **7,** 71

Wapen, B. D. and Gutmann, L. (1974). 'Wound botulism. A case report.' *Journal of the American medical Association,* **227,** 1416

Warren, C. P. W. and Mason, B. J. (1970). '*Clostridium septicum* infection of the thyroid gland.' *Postgraduate medical Journal,* **46,** 586

Weekly Report (1974). 'Wound botulism – Idaho, Utah, California.' *Morbidity and Mortality Weekly Report,* **23,** 246 and 251

Weekly Report (1975). 'Botulism – United States, 1974.' *Morbidity and Mortality Weekly Report,* **24,** 39

Weinberg, M., Prévot, A. R., Davesne, J. and Renard, C. (1928a). 'Studies on the bacteriology and serotherapy of appendicitis.' *Annales de L'Institut Pasteur,* **42,** 1167 (French)

Weinberg, M., Prévot, A. R., Davesne, J. and Renard, C. (1928b). 'The microbial flora of appendicitis.' *Comptes Rendus de la Société de Biologie, Paris,* **98,** 749 (French)

Weinberg, M., Renard, C. and Davesne, J. (1926). 'The presence of gas gangrene anaerobes in the microbial flora of appendicits.' *Comptes Rendus de la Société de Biologie, Paris,* **94,** 813 (French)

Weinstein, B. B. and Beacham, W. D. (1941). 'Postabortal tetanus. A review of the literature, and a report of 14 additional cases.' *American Journal of Obstetrics and Gynecology,* **42,** 1031

Wentzel, L. M., Sterne, M. and Polson, A. (1950). 'High toxicity of pure botulinum type D toxin.' *Nature, London,* **166,** 739

Werner, S. B. and Chin, J. (1973). 'Botulism – Diagnosis management and public health considerations.' *California Medicine,* **118,** 84

Wheatley, P. R. (1967). 'Research on missile wounds – the Borneo operations January 1963 – June 1965. '*Journal of the Royal Army medical Corps,* **113,** 18

White, W. G., Barnes, G. M., Barker, E., Gall, D., Knight, P., Griffith, A. H., Morris-Owen, R. M. and Smith, J. W. G. (1973). 'Reactions to tetanus toxoid.' *Journal of Hygiene, Cambridge,* **71,** 283

White, W. G., Gall, D., Barnes, G. M., Barker, E., Griffity, A. H. and Smith, J. W. G. (1969). 'Duration of immunity after active immunization against tetanus.' *Lancet,* **2,** 95

Wilkinson, J. L. (1961). 'Neonatal tetanus in Sierra Leone.' *British medical Journal,* **1,** 1721

Willis, A. T. (1969). *Clostridia of Wound Infection*. London; Butterworths

Willis, A. T. (1972–73). 'Tetanus.' In *Medicine, 5, Infectious Diseases 1*, p. 413. Ed. by H. Smith, London; Medical Education (International)

Willis, A. T. (1975). 'Tetanus.' In *Medicine 3, Infectious Diseases Part 3* p. 133 Ed. by H. Smith and R. T. D. Emond. London; Medical Education (International)

Willis, A. T. and Jacobs, S. I. (1964). 'Meningitis due to *Clostridium welchii.*' *Journal of Pathology and Bacteriology,* **88,** 312

Wilson, V. J., Diecke, P. J. and Talbot, W. H. (1960). 'Action of tetanus toxin on conditioning of spinal motoneurones.' *Journal of Neurophysiology,* **23,** 659

Wood, L. W. (1973). 'Botulinal toxin in commercially canned mushrooms.' *Morbidity and Mortality Weekly Report,* **22,** 115

Woodward, W. W. (1960). 'Tetanus eight days after administration of anti-tetanus serum.' *British medical Journal,* **2,** 961

World Health Organization (1950). 'Active immunization against common communicable diseases in childhood.' *Technical Report Series,* **6,** 15

Wright, G. P. (1955). 'The neurotoxins of *Clostridium botulinum and Clostridium tetani.*' *Pharmacological Reviews,* **7,** 413

Wright, G. P. (1959). 'Movements of neurotoxins and neuroviruses in the nervous system.' In *Modern Trends in Pathology,* p. 212. Ed. by Collins, D. H. London; Butterworths

Young, M. (1927). 'Tetanus IV. The geographical incidence of tetanus in England and Wales.' *British Journal of experimental Pathology,* **8,** 226

Yudis, M. and Zucker, S. (1967). '*Clostridium welchii* bacteraemia: A case report with survival and review of the literature.' *Postgraduate medical Journal,* **43,** 487

Zeissler, J. and Rassfeld-Sternberg, L. (1949). 'Enteritis necroticans due to *Clostridium welchii* type F.' *British medical Journal,* **1,** 267

Index